The Global Genome

LEONARDO

Roger F. Malina, series editor

The Visual Mind, edited by Michele Emmer, 1993

Leonardo Almanac, edited by Craig Harris, 1994

Designing Information Technology, Richard Coyne, 1995

Immersed in Technology: Art and Virtual Environments, edited by Mary Anne Moser with Douglas MacLeod, 1996

Technoromanticism: Digital Narrative, Holism, and the Romance of the Real, Richard Coyne, 1999

Art and Innovation: The Xerox PARC Artist-in-Residence Program, edited by Craig Harris, 1999

The Digital Dialectic: New Essays on New Media, edited by Peter Lunenfeld, 1999

The Robot in the Garden: Telerobotics and Telepistemology in the Age of the Internet, edited by Ken Goldberg, 2000

The Language of New Media, Lev Manovich, 2000

Metal and Flesh: The Evolution of Man: Technology Takes Over, Ollivier Dyens, 2001

Uncanny Networks: Dialogues with the Virtual Intelligentsia, Geert Lovink, 2002

Information Arts: Intersections of Art, Science, and Technology, Stephen Wilson, 2002

Virtual Art: From Illusion to Immersion, Oliver Grau, 2003

Women, Art, and Technology, edited by Judy Malloy, 2003

Protocol: How Control Exists after Decentralization, Alexander R. Galloway, 2004

At a Distance: Precursors to Art and Activism on the Internet, edited by Annmarie Chandler and Norie Neumark, 2005

The Visual Mind II, edited by Michele Emmer, 2005

CODE: Collaborative Ownership and the Digital Economy, edited by Rishab Aiyer Ghosh, 2005

The Global Genome: Biotechnology, Politics, and Culture, Eugene Thacker, 2005

Media Ecologies: Materialist Energies in Art and Technoculture, Matthew Fuller, 2005

From Technological to Virtual Art, Frank Popper, 2005

The Global Genome

Biotechnology, Politics, and Culture

Eugene Thacker

The MIT Press
Cambridge, Massachusetts
London, England

MIT Press books may be purchased at special quantity discounts for business or sales promotional use. For information, please email special_sales@mitpress.mit.edu or write to Special Sales Department, The MIT Press, 5 Cambridge Center, Cambridge, MA 02142.

This book was set in Garamond 3 and Bell Gothic by SNP Best-set Typesetter Ltd., Hong Kong. Printed and bound in the United States of America.

Library of Congress Cataloging-in-Publication Data
Thacker, Eugene.
The global genome : biotechnology, politics, and culture / Eugene Thacker.
p. cm. – (Leonardo)
Includes bibliographical references and index.
ISBN 0-262-20155-0 (alk. paper)
1. Biotechnology industries. 2. Bioinformatics. 3. Globalization. I. Title. II. Leonardo (Series) (Cambridge, Mass.)

HD9999.B442T453 2005
338.4′76606–dc22

2004061148

10 9 8 7 6 5 4 3 2 1

Contents

Series Foreword

The cultural convergence of art, science, and technology provides ample opportunity for artists to challenge the very notion of how art is produced and to call into question its subject matter and its function in society. The mission of the Leonardo book series, published by The MIT Press, is to publish texts by artists, scientists, researchers, and scholars that present innovative discourse on the convergence of art, science, and technology.

Envisioned as a catalyst for enterprise, research, and creative and scholarly experimentation, the book series enables diverse intellectual communities to explore common grounds of expertise. It provides a context for the discussion of contemporary practice, ideas, and frameworks in this rapidly evolving arena where art and science connect.

To find more information about Leonardo/ISAST and to order our publications, go to Leonardo Online at <http://mitpress.mit.edu/e-journals/Leonardo/isast/leobooks.html> or send e-mail to <leonardobooks.mitpress.mit.edu>.

Joel Slayton
Chair, Leonardo Book Series

Leonardo/International Society for the Arts, Sciences, and Technology (ISAST)

Leonardo, the International Society for the Arts, Sciences, and Technology, and the affiliated French organization Association Leonardo have two very simple goals:

1. to document and make known the work of artists, researchers, and scholars interested in the ways that the contemporary arts interact with science and technology, and
2. to create a forum and meeting places where artists, scientists, and engineers can meet, exchange ideas, and, if appropriate, collaborate.

When the journal *Leonardo* was started some 35 years ago, these creative disciplines existed in segregated institutional and social networks, a situation dramatized at that time by the "Two Cultures" debates initiated by C. P. Snow. Today we live in a different time of cross-disciplinary ferment, collaboration, and intellectual confrontation enabled by new hybrid organizations, new funding sponsors, and the shared tools of computers and the Internet. Above all, new generations of artist-researchers and researcher-artists are now at work individually and in collaborative teams bridging the art, science, and technology disciplines. Perhaps in our lifetime we will see the emergence of "new Leonardos," creative individuals or teams who will not only develop a meaningful art for our times but also drive new agendas in science and stimulate technological innovation that addresses today's human needs.

For more information on the activities of the Leonardo organizations and networks, please visit our Web site at <http://mitpress.mit.edu/Leonardo>.

Roger F. Malina
Chair, Leonardo/ISAST

Foreword

April 2003 marked the fiftieth anniversary of the discovery of the structure of DNA (deoxyribonucleic acid). Discovery of the double-helical structure by James Dewey Watson and Francis Crick along with Maurice Wilkins and Rosalind Elsie Franklin at the University of Cambridge in 1953 redefined our fundamental understanding of the biological structure and function of life. Passing decades of research have continued in a rapid evolution of science and engineering resulting in many new disciplines ranging from pharmacogenomics to the neural correlates of consciousness. This dramatic transformation of biology is represented by a trajectory of codification and informatics. Francis Crick died, July 29, 2004. His proclaimed announcement, "We have discovered the secret of life," rings in our ears.

The Global Genome: Biotechnology, Politics, and Culture by Eugene Thacker explores the intersection of biology and informatics. It is a book about the mutability of biology and the fluidity of values, signs and power that exist across scientific disciplines and political-economic culture. Analysis informed by Foucault's biopolitics; Marx's economica (species being and inorganic body); Shannon's theory of information; von Neumann's cybernetic reasoning; Canquilhem's study on the ontology of organization; and the reality of informational excess represented by Luca Cavalli-Sforza and Allan Wilson's Human Genome Diversity Project establish a context for consideration. It is a context Eugene Thacker builds upon and expands within a central theme: clarification of the ontological struggle with simplistic notions of biology as information.

Eugene Thacker speaks directly to the tension of convergence involving biology and informatics. His is a concern with the implication of new modes of regulation management and control determined by a social system that values increasingly more effective forms of biological prediction, analysis, accuracy, efficiency, probability, and optimization. It is a description of a cultural system that is politically driven toward opportunities and discovery of economic benefit. It is a view of genomics as a political economy of the genetic body, a view that understands life as reprogrammable, instrumentalized, and networked. For Thacker, the network is the dominant paradigm of globalization and the terms of globalization and mobility are complimentary.

The Global Genome: Biotechnology, Politics, and Culture is also an address on the mobility of biology across media in which production, distribution, and consumption define the contemporary model of biology and life itself. According to Eugene Thacker, encoding, recoding, and decoding represent the primary activities of biotechnology today. Interlaced production, distribution, and consumption shape the essentiality of biological being as material and immaterial, property and information. Genes, cell lines, plants, and animals exist as semiotic deployments of life that dynamically shift from body to body, body to code and code to body.

The Global Genome: Biotechnology, Politics, and Culture examines the nature of a biological mobility that is deeply embedded in a biologicalness that necessarily integrates the goo of life with code and network.

Reconciliation of the artificial with the natural continues to be a central issue of our times. In a world where the DNA of an organism can be reshuffled and engineered, the struggle to formulate coherent theoretical constructs for the structure and function of life is both a technical and ethical question. Eugene Thacker provides a glimpse at the implications of a biological body that is both wet and dry, presence and pattern. The Leonardo Book Series is very pleased to include *The Global Genome: Biotechnology, Politics, and Culture* by Eugene Thacker.

Joel Slayton
Chair, Leonardo Book Series

Preface

For some reason, I have always felt that a preface should really be called a "confession." So here is mine: this is not only a book about genes, proteins, and codes, but a book of philosophy. In itself, this is not much of a confession (unless "doing philosophy" is something to hide). Perhaps what I should really say is that this is a book about how, in biotechnology, ontological questions immediately fold onto questions that are social, economic, and cultural.

For example, take three "objects." The first is your DNA in a test tube. Actually it is very easy—using supplies from your local grocery, you can extract your own DNA; all you need is a toothpick, detergent, salt, and alcohol. Your DNA does not look particularly "high tech" or sexy; it is more like stringy, semitransparent goo. So that is our first object. The second object is a computer file of a DNA sequence from a database. Anyone with an Internet connection can access GenBank, the main public repository for genome data. You can search for a particular sequence you are after, and the file will display all sorts of information, from research publications to the actual sequence of As, Ts, Cs, and Gs. Now take a third object, a computer file of the patent for the DNA sequence file I just mentioned. Again, this is easily accessible via the U.S. Patent and Trademark Office Web site. This file is a bit different, containing not only scientific information, but technical, economic, and legal information as well. So we have three kinds of DNA: "wet" DNA in a test tube, "dry" DNA from a computer database, and valuable DNA as part of a patent.

This book is about the relationship between these three entities. Between the test tube DNA, the GenBank file, and the patent record, there are a number of relationships that have to do with how the concept of "life itself" is being fundamentally transformed in the era of biotechnology. The argument I make is that these changes are not merely additive changes to already-existing notions of life, labor, and property, but that, in their very existence, these unique entities—genome databases, DNA synthesizers, regenerative tissues—are ontologically redefining the notion of biological "life itself," a notion that has always been at the center of biological thought.

But, for all this talk about ontology, there is little talk of the niceties of ontological categories or philosophical logic. Instead, many of the pages are filled with discussions on genome databases, drug discovery informatics, lab-grown tissues, and plasmid libraries. Not only that, but there is also talk of international genome sequencing organizations, national biobank initiatives, and, of course, the private sector emphasis on fields such as bioinformatics and pharmacogenomics.

So, if this is a book of philosophy, then it is a strange one, one that talks as much about reprogrammed stem cells as it does about Aristotelian biology or Marxian "species being." My hope is that between this triangulated relationship of "real" biology (test tube DNA), computer codes (the GenBank file), and biological property (the patent record), we can develop a more complex awareness of the ways in which "life itself" is deployed in a range of areas, from the creation of new medicines to the transformations in food production, national security, and the cultural understanding of biological life.

Acknowledgments

This book has had a long gestation period. A number of the concepts have previously appeared in publication, either in different essays or as earlier versions of the chapters presented here. Chapter 3 is an expanded and modified version of an essay commissioned by the Walker Arts Center's Gallery 9. The sections of chapter 4 on recombinant capital are derived from an essay published in *Dialectical Anthropology*. Chapter 5 grew out of an essay commissioned by Locus+ for a catalog. It also incorporates arguments from a short piece entitled "The Anxieties of Biopolitics," commissioned in 2001 by the Information, War, and Peace Project at Brown University. Chapter 6 is a modified version of an essay published in *Theory & Event* and republished later in the *Ars Electronica* '99 catalog. Chapter 7 grew out of two talks I gave, one at the Dutch Electronic Art Festival (DEAF) and another at the Society for Literature and Science (SLS) conference, both in 2000. Portions of these talks were subsequently published in the *Journal of Medical Humanities*.

There are many to thank. In the earliest stages of this book, Ed Cohen, Samira Kawash, and especially Josephine Diamond provided a supportive environment for me to explore some of its themes.

Over the years, many colleagues provided helpful comments and conversation, both in person and via email, including Oron Catts and Ionat Zurr, Joe Davis, Nick Dyer-Witheford, Michael Fortun, Alex Galloway, Diane Gromala, Natalie Jeremijenko, Ken Knoespel, Eduardo Kac, Steve Kurtz, Lisa Lynch, Wendy Newstetter, Kavita Philip, Steve Potter, Birgit Richard, Steve Shaviro, and Cathy Waldby. The editorial team at MIT Press, especially Doug

Sery and Joel Slayton, have been instrumental in seeing this book through to publication. Online communities such as Nettime, Rhizome, and the Thing have also provided a forum in which to experiment with new ideas.

Certain things go without saying, but should be said nevertheless. Without Edgar Thacker, Moonsoon Thacker, Marie Thacker, and Prema Murthy, undertaking a project such as this would have been difficult. Finally, I would like to dedicate this book to my grandfather, Dr. E. A. Thacker (1905–99).

Introduction

"Templates broken, membranes burst"

Bioscience research and the biotech industry are increasingly organized on a global level, bringing together novel, hybrid artifacts (such as genome databases and DNA chips), with new means of distribution and exchange (most notably, the use of the Internet in exchanging biological data).[1] Indeed, a glance at contemporary biotech-related events reveals novel artifacts and practices that are in their very constitution networked and global: international genome sequencing efforts such as the International Human Genome Sequencing Consortium (IHGSC), the proliferation of genomic databases (GenBank in the United States, the European Molecular Biology Laboratory in Europe, the DNA Data Bank of Japan in Japan), continued developments in national and international property laws (World Intellectual Property policies [WIPs]), the "borderless" business of biotech start-ups, spin-offs, and subsidiaries, and the various efforts to integrate health care and medicine with the concerns of national security (the World Health Organization [WHO] Global Outbreak and Response Network, the U.S. government's Project BioShield). Biotechnology, it seems, takes place on a global level, be it in terms of exchanging biological information, controlling epidemics, deterring biological attacks, or standardizing intellectual property laws.

Yet, in another sense, none of this is new, for modern biotechnologies have always been global, whether they involve the development of worldwide markets for pharmaceuticals or the collection of biological materials from various cultures around the globe. In a sense, the degree to which the biotech industry is globalized is proportional to the degree to which the business of health care is seen as financially independent of particular governmental regulations. Although the division between the private and public sectors in bioscience research and health care is today more complex that it has ever been, there is also a sense in which the very notion of a semiautonomous, biotech "industry" implies the active development of new alliances between government regulation, federally funded research, biotech start-ups, information technology companies, and "Big Pharma."

What is new, however, is the extent to which information technologies—and an informatic worldview—are increasingly becoming part and parcel of our understanding of biological "life itself." Certainly, the discursive intersections between genetics and cybernetics, biology and information theory, are in themselves not new. Yet in an era in which it is commonplace to move from DNA in a test tube to DNA in an online database—and back again—the metaphorical relationship between biology and information is raised to a new level. The various efforts to map the genomes of a range of organisms, from the roundworm to human beings, is only the most recent sign of this intersection of biology and informatics, of genetic "codes" and computer "codes." The combination of biology and informatics is arguably affecting nearly every field within the biotech and health care industries, from the quotidian use of bioinformatics software, to the ongoing transition to "infomedicine."

Yet this integration of biology and informatics brings with it a host of difficult questions, questions that are at once philosophical, economic, and political. If the emphasis on informatics is indeed a constituent part of biotech today, how does this ease of mobility between the material and the immaterial, the "wet" and the "dry," affect the biological and medical understanding of the body (and should biotech become "moist")? If biology is really just pattern, code, or sequence, then how does this fact change the conventional property relations surrounding biological materiality? How has our understanding of the relationship between organism and environment changed in light of recent network forms of organization (e.g., disease surveillance networks, bioinformatics and genomics databases)? How might the economic, political, and cultural dimensions of globalization affect the current informatic paradigm of molecular biology?

Thus, not only do we see biology being networked and distributed across the globe, but we are also seeing a hegemonic understanding of biological "life itself" that ceases to make a hard distinction between the natural and the artificial, the biological and the informatic. This is the twofold aspect of biotechnology that this book intends to explore. On the one hand, we witness a range of current events that readily display features of a globalizing industry. On the other hand, we also witness an ongoing integration of biology and informatics, genetics and computers, DNA chips and gene-finding software. The aim of *The Global Genome* is, then, to comprehend these two developments: to situate changes in the biotech industry within the larger context of globalization and political economy, and, conversely, to understand globalization as a core part of the practices and concepts of the biotech industry.

"Globalization" is a phenomenon that has been widely discussed in a number of contexts: economic (e.g., international organizations such as the World Trade Organization [WTO], the International Monetary Fund [IMF], and the World Bank), political (e.g., the so-called withering of the nation-state and the limits of geopolitical borders), and cultural (e.g., the hegemony of American culture or what some call "cultural imperialism"). But little attention has been given to the consideration of globalization as a biological phenomenon—that is, globalization in the context of biotechnology. The emergence of a biotech "industry" in the 1970s and the continued expansion of the pharmaceutical industry have arguably been global endeavors from the beginning. In this sense, biotechnology is coextensive with globalization. Similarly, "big science" projects such as the mapping of the human genome and events such as the emergence of new infectious diseases (the 2003 Severe Acute Respiratory Syndrome [SARS] outbreak) take place on a global level that includes networks of all kinds. Biological networks of infection (a novel virus strain) are contextualized by networks of transportation (the air-travel industry), which are then affected by governmental modes of regulation (travel restrictions, quarantine), which are countered by medical response efforts (disease surveillance networks), which are linked to medical-economic interventions (new vaccines, "fast-track" U.S. Food and Drug Administration [FDA] drug approval), all of which have a palpable effect in terms of the mass media and a larger cultural impact (science fiction films or television programs featuring bioterrorist attacks or dirty bombs).

Thus, the primary aim of this book is to understand how the processes of globalization form a core component of biological knowledge and practice, without, however, simply determining it. In our contemporary context, biotechnology and globalization will be seen as indissociable, while not simply being identical. Indeed, it can even be stated that biotechnology and, by extension, a biotech "industry" are unthinkable without a globalizing context. The intersection of biology and informatics in this case is instructive. A large-scale scientific endeavor such as the mapping of the human genome integrates genetic and computer codes at many different levels, from the agglomeration of information in databases, to the use of diagnostic technologies, to the development of novel genetic pharmaceuticals. In this increasing globalization of biotechnology, we witness the exchange of not only information, but, specifically, many types of genetic information. The "global genome" is the result of what happens when biotechnology is globalized, when economic exchanges, political exchanges, and semiotic exchanges are coupled with biological exchanges.

But the biological exchanges that characterize the global genome are not simply just another type of information. Modern biological thought always makes two demands of "life itself": that it be essentially information (or pattern) and that it also be essentially matter (or presence). The tensions in this dual demand often lead to apparently contradictory positions within the biotech industry. For example, in the ongoing debates over the patenting of biological life (genes, cell lines, plants, animals), biotech and pharmaceutical corporations have often put forth an awkward position. They make the claim that a given gene or genetically modified organism (GMO) is fully artificial and not something already found in nature. This claim is the foundation for fulfilling the patentability criteria in the United States and the European Union: that patents be "new, useful, and nonobvious." Those against genetic patents argue that by definition "life itself" cannot be subject to patent laws because "life itself" is synonymous with "nature," or something that already preexists human intervention. The claim that a genetic sequence or GMO is artificial underscores the "tech" part of the biotech: it is in some minimal way the result of human intervention, industry, and technology.

But we also see another, contradictory claim from many of the very same companies that are ostensibly in the business of patenting and licensing. This claim is the exact opposite of the first: that these new, useful, and nonobvious inventions are "natural" and thus safe for the environment, for the human body, for agriculture, and for medical application. In other words, when discussing a GMO or genetically modified (GM) foods or a new drug, the claim often made is that the GMO, GM food, or drug does nothing that "nature itself" does not already do. This argument has especially been put forth in debates over the safety of GM foods, but the same rhetoric is in the research, advertising, and promotion of new pharmaceuticals. The logic here is that biotechnology is not the opposite of nature, but rather that which compliments how the environment, nutrition or human health normally operates. In fact, much biotechnology research takes its cue from nature or from the normal functioning of the body. This is the basic approach of fields such as regenerative medicine or gene-based therapies. The "technology" in these practices is seen as nothing more than biology itself, or "life itself."

The very concept of a biotechnology is thus fraught with internal tensions. On the one hand, the products and techniques of biotech are more "tech" than "bio"; biology is harnessed from its "natural" state and utilized in a range of industrial and medical applications. On the other hand, there is no "tech", only "bio"; the unique character of the technology is that it is fully biological, composed of the workings of genes, proteins, cells, and tissues. On the one hand, biotechnology appears not to be a technology at all, but only "life itself" rearranged or recontextualized, but nevertheless performing the same functions it always has. On the other hand, biotechnology appears to be the new nature, the promise of a healthy and optimized body without cyborglike accoutrements of artificial organs, pacemakers, prosthetics, or invasive surgery. The advantage claimed for biotechnology is that it is more natural, a direct working with "life itself." In its ideal guise, biotechnology promises to bypass technology altogether, a biology working upon itself.

On the surface, it appears that the tensions inherent in the concept of biotechnology have to do with the relation between biology and technology, between nature and artifice, and so on. But I suggest something further: the core tension in the concept of biotechnology is not that between biology and technology or that between the natural and the artificial, but rather a tension between biology and political economy. The aim of this book is to present a set of concepts for understanding this twofold tendency within biotechnology—its globalizing tendency and its tendency to integrate biology and informatics. Another aim is to develop a better understanding of the contradictory logic of the biotech industry, its claims to be at once natural and artificial, biological and technological. With regard to the global genome, the central issues are those of biological exchanges, genetic and computer "codes," and the network effects of health and medicine.

A Brief Overview of the Chapters

Again, a point repeated frequently throughout the book: the mere existence of economic models in relation to biotechnology research and application is not necessarily problematic in itself. However, when different value systems conflict—conflicts between medical and economic value—the ethical quandaries and scientific controversies emerge. *The Global Genome* can be considered an attempt to trace and outline some of these points of tension. As previously stated, one of the general goals of the book is to comprehend the twofold tendency that currently characterizes the biotech industry: its globalization and its integration of biology and informatics. Thus, each chapter can be seen as a more focused exploration of some of the ontological, political, and economic facets of biotechnology. In fact, a better, more methodological description of this book might be a "political economy of molecular biology." Today, the term *political economy* has largely been replaced by the more familiar term *economics*, but this change in a way provides the occasion to revitalize the philosophical and political dimensions of the former term.

A glance at the table of contents reveals a larger organizing principle of this book as well. The three sections—encoding, recoding, and decoding—are meant to cover the three primary activities of biotechnology today: the encoding of biological materials into digital form, the recoding of that digital form in various ways, and the decoding of that digital form (back) into biological materiality. Together, these three activities form an incomplete circle—or, rather, a spiral. The biology produced is not necessarily the biology with which one begins. Whereas in more straightforward techniques such as DNA synthesis, the input and output do tend to be identical, in other instances, such as the lab-based regeneration of skin or the engineering of a patient's stem cells, the result may not be entirely identical, though it may be identical enough to ensure biocompatibility. In still other instances, such as genetic drugs or gene therapies, the output can radically diverge from the input, and compounds taken in can often have drastic side effects. Thus, the overall tripartite organization of encoding, recoding, and decoding is meant to underscore the different ways in which biotechnology mediates between the biological and the informatic.

This same tripartite division is also a political-economic one as well. In a sense, *encoding* is synonymous with *production*, for it is in the process of encoding the biological that the biotech industry is able to accrue profits (as intellectual property, as a proprietary database or software). *Recoding* is then synonymous with *distribution* (and its related term *circulation*), for the practices of bioinformatics, database management, and computer networking are predicated on the ability of biological information to be widely distributed and circulated. Finally, *decoding* is synonymous with *consumption* in that, in a medical sense at least, it is in the final output or rematerialization of biology that biological information is used, consumed, or incorporated into the body.

However, this separation of production, distribution, and consumption is only heuristic. As Marx notes in the *Grundrisse*, "production, then, is also immediately consumption, consumption is also immediately production," and furthermore, "distribution is not structured and determined by production, but rather the opposite, production by distribution."[2] There are many ways in which production is consumption (in the use of raw materials), in which consumption is production (in the creation of need), and in which distribution conditions both production and consumption (exchange dictates what is made and used). This confusion of categories is further illustrated in the way in which the practices of biotechnology incorporate information technologies at the same time that they insist on the referent of biological materiality.

Given this political-economic organization, each of the chapters within these three sections takes up a particular field or set of practices within the biotech industry. Chapters 2 and 3 consider the fields of bioinformatics and genomics, respectively. In particular, chapter 2 focuses on the way that bioinformatics—as a practice and as an industry—redefines the problem of "life itself" that has been at the heart of biological thinking for some time. Georges Canguilhem's work in the history of biology is useful in this regard, for it connects the modern, genetic view of the body with a rich tradition that extends from Aristotle through molecular biology. This problem of "life itself" is also understood as inseparable from the question of economic value, and it is in chapter 3 that the notion of "intellectual property" is explored in greater depth. This topic links chapter 2 to chapter 3 by considering genomics from a political-economic perspective, but one based on "excess" as elaborated by Georges Bataille. Placing the economic critique of Marx next to Bataille allows us to see the way in which a field such as genomics manages all that "junk DNA" with databases and Gene Expression Markup Language (GEML).

The section on recoding contains three chapters, each of which takes up a field that depends on or makes use of the results of bioinformatics and genomics (chapters 2 and 3). Chapter 4 considers the recent efforts to map the genomes of genetically isolated ethnic populations, what some have critically referred to as *biocolonialism*. Michel Foucault's work on the historical emergence of biopolitics is helpful here in talking about how biocolonialism makes the transition from "territory" to the "population." Chapter 5 considers the financial strata of the biotech and pharmaceutical industries—not from a purely economic standpoint, but rather as a layer connected to scientific and technical advance. The concept of biomaterial labor makes a reappearance here, in dialogue with the work of Marx as well as with the Italian autonomist tradition. Chapter 6 takes a look at the areas of biological warfare and bioterrorism, areas that are at once medical and nonmedical in their application. Foucault's Collège de France lectures, which position biopolitics in relation to the "war between races," and Paul Virilio's notion of the "genome bomb" (which I argue against) help to situate biowarfare within the current concerns over the weaponizing of biology. At the core of these topics is a pervasive concern over what I call "biological security."

The final section, on decoding, contains two chapters that deal with the overlapping fields of tissue engineering and regenerative medicine. Chapter 7 revisits Canguilhem's theories of normativity as a way of elaborating the philosophical and ethical challenges brought about by tissue engineering. The confusion of the boundaries between biology and technology in tissue engineering leads to a strange type of body that is at once restored and "natural," yet augmented, improved, and technically "optimized." Similarly, regenerative medicine research (including telomere research and stem cell research), discussed in chapter 8, promises a great deal, but is ultimately compromised by the question of normativity and the Aristotelian theories of morphogenesis. Here Virilio's philosophy of speed helps to highlight the aporias in the "biotechnical time" of the regenerative body, a particular kind of time that is, as Antonio Negri points out, also a value time.

Finally, the book closes with a chapter on possible models through which critical questions can be raised and sustained. A consideration of the discourse of "tactical media" and "post-media" is juxtaposed to selected groups and individuals who practice what some have called "bioart," or art projects that make use of biological techniques, knowledges, and practices. Whether such bioart practices can offer a viable, alternative venue for questioning biotechnology is left an open issue for the reader.

A closing note concerning what is not in this book. For various reasons, I have not included a number of equally important fields here, either because they are outside the scope of this book or because they have been dealt with thoroughly elsewhere. For instance, I do not consider agricultural biotechnology or the industrial use of GMOs, largely because the focus of *The Global Genome* is exclusively biomedical and genetic research in the biotech industry. More popular topics, such as human cloning or gene patents, have been dealt with elsewhere in more detail. The relationship between biotechnology and popular culture is also a fascinating one that deserves more attention in a political context, and appendix D briefly offers some informal thoughts on this issue and several references.

I

Encoding / Production

If there is no production in general, then there is also no general production . . . production is also not only a particular production. Rather, it is always a certain social body.

—KARL MARX, *GRUNDRISSE*

1

The Global Genome

Biotechnology and Globalization

To speak of "biological exchanges" raises all sorts of odd images. In an era where AIDS and bioterrorism dominate the headlines, most of us may tend to avoid biological exchanges in favor of various defensive or prophylactic practices. Yet biological exchanges take place all the time within the health-care and biotechnology sectors. Various "banks" that house biological materials (blood, sperm, ova) regularly participate in the circulation of bodily fluids. Likewise, tissues and organs are regularly distributed (legally or illegally) for the purposes of transplantation. On a more quotidian level, a visit to the doctor often involves any number of biological exchanges, from blood or urine samples to the biological data extracted and abstracted in X rays or blood pressure readings, all of which make their way into medical files on a computer.

The dual biological and economic meanings of the term *biological exchange* reflect the dual nature of such practices. Biological materials literally move from one body to another via a set of techniques and technologies (transfusion, insemination, transplantation). In this sense, they form a kind of network wherein biological materials flow between nodes that may be individual bodies or containment systems ("banks"). But such a network is not purely biological, for it is aided by technical, medical, and legal systems that mediate the bodies and the biological materials. Hospitals, medical technologies, and the particular laws surrounding the acquisition and donation of materials participate in this biological network. This means, among other things, that a set of protocols governs this network, protocols that ensure that the exchanges benefit both individual nodes (patients, hospitals, insurance companies) and the network as a whole.

The protocols governing these biological networks are often economic and juridical in nature, and, indeed, they are often in conflict when it comes to the controversial issues of property and consent. The controversies surrounding the patenting of cell lines and genes, genetic screening in health insurance, preimplantation diagnostics in in vitro fertilization, the marketing of bioartificial tissues and organs, and a host of other topics raise fundamental questions: Does an individual "own" his or her own body, and if so, does this ownership constitute his or her biology as "property"? Under what conditions are individuals authorized to sell their organs, cells, or DNA? Under what conditions are companies or government agencies authorized to collect biological materials? Is the collected biological "information," such as a DNA sequence, different from "the thing itself"? Can such information be circulated within an economy, totally separate from the thing itself? Such questions lie at the heart of our contemporary debates over the circulation and exchange of biological materials.

However, these questions still only begin to address the complexities of biological exchange. Certainly, at some level, biological exchanges have existed for some time, from the circulation of cadavers in Renaissance anatomy theaters to the collection of exotic artifacts in natural history museums, to the trade in slaves during the long history of European colonial expansion. Indeed, in this more general sense, the very development of agriculture, animal breeding, and fermentation can be seen as a kind of biological exchange. If this is the case, then biotechnology is as old as civilization itself. To gain an understanding of our current condition we need a way of specifying this term *biological exchange* beyond its more general connotations.

One way of doing this is to consider biological exchange within the context of globalization. Although this is not the place to summarize recent theories of globalization, we can take a few definitions of the term as a kind of shorthand to help us situate biotechnology in relation to globalization. For instance, consider Anthony Giddens's definition: "Globalisation can . . . be defined as the intensification of world-wide social relations which link distant localities in such a way that local happenings are shaped by events occurring many miles away and vice versa."[1] This definition can be contrasted to the description given by Saskia Sassien, who in her analysis of "global cities," immigration, and labor emphasizes the relevance of place in globalization: "The global economy materializes in a worldwide grid of strategic places, from export-processing zones to major international business and financial centers. We can think of this global grid as constituting a new economic geography of centrality, one that cuts across national boundaries and across the old North-South divide."[2] The emphasis on the local-global relationship is further extended by Malcolm Waters, for whom globalization is "a social process in which the constraints of geography on economic, political, social and cultural arrangements recede, in which people become increasingly aware that they are receding, and in which people act accordingly."[3]

We should also note that theories of globalization often tend to emphasize one or more factors in the globalizing process, be it economics, technology, culture, or politics. For instance, Immanuel Wallerstein's multivolume study of "world systems" places an emphasis on the historical development of capitalism and its subsequent modifications. For Wallerstein, one of the defining characteristics of modern world systems is not their exhaustiveness, but rather the extent to which they supercede traditional nation-states. What he calls the "European world-economy" of the fifteenth and sixteenth centuries is "a 'world' system, not because it encompasses the whole world, but because it is larger than any juridically defined political unit. And it is a 'world-economy' because the basic linkage between the parts of the system is economic."[4]

By contrast, sociologist Manuel Castells emphasizes the ways in which information technologies have transformed economics, culture, and political relations. Castells's model is the "network," a structure of interconnections that stands in direct contrast to the monadic, semiautonomous subject. At the root of any network—be it financial, political, or consumer based—is an infrastructure for the "production of information." As Castells notes, "[t]he emergence of a new technological paradigm organized around new, more powerful, and more flexible information technologies makes it possible for information itself to become the product of the production process."[5] Such observations are also echoed by Michael Hardt and Antonio Negri, who define "immaterial labor" as the labor of informational problem solving (calculation), symbol analysis (communications), and the production of affects (entertainment).[6]

Theorists such as Giddens, although not ignoring the economic and technological aspects of globalization, also raise a set of issues that are primarily social and cultural. For Giddens, there is an equal focus on the experiential dimensions of the economic and technological aspects of globalization. In particular, he identifies three main affects that are part of the globalizing or modernizing process. The first is a "universalizing tendency," which is really the simultaneous fragmentation and coordination of world cultures. The second is time-space distanciation, the process of coordinating human activities across temporal and spatial locales, across time zones and national boundaries. The third is the process of "disembedding," which is "the 'lifting out' of social relations from local contexts of interaction and their restructuring across time and space."[7]

Waters gives what is perhaps the most helpful overview of the various theories of globalization and their commonalities. He suggests a tripartite analysis of globalization based on the nature of the exchanges that take place. He identifies three main types of exchanges that are constitutive of globalization: *economic exchanges*, *political exchanges*, and *cultural exchanges*. Economics, politics, and culture can indeed be found in many theories of globalization, though their precise relations differ from one theory to the next. Each type of exchange involves a certain unit: economic value, political power, and cultural signs. Furthermore, Waters notes, each type of exchange has a different effect on the globalizing process as a whole: economic exchanges "localize" (purchasing-selling activities), political exchanges "internationalize" (international organizations such as the United Nations), and cultural exchanges "globalize" (circulation of mass media images).[8]

Although definitions of what is or is not an instance of globalization vary, one trend worth pointing to is the increasing emphasis on information technologies and information networks as the paradigm for globalizing tendencies.[9] Nearly all accounts of globalization do attest to the advantages of information technologies within a globalizing context. As the logic goes, when everything, including production and services, can be accounted for by information or reconstituted as information, the resulting fluidity and mobility leads to a condition in which the obstacles of place, nation, and geopolitical distance are in effect transcended.[10]

Theories of globalization are helpful in understanding our contemporary context, but they often leave out an additional type of exchange: *biological exchanges*, or the exchange, circulation, and distribution of biological information and materials. Can we add biological exchange to the theories presented by Waters, Castells, Giddens, and others? Within Waters's model, biological exchange would, at first, seem to fit quite well. If economic exchanges have to do with value, political exchanges with power, and cultural exchanges with signs, then biological exchanges would be the exchanges, simply, of biological "life itself." If economic exchanges localize, political exchanges internationalize, and cultural exchanges globalize, then biological exchanges would rematerialize, via the concept of a genetic "code." However, the matter is more complicated than that, for there are a number of apparent contradictions in biotechnology and the notion of biological exchange. For instance, even though biotech is dominated by the concept of a genetic code, biotech by definition deals with the realm of biology—the material, the "wet," messy stuff of biological science.

What, then, can we make of biological exchanges in the context of globalization? It would seem that biological exchanges have very little to do with the three types of global exchanges outlined earlier (economic exchanges, cultural exchanges, political exchanges). The fluidity, mutability, and mobility of values, signs, and power seem to be in stark contrast with the inert, messy "stuff" of biology. It would seem that immateriality is in some way a prerequisite for the mobility of globalizing exchanges. Indeed, it is difficult to talk about networks at all without immediately bringing up information networks, be they LANs, PANs, or WANs. And when we further consider the emergence of the network as the dominant paradigm of globalization, biology seems to be at odds with the light-speed activity of the Internet or wireless networks. Biology seems all but left behind in the race toward digital, wireless, mobile networks.

What Is Biological Exchange?

However, it is precisely in the network—the structure that would seem to exclude biology categorically—that we discover several important characteristics of biological exchange as it stands today. Within the context of globalization, biological exchange can be defined as *the circulation and distribution of biological information, be it in a material or immaterial instantiation, that is mediated by one or more value systems*. Biological exchange is the ability to render the biological not only as information, but as mobile, distributive, networked information. The criteria that dictate which kinds of information are exchanged, how they are exchanged, and between which parties have traditionally been defined by standards that are at once scientific, economic, and juridical. For instance, GenBank, one of the main data repositories for information pertaining to the Human Genome Project (HGP), is supported by federal funds, housed within the U.S. National Institutes of Health (NIH), is managed by the National Center for Biotechnology Information (NCBI, a subunit of the NIH), and offers free access to anyone who wishes to view the database online via its Web site. Other genome databases, such as those managed by Celera Genomics, Structural Genomix, or Incyte Genomics, are proprietary databases and offer access based on a substantial subscription (mostly from institutions or corporations). Each of these databases has particular guidelines surrounding the patentability of novel compounds or processes discovered using their systems. In addition, the organizations housing and managing these databases—public and private sector alike—form alliances and partnerships with pharmaceutical corporations for drug development. The convergence of information technology with biotechnology has spawned a new generation of biotech businesses not only in the United States and Europe, but also in Japan, Latin America, and India.[11] Finally, databases such as GenBank are open-ended databases; they are built and designed to be constantly updated and modified and to accommodate many simultaneous database searches. They are databases built for exchange.

Already we can see several characteristics that distinguish biological exchange in its modern context from earlier examples such as agriculture or animal breeding. First and foremost among these characteristics is the fact that what is exchanged in biological exchange is *information*. That is, the "thing" that is the unit of exchange is biological information. Now, this biological information need not be exclusively in a digital or immaterial format; it can also exist in very material ways. Take, for example, DNA. DNA exists, certainly, in the nucleus of every cell in our living bodies. DNA can be extracted from our cells and kept in a test tube for diagnostics and analysis. DNA in this in vitro state can also be transferred from one cell to another or from one organism to another. This is the principle behind basic genetic engineering. DNA can also be encoded or digitized and stored as a sequence on a computer database. This process is, ostensibly, what the various genome sequencing efforts are all about. DNA can also be synthesized, and, in fact, it can be synthesized using the prior, digital form of DNA in databases. Finally, DNA can be materialized (or rematerialized) in the form of genetic therapies, genetic drugs, or GMOs. In such procedures, we move across many boundaries, and the thread that connects them all is this notion of biological information. Whether it is in a living cell, a database, or a test tube, what remains constant throughout (or so it is said) is biological information.

But biological exchange is more than just information. In other words, the aim of biological exchange is not to render everything digital and immaterial, despite the industry hype over fields such as bioinformatics and genomics. Rather, the aim of biological exchange is to enable a more labile, fluid *mobility across media*—to the extent that it is literally immaterial whether the DNA is in a database or in a test tube. This point cannot be stressed enough. The aim of biological exchange—and by extension the aim of the current intersection between biology and computers in genetics and biotechnology—is to define biology as information while at the same time asserting the materiality of biology. Biological fields such as genetics and biotech are unique in this respect, for, like the economic categories of production and labor, they always require material interactions at some level. Biology can never totally relinquish its reliance on the concepts of biological matter, for, were it to do so, it would be indistinguishable from computer science fields such as a-life or artificial intelligence. Even when biological theories want to make claim to a transcendent quality (as in theories of vitalism or animism), there is still a minimal acceptance of the "essential" material condition of biology. Indeed, our very cultural associations with biology evoke its material, even visceral basis: dissections, lab rats, microbes, microscopes, digestion, disease, reproduction, and decay. Thus, biological exchange is not simply the "digitization" of biology, for biology has arguably always been enamored of the "stuff" of biological life.

What we are witnessing in biological exchange is an emphasis on *the network properties of biology* through this formulation of biology equaling information. If biology is considered to be an abstract pattern or form—not just a formation but an in-formation—then the material substrates through which this information is distributed are of secondary concern, and that which is seen to be at the core of biology (pattern, information, sequence) can be seen to exist apart from the material substrate (cell, test tube, computer). The consequences of this view, which has had a long and diverse history, are manifold. On a scientific level, the development of biological exchange has made research more efficient and has opened more opportunities for the distribution and sharing of knowledge among researchers—though disputes over patents, proprietary "discovery systems," and privacy have often impeded as much as facilitated this potential. On a medical level, biological exchange has enabled more rapid and accurate diagnostics and has further formalized the relationship between medicine and drug treatments in many First World health-care systems—though public concerns over genetic discrimination, testing, and health insurance continue to be major obstacles to a fully realized genetic medicine. On the economic level, biological exchange has had perhaps the greatest impact, allowing the biotech industry to grow and diversify, especially in the pharmaceutical and agricultural sectors, though the vast inequalities between developed and developing nations limits the promised benefits of such applications. In each of these cases, a single DNA sequence may be what is exchanged, but, depending on the context, that DNA sequence can be seen as a disease-related gene, a symptom for the prescription of a certain drug, or a patent for a novel discovery. In each case, we see a DNA molecule, the very stuff of life; but in each case we also see how biological exchange transforms DNA into much more than just the stuff of life.

Thus, the bases for our modern version of biological exchange are that (1) biology be understood as information and that (2) information, in this case, be understood as being both material and immaterial, with (3) the aim of networking that information. What are the results of establishing these three foundations? One result is that biological exchange creates the conditions for networks of biological exchange that operate irrespective of the traditional boundaries between the immaterial and the material, the natural and the artificial, the medium and the message. Biological exchange, in conceiving of biology as information that exists—and persists—across media, radically widens the possibility of what can be exchanged within the biological domain. Not only is the biological commensurate with the economic (e.g., microbes, cells, or DNA that is patented or purchased for research), but the biological can be internally exchanged in ways that are not limited by the division between the material and the immaterial. A DNA sequence, as previously stated, can be extracted, used for wet-lab analysis, encoded into a computer database, and then synthesized for further testing (see figure 1.1). DNA's mobility across the in vivo, in vitro, and in silico contexts enables exchanges across media, a condition I have previously called *biomedia*.[12] Such a mobility to biological exchange enables a DNA sequence downloaded from GenBank's database to result in a synthesized DNA strand, which may then be inserted into a bacterial plasmid for culturing. Biological molecules can be encoded into strings of data in a computer file (in the case of genomics); digital computer files can be exchanged for biological molecules (in the case of DNA synthesis); biological molecules can be rendered as intellectual property (in the

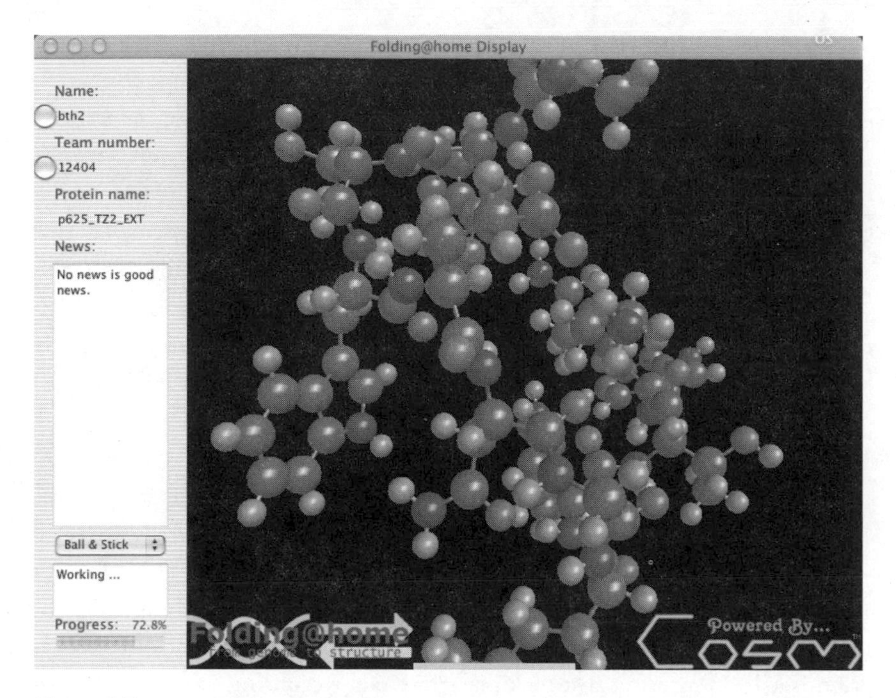

Figure 1.1 Computer networks and biological networks: the Folding@home project at Stanford University makes use of distributed computing over the Internet to analyze the complex phenomena of protein folding.

case of gene patents); and intellectual property can be transformed over time into genetic medicines tested in human clinical trials.

Thus, in contemporary molecular genetics and biotechnology, biology is information, and information is both immaterial and material. Furthermore, this biological information is understood to be operative only within a network. Yet much of this seems contradictory: material information, immaterial biology, and networks that seem to be indissociable from biology itself. I can thus state a variant of the definition given previously: biological exchanges conceive of "life itself" as informatic, and in doing so *biological exchanges informationalize without dematerializing*.

Biological Exchange in the Human Genome Project

These characteristics of biological exchange can be further illustrated by an example. Consider the international effort to sequence and map the human genome, which currently goes under the cumbersome acronym IHGSC. Histories of the "origin" of the HGP are many; and, as may be guessed, there is a great deal of controversy about the details—both past and future. As Michael Fortun notes, the HGP has been as concerned with its "promissory" aspects as with its historical origins. Fortun frames the HGP's convoluted history as taking place in a "future anterior," a promise of DNA sequencing as that which will have been the foundation for a future scientific revolution. In this sense, the argument that the HGP will transform "all of biology" (to use biologist Walter Gilbert's term) can be proved only by actually doing the work. As Fortun notes, "one does not know what 'all of biology' is, yet one presumes, in advance, that DNA sequences are 'fundamental' to it—or rather, that DNA sequences will have been fundamental to it."[13]

For my purposes here, I will not reiterate the many-sided history of the HGP. Rather, I will pick up on some of the more disputed and controversial points, especially concerning the institutional backdrop of the HGP, for this backdrop reveals that, already, in the "beginning," there was an implied focus on the integration of informatics with biology. One such point of contestation is a 1984 meeting that took place in Utah. According to some accounts, scientists such as Charles DiLisi were gathered to discuss the future prospects of DNA sequence analysis. Funded by the U.S. Department of Energy (DoE), the meeting turned on issues pertaining to new techniques for the analysis of genetic mutations resulting from radiation exposure. The DoE had already been studying such phenomena, much of the data from Hiroshima and Nagasaki survivors. In the mid-1980s, research scientists at the DoE began linking their research on the biological effects of radiation to the then-emerging techniques of genetic sequencing.[14] But, as Fortun notes, the DoE was less interested in national security than in the rising economic competition from Japan in the microelectronics and computing industries. However, a transition can be detected between this meeting and another one a year later. In 1985, Robert Sinsheimer organized an initial meeting on the sequencing of the human genome at the University of California, Santa Cruz. The discussion at this meeting revolved around the possibility of sequencing the DNA of an entire genome. This meeting and subsequent meetings would include prominent figures in the scientific community, such as James Watson, Paul Berg, Walter Gilbert, and Leroy Hood (Hood is an important figure in this context because he is one of the developers of the automated gene sequencing machines now commonly used in genomics labs). In addition, numerous accounts of the history of the HGP note that the issue of economics was also raised at subsequent meetings, particularly on the issues of patenting and possible public-private affiliations.[15] Thus, if the 1984 meeting with DeLisi was framed by the initiatives of Cold War science, the 1985 meeting organized by Sinsheimer was framed by the emergence of new, computer-based technologies and the possibility of an industry specific to what would later be called genomics.

The last point is significant because, from the beginning, the idea of mapping the human genome was indissociable from the development of new technologies for doing so. From the beginning, the DoE developed scientific and technological plans for its proposed mapping efforts, and informatics was a core part of this planning. Recall that this is 1986, some five years following the introduction of the first IBM personal computers (PCs) and just two years following the popularity of Apple's Macintosh. Computing was not only becoming less expensive, but also becoming more multipurpose. In addition, the Internet was growing and diversifying. What had begun as a military research project in the late 1960s had by the mid-1980s become a largely university-based, American project. The network infrastructural basics had been in place by the mid-1970s (TCP and IP protocols), and the 1980s saw the emergence of USENET, the WELL, the Domain Name System (DNS), and of course a host of network-based computer viruses. These parallel developments in the computer and information technology industries had a significant impact on the original Human Genome Initiative. Without the promise of a future integration of informatics and biology, it is likely that the idea of genome mapping would have been much slower to start. By 1987, the Human Genome Initiative had obtained its first budgets from the U.S. House and Senate Appropriations Committee, and the project now became a joint DoE and NIH endeavor. The HGP—the initiative's new name—was officially launched in 1990. It included a multi-million-dollar budget, a host of labs throughout the United States, and other special provisions, such as a working group on ethical, legal, and social issues. In 1993, Watson was replaced by Francis Collins as head of the HGP, and by the late 1990s the project was internationalized in its scope, including major sequencing labs in the United Kingdom, France, Germany, Japan, and China. At this point, the HGP became the IHGSC (see table 1.1).

At the same time that the original ideas for the HGP were being organized, J. Craig Venter, an NIH research scientist, became interested in the

Table 1.1 The "Global" Organization of the IHGSC

Institute	Location	Total human DNA (in Kilobases)
Whitehead Institute for Biomedical Research, Center for Genome Research	Cambridge, Mass.	1,196,888 kb
The Sanger Centre	Cambridge, United Kingdom	970,789 kb
Washington University Genome Sequencing Center	St. Louis, Mo.	765,898 kb
U.S. DoE Joint Genome Institute	Walnut Creek, Calif.	377,998 kb
Baylor College of Medicine Human Genome Sequencing Center	Houston, Tex.	345,125 kb
RIKEN Genomic Sequencing Center	Yokohama, Japan	203,166 kb
Genoscope	Evry Cedex, France	85,995 kb
GTC Sequencing Center	Waltham, Mass.	71,357 kb
Department of Genome Analysis, Institute of Molecular Biotechnology	Jena, Germany	49,865 kb
Beijing Genomics Institute and Human Genome Center	Beijing	42,865 kb
Multimegabase Sequencing Center, Institute for Systems Biology	Seattle, Wash.	31,241 kb
Stanford Genome Technology Center	Palo Alto, Calif.	29,728 kb
Stanford Human Genome Center	Stanford, Calif.	28,162 kb
University of Washington Genome Center	Seattle, Wash.	24,115 kb
Department of Molecular Biology, Keio University School of Medicine	Tokyo	17,364 kb
University of Texas Southwestern Medical Center	Dallas	11,670 kb
University of Oklahoma Advanced Center for Genome Technology	Norman, Okla.	10,071 kb
Max Planck Institute for Molecular Genetics	Berlin	7,650 kb
Cold Spring Harbor Laboratory	Cold Spring Harbor, N.Y.	4,338 kb
German Research Center for Biotechnology	Braunschweig, Germany	4,639 kb
Genoma Analysis Group/other	various	59,574 kb

Source: *Nature* 409 (15 February 2001): 860–921.

technology of automated genome sequencing. In the mid-1980s, he met with Michael Hunkapiler at Applied Biosystems, a company formed by Leroy Hood for the manufacture of an automated genome sequencing machine. Venter obtained a prototype sequencer, and in 1991 he and his colleagues filed a patent claim for approximately 2,000 human gene fragments (which were later refuted by the U.S. Patent and Trademark Office [PTO]). The patent claim set off a storm of controversy and was quickly derided by Watson, then the head of the HGP. Venter, after leaving the NIH, went on to start an independent genomics company—the Institute for Genomics Research (TIGR)—where he codeveloped the "shotgun sequencing" technique now used by many genomics labs. In 1998, he formed Celera Genomics, and announced its intentions to sequence the human genome in less time and for less money than the federally funded HGP. The rest of this story is by now familiar, with the tensions and competition between the public- and private-sector projects. The competition resulted in a 2000 press campaign that announced, with great fanfare, the completion of a "rough draft" of the genome and that featured front-page images of President Clinton flanked by Collins and Venter, shaking hands.

However, such media images do not mean that the two projects—public and private, government and corporate—have become one, unified project. Instructive differences can be noted between them on an institutional level. Whereas the HGP emerged out of concerns of military science, Celera emerged out of an interest in increasing the market for biotechnology, independent of government regulation. However, both projects are united in their commitment to developments in high technology: informatics, sequencing, and data management systems. In fact, Perkin-Elmer, a major manufacturer of genome sequencing machines, supplied the tools needed for both projects. Indeed, informatics has become one of the central meeting points in the varied tensions between the public and private sectors. The government-funded HGP set up the Genome Data Base in 1991 and a year later established data access and sharing guidelines for its labs. In the mid-1990s, the Genome Data Base evolved into GenBank (managed by the NCBI at the NIH). A standard for molecular biology and genetics research, GenBank is accessed by more than 40,000 users daily. GenBank was also used by both the IHGSC and Celera in the error-detection phases of their sequencing efforts. Last, changes in government policy concerning patenting have further blurred the division between the public and private sector. Although the U.S. PTO guidelines prohibit the patenting of any "raw" or unaltered genetic sequence, a host of patents have been granted for gene sequences derived from or isolated by novel means, in effect a "back door" for patenting gene sequences. Other policies (such as the Bayh-Dole Act of 1980) sought to encourage the economic benefits of genome research by allowing labs to apply for patents based on research carried out with federal funds.

This brief account of the institutional history of the human genome reveals the ambiguous relationship between informatics and economics. In a sense, informatics is seen as that which enables economic development for genomics. The advent of new sequencing technology, combined with computer databases and networks, has ushered in an era of what Francis Collins, Michael Morgan, and Ari Patrinos have called "large-scale biology."[16] It is precisely the two developments I noted at the outset—globalizing production and integrating informatics and biology—that Collins Morgan, and Patrinos point to as the key factors in the field of genomics. It is difficult to imagine a genome project without the use of computers, databases, and networks (see figure 1.2).

Figure 1.2 Computer lab or biology lab? The Whitehead Genome Sequencing Center in Cambridge, Mass. Image courtesy of the National Human Genome Research Institute.

However, the foundation of informatics also creates new layers of economic valuation (in part, it is said, to cover the high costs of informatics and information technology). The conversion of DNA into "intellectual property" is unthinkable without an informatic infrastructure that organizes DNA into data strings, computer files, and information visualizations. It is almost tempting to see this conversion as a broader redefinition of DNA and biology as a concept, a concept that is also property (a phenomenon addressed in chapter 2).

But genomics is not only about transforming all of biology into informatics, for the assumptions of materiality that still inform biological science have their part to play as well. The IHGSC's efforts are instructive in this regard. Not only does the IHGSC distribute its production centers across a number of countries, but it also distributes biological information in two types of "banks" or databases. The first type, as stated, is the computer database. There are currently some 200 such databases online, with more than 500 commercial software packages for sale and a host of noncommercial freeware or open source projects as well.[17] Aside from genomics databases, there are databases for comparative genomics (genomes of different organisms), disease genome databases, and databases dedicated to specific DNA fragments such as single nucleotide polymorphisms (SNPs) or expressed sequence tags (ESTs). In addition to this more familiar informatic type of database, however, there are also a host of "wet" biological banks or "libraries," which have been the conventional way of archiving genetic material before the widespread introduction of computers into the lab. An example is the bacterial artificial chromosome (BAC). As its name indicates, the BAC involves using simple bacteria (which have a small, circular DNA or "plasmid" as their genome) as the host for a gene sequence from a human sample. Using the cut-and-splice techniques of genetic engineering, scientists can insert the human gene into the bacterial plasmid, thereby creating an in vitro database. As the bacteria replicates, so will the inserted human DNA, making for a kind of biological copy machine. Though more cumbersome than the manipulation of computer files, BACs have been extensively used by the IHGSC labs in tandem with the computational methods of sequencing and alignment.

The two types of databases employed by genomics—computer databases (UNIX) and biological databases (BACs)—demonstrate the extent to which biotechnology privileges biological information across material substrates. Furthermore, inasmuch as a database is not a database unless it is networked or accessed, these two types of databases form a complex, global genomics network that spans communications as well as geographical relations. In this way, the human genome is global in a technological sense: it is an online database, accessible all over the world, a database to which researchers in labs all across the world contribute and from which they retrieve data. The human genome is global in a scientific sense: the ease of access and distribution of data also means that scientific knowledge can be shared worldwide. The human genome is also global in an economic sense: data in the genome database are directly connected to data in the various U.S. and EU patent databases (a connection that, in turn, restricts the sharing of information). In addition, many pharmaceutical companies that do drug development based on genome data are "global" companies in that their administrative, production, distribution, sales, and research-and-development (R&D) departments may be spread across different regions of the globe. The human genome is global in a political sense as well: the IHGSC is composed of labs, institutes, and universities from the United States, Europe (United Kingdom, France, Germany), and Japan—a kind of United Nations of genetics. Finally, the human genome is global in a cultural sense: the pervasiveness of genetics themes in popular culture and the mass media in different cultures attests to the increasing popularity of the "cultural icons" of cloning, genetic engineering, greedy corporations, and "mad scientists."

We should also note a number of inconsistencies in this view of genomics as a "global" activity, however. For instance, the IHGSC is a selectively global endeavor; that is, only developed nations with the technological and economic infrastructure to support bioscience research are included in its membership. This exclusivity has meant, among other things, that the IHGSC has set both technical and political criteria for a globally accepted, modern science. In addition, despite the press that the IHGSC is a "public" endeavor (as opposed to the privatized genome project by Celera), the term *public* in this context means something very specific: that the IHGSC centers in the United States (for example) are federally funded and that the results of the genome mapping research are made available to the public via the Internet. Although these are positive steps toward making biological research and knowledge "open source," a number of questions follow: for example, has this public project generated any patents or public-private sector alliances that might inhibit the accessibility of information? At the very least, we can note that *public* in such "Big Science" projects does not necessarily mean free of economic interest.

Again, it is not the very fact of economics that is problematic in itself (for funding is always a reality), but whether certain kinds of economic interests may inhibit rather than facilitate scientific innovation. Finally, a point about tools: although much was made about the division between the IHGSC and Celera as a split over public and privatized values, a third group joined the two sides, Perkin-Elmer, the manufacturer of technologies for life science research, such as automated genome sequencing computers. If the mapping of the human genome has been a technology-driven quest, then it would appear that Perkin-Elmer (and not the IHGSC or Celera) is the real motor behind the genome mapping process. These and other characteristics serve to remind us that genomics is a selectively global industry, creating a specific map determined by Western science, technology, and governmental and economic interests.

If the IHGSC is an example of a globalizing, biological exchange, then we might ask a further question: Has the exchange of biology not been a part of classical biotechnologies as well (e.g., livestock, breeding, agriculture)? Earlier, I pointed out that, in a very general sense, biotechnology can be considered as old as civilization itself if we include such practices. Biological exchanges, in this case, would simply be any appropriation and instrumentalization of nature, a definition that cannot account for the specificities of globalization and the biotech industry. How can we distinguish classical modes of biological exchange from contemporary ones? One of the primary concepts that distinguishes modern genetics and biotechnology from earlier examples is the role that information plays in articulating our knowledge about biological "life itself." Scholars and historians of science have long since noted the close ties between cybernetics and biology in the development of molecular biology during the post–World War II era. Indeed, as early as 1963—tens years after the Watson-Crick publications on the structure of DNA—the philosopher of science Georges Canguilhem was situating the then-emerging field of molecular biology and genetics within the larger framework of the history of biology. As Canguilhem noted in one article, "messages, information, programs, code, instructions, decoding: these are the new concepts of the life sciences."[18] At the same time, molecular biologists such as Jacques Monod and François Jacob were fond of describing protein synthesis and gene expression as if they were computer systems. I explore this transformation of biology into information in chapter 2, but, for the time being, let me note that the influence of cybernetics and information theory in molecular biology has had profound consequences in the development of a biotech industry. Online genome-databases, gene-finding software, and genome-sequencing computers can be seen as the technical outcome of this intersection between biology and informatics. In a historical sense and in a contemporary, technical context, the ongoing motto of molecular genetics and biotechnology has been "biology is information."

However, it is one thing to note how the metaphor of information has been employed within molecular biology, but quite another to suggest that that metaphor is instrumentalized in artifacts such as DNA chips or genome databases. As Lily Kay and Evelyn Fox Keller have noted, when the concepts from cybernetics and information theory found their way into molecular biology, they inadvertently raised fundamental questions about the substance of "life itself."[19] The concept of information is actually a nonconcept, it holds no content, only data. In the late 1940s, the mathematician and engineer Claude Shannon was explicit on this point: information had nothing to do with content or the message. All that was important was that a certain quantity of information sent at point A arrives at point B.[20] By contrast, molecular biologists working during the same period were equally explicit about the content of DNA and proteins. For Watson and Crick, the structure of DNA was intimately tied to its sequence, or its "content." Indeed, by the time molecular biologists of the late 1960s were at work on "cracking the genetic code," all that mattered was the content. By the time the HGP was first proposed by the U.S. DoE and NIH in the late 1980s, the content or sequence was the primary goal, with structure (e.g., proteins) being of secondary concern.

In short, molecular biology has continually dealt with the tensions between two views of biological life: one view is that biology is indissociable from biological materiality, from the very "stuff" of life (molecules, cells, tissues, and so on). Another, opposite view is that biological life is a kind of immaterial pattern or sequence, information that is separate from its material instantiation. These tensions between content and form, quality and quantity, sequence and structure revolve around the basic premise that biology is information. However, the key to understanding the complexities of genetics and biotechnology is in the realization of a paradox at the core of the concept of biological exchange: that biology is information, and, crucially, that information is both material and immaterial. One of the major tensions that resulted from the introduction of information theory and cybernetics into molecular biology is this tension between biology and informatics, the material and immaterial views of biological life.

What do such tensions say about the concept of biological exchange? For one, they point to the *operative* tension within molecular genetics and biotechnology: the domain of the biological is at once the living body, an in vitro cell culture, a sequence in a computer database, a process of genetic recombination or modification, even a file that describes a patent on a DNA sequence. Again, however, the fundamental tension within molecular genetics and biotechnology in no way implies the unfeasibility of these fields and their practices. In a sense, it can be suggested that molecular biology and its related fields are constituted by this tension between the biological and the informatic. Biological exchange, then, can be understood as being based on two seemingly contradictory statements: *that biology is information, and that information is both immaterial and material.* This foundation enables biological exchange to take place on a level akin to the economic, cultural, and political exchanges we see in globalization.

Biopolitics in the Global Genome

If biological exchange is constituted by the contradictory—yet operative—relation between the material and the immaterial, the biological and the informatic, what is needed is a way of accounting for the operative nature of biological exchanges. That is, despite (or, perhaps, because of) these tensions, biological exchanges happen daily, as we saw in the example of the HGP. We can continue in this direction by bringing together two further concepts that will help to emphasize the social, political, and economic aspects of biological exchange, as well as the scientific and technological aspects. These two concepts are Michel Foucault's notion of "biopolitics" and, later, the notion of "species being" formulated by Karl Marx.

First, however, we can begin by discussing the Foucauldian concept of biopolitics. Much of Foucault's interest in the life sciences can be understood as an interest in the manifold ways in which "power takes hold of life." For Foucault, the nature of political power underwent a gradual change during the seventeenth and eighteenth centuries, particularly in the way in which political power related to the life of its subjects through the concept of the "population." Yet Foucault's context for his discussion of politics was not that of contemporary biotech, but rather the eighteenth-century practices of demographics and population control. Thus, our question: How do we understand Foucault's concept of "biopolitics" after Foucault? That is, how has the concept of biopolitics been transformed in the current context of the biotech industry? What does biopolitics mean in an era in which genetic engineering is a routine practice, in which biotechnologies are used in both medical and nonmedical areas (law, engineering, computer science, warfare), in an era in which new, hybrid technologies such as genome databases and genetic drugs regularly appear?

Immediately, I should reiterate the twofold observation with which I started: on the one hand, the globalizing tendencies of the biotech industry and, on the other hand, the increasing integration of biology and informatics, molecular biology and computer science, DNA and data. Yet, we can already see the nascent evidence of this contemporary trend in the way Foucault talks about biopolitics. For instance, in his inaugural lecture at the Collège de France, he outlined as one of his future concerns the production of "the knowledge of heredity," or the role of genetics in the construction of the population.[21] Likewise, one of the defining characteristics of Foucault's concept of biopolitics is the role that the mathematically driven science of statistics plays in defining the population as an entity amenable to political control. Foucault links statistics and its related practices to the emergence of the field of political economy; in such a context, statistics, demographics, and the census become "sciences of the state."

It can be said that Foucault's emphasis on the emergence of the political concept of "the population" stems from these two innovations in the eighteenth and nineteenth centuries: on the one hand, the development of biological fields such as natural history, nosology, evolutionary and contagion theories and, on the other hand, the application of mathematical and statistical techniques toward those living beings, the "living biomass," that constitutes the state. We can, then, take a new look at Foucault's concept of "biopolitics" with the twofold tendencies of biotechnology in mind.

Foucault calls "biopolitics" that mode of organizing, managing, and above all regulating "the population," considered as a biological, species entity. This concept is differentiated, but not opposed to, "anatomo-politics," in which a range of techniques, gestures, habits, and movements—most often situated within social institutions such as the prison, the hospital, the military, the school—collectively act on the individual and individualized body of the subject. In a 1976 lecture at the Collège de France, Foucault summarized the distinction between these two types of biopower:

> Unlike discipline, which is addressed to bodies, the new nondisciplinary power is applied not to man-as-body but to the living man, to man-as-living-being; ultimately, if you like, to man-as-species. . . . And this new technology that is being established is addressed to a multiplicity of men, not to the extent that they are nothing more than their individual bodies, but to the extent that they form, on the contrary, a global mass that is affected by overall processes characteristic of birth, death, production, illness, and so on. So after a first seizure of power over the body in an individualizing mode, we have a second seizure of power that is not individualizing, but, if you like, massifying, that is directed not at man-as-body but at man-as-species. After the anatomo-politics of the human body established in the course of the eighteenth century, we have, at the end of that century, the emergence of something that is no longer an anatomo-politics of the human body, but what I would call a "biopolitics" of the human race.[22]

Foucault mentions the regulation of birth and death rates, disease control and patient monitoring in hospitals, as well as more contemporary examples of consumer data, individual identification forms, health insurance, health data related to sexuality and psychology, and institutional surveillance of subjects. Principle among these and other examples is the accumulation and ordering of different types of information. The fields of statistics and demography are thus key developments in the modern forms of biopolitics. The population is articulated in such a way that its nontotalization may be rerouted via methods or techniques of statistical quantification. This is not only a mode of efficiency, but also a political economy of subjects and bodies. The result—or rather the aim—is the development of "a power that exerts a positive influence on life, that endeavors to administer, optimize, and multiply it, subjecting it to precise controls and comprehensive regulations."[23]

However, what makes Foucault's concept of biopolitics useful for a critical understanding of the biotech industry today is that it highlights two tendencies that are a core part of the practices of biotechnology. First, biopolitics defines an object of governing, which, as the quote earlier states, is put in specifically biological, species terms: this is the modern concept of "population." The concept of the population defines, for Foucault, "a new body, a multiple body, a body with so many heads that, while they might not be infinite in number, cannot necessarily be counted. Biopolitics deals with the population, with the population as political problem, as a problem that is at once scientific and political, as a biological problem and as power's problem. And I think that biopolitics emerges at this time."[24]

Foucault notes a gradual historical shift from the governance of territories to the governance of populations. *Population* does not just mean the masses or groups of people geographically bound (that is, *population* is not the same as *territory*). Rather, the population is a flexible articulation of individualizing and collectivizing tendencies—"the endeavor, begun in the eighteenth century, to rationalize the problems presented to governmental practice by the phenomena characteristic of a group of living human beings constituted as a population: health, sanitation, birthrate, longevity, race."[25] In this sense, eighteenth-century medical interventions—programs such as the "medical police" programs in Prussia, urban hygiene regulations in Paris, and the English Poor Laws—had as their aim the correlation between political-economic health and biological or species health. That is, biopolitics has as one of its principle aims the nuanced coordination between individual and mass bodies in which the political existence of the state is consonant with the "health" of the state.

Along with this, biopolitics defines a means of production of data surrounding its object, the population. This includes asking how data pertinent to the regulation of the health of the population can be extracted from the biological-political population.

> The mechanisms introduced by biopolitics include forecasts, statistical estimates, and overall measures. And their purpose it not to modify any given phenomenon as such, or to modify a given individual insofar as he is an individual, but, essentially, to intervene at the level at which these general phenomena are determined, to intervene at the level of generality. . . . And most important of all, regulatory mechanisms must be established to establish an equilibrium, maintain an average, establish a sort of homeostasis, and compensate for variations within this general population and its aleatory field.[26]

Within biomedicine, the "vital statistics" of reproduction, mortality rates, the spread of diseases, and other factors become means of monitoring the population as a single, dynamic entity.[27] As a defined unit, the population-species can not only be studied and analyzed, but also be extrapolated, its characteristic behaviors projected into plausible futures (growth, development, etc.). The proto–information sciences of demographics and, later, statistics provided a technical ground for a more refined, mathematically based regulation and monitoring of the population. As Ian Hacking notes, "[o]ne can tell the story of biopolitics as the transition from the counting of hearths to the counting of bodies. The subversive effect of this transition was to create new categories into which people had to fall and so to create and to render rigid new conceptualizations of the human being."[28] In a historical analysis of the "statistical enthusiasm" that swept English bureaucratic practices in the nineteenth century, Hacking notes a number of significant developments in Western biopolitics: the immense agglomeration of data concerning sicknesses of all types and kinds, the incorporation of the biological processes of sickness into statistical calculation, and, finally, a number of governmental and bureaucratic offices whose express purpose was to collect and categorize these data, thereby forming a system in which each individual was accounted for via the science of vital statistics. "Disease, madness, and the state of the threatening underworld, les misèrables, created a morbid and fearful fascination for numbers upon which the bureaucracies fed."[29]

It is important to note that biopolitics is not simply a technique employed by governments for homogenizing the mass of its citizens. A key point stressed by Foucault is that the "governmentality" of biopolitical practices is predicated on a dual approach that both universalizes and individualizes the population. Biopolitics accounts for "each and every" element of the population, the individual and the group, and the groups within the group (the poor, the unemployed, the resident alien, the chronically ill). In this gradated approach, populations can exist in a variety of contexts (defined by territory, economic/class groupings, ethnic groupings, gender-based divisions, or social factors)—all within a framework analyzing the fluxes of biological activity characteristic of the population. Such groupings do not make sense, however, without a means of defining the criteria of grouping—not just the individual subject (and in Foucault's terminologies *subject* is also a verb), but a subject that can be defined in a variety of ways, marking out the definitional boundaries of each grouping. As individuated subjects, some may form the homogenous core of a group, others may form its boundaries, its limit cases. The methodology of biopolitics is therefore informatics, but a use of informatics in a way that reconfigures biology as an information resource. In biopolitics, the body is a database, and informatics is the search engine.[30]

In the context of the contemporary biotech industry, what is helpful about Foucault's concept of biopolitics is the way in which it historically identifies two elements central today to the ongoing development of biotechnology: the role of biological knowledge and the techniques of informatic data management and agglomeration. *In the integration of biology and informatics, biopolitical practices establish novel modes of regulation, management, and control.* In fact, the effects of biopolitics—medical normativity, an emphasis on the security of the population, and the development of a political economy—can be seen as issuing from the particular relationship between biology and informatics in the biopolitical viewpoint.

After defining its object (the population) and its method (informatics), biopolitics reformulates power as a positive, generative function and the role of governance in terms of "security." Here it is important to reiterate that, for Foucault, biopolitics is most sharply differentiated from the sovereign model by its generative characteristics. If the sovereign was defined in part by the right to condemn to death, biopolitics is defined by the right to foster life: "One might say that the ancient right to take life or let live was replaced by a power to foster life or disallow it to the point of death."[31] For the classical political economists—Malthus, Smith, Ricardo—the population was an entity at once volatile and yet inscribed by artificial boundaries, by the knowledge of natural history, biology, and, later, evolution. The population was therefore characterized by a number of problems that defined it: it could be unstable, vulnerable to the collective effects of biological processes (disease, infection, sanitation). Not only was the population biological, but its fluctuations affected the body politic—the laboring body, the consuming body, the military body, and so forth.

Thus, the imperative of security in biopolitics is at once biological and political. For Foucault, the more familiar security of territory is further elaborated in a security of the population, "in a word, a matter of taking control of life and the biological processes of man-as-species and of ensuring that they are not disciplined, but regularized."[32] The security of biopolitics thus "has to do with the preservation, upkeep, and conservation of the 'labor force.' No doubt, though, the problem is a wider one. It arguably concerns the economico-political effects of the accumulation of men."[33]

Despite this imperative of security in biopolitics, the relation between life and politics within a technoscientific frame is never obvious or stable. In both the sovereign and liberal-democratic formations, we find what Georgio

Agamben calls a "zone of indistinction" between "bare life" and the qualified life of the citizen.[34] In more contemporary situations, we find what Hardt and Negri refer to as "living immaterial labor" (the "intellectual labor" of the information technology industries, technoscience, and media industries).[35] Biopolitics is therefore never an ideology, but rather a particular problem emerging within protocological relations, about how and whether to distinguish "life" from "politics."

The security of biopolitics is precisely this challenge of managing a network of bodies, data, and their interlinkages—travel advisories, global health alerts, emergency-response protocols, selective quarantines, high-tech diagnostics, and the medical and economic assertion of newer and better prescription drugs. The problem of security for biopolitics is the problem of creating boundaries that are selectively permeable. Whereas certain transactions and transgressions are fostered (trade, commerce, tourism), others are blockaded or diverted (information sharing, a biological commons, migrating). All of these network activities, many of which have become routine in technologically advanced sectors of the world, stem from a foundation that is at once biological and technological.

To summarize briefly, in Foucault's concept of "biopolitics" there is an implicit connection between two modes of articulating biological "life itself." One is an informatic view of the biological domain, its dynamisms, its temporal processes, its manifest characteristics. Foucault locates the emergence of this view in the eighteenth century, with the birth of demographics, political economy, and statistics. This informatic view of biological processes—birth and death rates, public health and disease, the formation of state hospitals—has its roots in a set of techniques for quantifying and thus, in a more refined manner, for articulating "life itself." Along with this, Foucault also points to the shift in modern forms of statecraft, from a concern with territories to a concern over "the population." The emergence of the concept of the population within state-management forces is significant for Foucault, for it defines the health of the population as the foundation of the health of the body politic. This concept of the population is, as Foucault notes, one grounded in the developing sciences of natural history, biology, and, later, evolutionary biology. The population is not only a political matter, but also a biological matter—and today, a genetic matter.

Thus, we have a constellation of a number of issues central to the concept of biopolitics: race-population, biology-genetics, statistics-informatics, variation-polymorphism. One is tempted to see a secret link between the two methodological (and ontological) shifts Foucault notes: an informatic view of life and a genetic approach to life. Despite the uncanny intimacy between these two threads—the informatic and the biological—Foucault never makes an explicit link between them, except by a general inference that both the science of statistics and the science of life (*bio-logy*) constitute a biopolitics, a bioscience of the state.[36] Although these sciences, taken individually, are explicit in biopolitical practices, the precise link between them is elusive, and Foucault rarely addresses their exact relation. One reason for this is that neither perspective, in itself, can fully address the transformations in both the nascent life sciences and the equally nascent approach of informatics, for it is not that a changing informatic approach addresses an unchanging set of concepts surrounding biological life, or vice versa. Both undergo significant changes. And it can be argued that those changes do not happen concurrently, but separately, although in coextension.[37]

To return to my initial question: How has biopolitics changed in the current context of the biotech industry? What is the specific biopolitics of biotechnologies? I can begin answering these questions by noting that the nascent relationship between population and statistics in Foucault's discussion of biopolitics is magnified and developed in the biotech industry, which results in the current integration and management of the relationship between biology and computers, genetics and informatics. Fields as diverse as genomics, patenting, GM foods, and pharmaceuticals share this basic commitment to the relationship between genetics and informatics. *Biopolitics can be understood today as the ongoing regulation of the bioinformatic inclusion of "life itself" into the political domain.* What results is a continuous production and permeation of "life itself" in contexts that are medical, social, economic, and political. Biopolitics mediates between genetics and informatics. The point of mediation is a certain notion of "information," which derives from communications, cybernetics, and information theory. In biopolitics, however, a twist is added: information is not exclusively immaterial or disembodied; information in biopolitics is precisely that which can account for the material and embodied and, furthermore, that which can produce the material, the embodied, the biological, the living—"life itself." Information is the point of mediation through which biopolitics regulates genetics and informatics into a sophisticated mode of governmentality or control.

This notion of "life itself" that is produced by and drives biopolitics is understood to be consonant with an informatic view of the world. Genome databases, biological "libraries" of cell lines, patient databases at hospitals and clinics, prescription databases, insurance databases, online medical services, and a host of other innovations are transforming the understanding of "life itself" into an understanding of informatics. Again, it is important to stress that this transformation does not have a dematerializing, disembodied effect. Quite the contrary. The pills, therapies, test results, diagnostic measures, insurance rates, and foods are the material output of this informatic view. In rarer cases, cell therapies, in vitro fertilization, genetic screening, and tissue engineering are literal instances of this biopolitical condition, in which data is made flesh.

Biopolitics continually regulates the relation between biology and informatics within the context of political and economic concerns. It does this in order to reconfigure the biological domain and "life itself" as open to intervention, control, and governance. At a very specific level, what is produced from this process are unique artifacts such as genome databases, prescription drugs, medical therapies, test results, insurance rates, patents, foods, and networks of uneven, asymmetrical production, distribution, and exchange. At a general level, what is produced from this process is the very same concept of "life itself" from which biopolitics begins. "Life itself" is the product of biopolitics, and there is nothing more *natural* than "life itself." If politics is biopolitics, and if biopolitics is fundamentally productive, then what is produced is a notion of "life itself" that is amenable to governmentality through the lens of informatics. It is in this sense that "life itself" is both material and informatic.

The Species Being in the Global Genome

If we can indeed "upgrade" Foucault's notion of biopolitics to contemporary biotechnology, we are able to begin to understand how the operative tension (of the material and immaterial, the biological and technological) in biotechnology works. We have seen how a biopolitical viewpoint has as its aim the ongoing management and regulation of both the biological and the informatic domains. However, this view leaves the relation between biological "life itself" and economics in the background, or, rather, as a derivation of the aims of "governmentality." Although Foucault was careful to articulate the economic dimensions of biology within the concept of biopolitics, we nevertheless need a way of bringing to the forefront the manifold economic characteristics of the biotech industry. Be it in terms of genetic patents, government-funded research, drug development, or health-care policy, the way that biopolitics manages the relation between "population" (its biological element) and "statistics" (its informatic element) is in many ways mediated by one or more systems of value. Thus, in addition to Foucault's concept of biopolitics, we can add another concept to our arsenal, one that places a greater emphasis on the biology-economics relationship. This concept is what Marx called the "species being" in his early writings.

What Marx brings to our current understanding of the biotech industry is a way of seeing biology, economics, and politics as inseparable. Following Marx's lead, this shift in viewpoint requires not just philosophy, but also a consideration of the "hidden abode of production" in biotechnology, the techniques and technologies that constitute biology as a technology, bioinformatics as in industry, and genetics as an economy. It requires a particular mode of reading. It requires the reader to read someone such as Marx through the lens of biotechnology. Although Marx's earlier works are often criticized for their tendency to idealize the human being's relationship with nature, there is much to be gained by a reconsideration of the early Marx—but through the lens of biotechnology. In particular, it is Marx's concept of the "species being," found in its most elaborated form in the *Economic and Philosophic Manuscripts of 1844*, that brings together biology and economics in a way that helps us understand the approach of the contemporary biotech industry.

Marx's 1844 manuscripts are most often read for the passages concerning alienation and "estranged labor"—the condition in which "the object which labor produces . . . confronts it as *something alien*, as a power *independent* of the producer."[38] In a capitalist system, the human subject, having nothing to sell, must sell his or her capacity for work (what Marx would later call "labor power"). But this then involves transforming the capacity for work into a "thing," an object for purchase and for sale. The result is that neither the products of labor nor the laboring activity itself belongs to the human subject, for they have been sold in return for wages. The process of laboring and the products of labor do not directly benefit the human subject doing the laboring. These products, which enter the market, become so many "alien objects," or "labor which has been congealed in an object."[39] Your actions are not your own, nor are the things that you produce, which are the results of your actions. In Marx's paradoxical formulation, the "*increasing value* of the world of things proceeds in direct proportion to the *devaluation* of the world of men."[40] The more you produce, the less you are. Marx emphasizes this point (in some ways a phenomenological point) by making reference to the relation between labor and "life": "The worker puts his life into the object; but now his life no longer belongs to him but to the object. . . . Whatever the product of his labor is, he is not."[41]

However, Marx's points concerning life, labor, and estrangement make sense only if the act of production or labor is seen to be inimical to being "human." That is, the human subject can become alienated and disenfranchised as a laborer only if there is some foundational link between productive activity and the "life," which is then pulled apart and separated into purchased labor power, and a "life" driven by immediate necessities. This, in turn, requires that we think of production or productive activity as a core part of the human condition. Indeed, Marx stresses this point both in early writings and in the *Grundrisse* and *Capital*, although it takes different shapes and forms in the later works. "[T]he productive life is the life of the species."[42] "The only thing distinct from *objectified* labor, is *non-objectified* labor, labor which is still objectifying itself, *labor* as subjectivity."[43] "The labor process . . . is purposeful activity aimed at the production of use-values. . . . It is the universal condition for the metabolic interaction between man and nature, the everlasting nature-imposed condition of human existence . . . it is common to all forms of society in which human beings live."[44] In general, productive activity—but not necessarily "work"—is, for Marx, the core of what it means to be human, both in the sense that production qualifies the human subject and in the sense that the human subject becomes human through productive activity.[45] But this fundamental link between life and labor also provides the conditions for the estrangement of labor from life, production from the species. In this way, "private property is thus the product, the result, the necessary consequence, of alienated labor, of the external relation of the worker to nature and to himself."[46]

But, aside from the points Marx makes concerning production, what is often left undermentioned is the fact that he frames his discussion of estranged labor within the context of biology. The 1844 manuscripts articulate three types of estranged labor that result when the human subject treats his or her capacity for labor as a thing for purchase and sale. There is, first, the estrangement of the human subject from the products of labor, from what he or she makes. The products of labor are not direct use values and do not directly benefit the human subject making them. Second, there is an estrangement from the very process of production, from the very productive activity itself. The capacity to produce is dissociated from the human subject's experience of being a subject and is externalized in commodity production.[47] Thus, both what is made and the act of making are rendered into something outside of the human subject. Finally, Marx notes that estranged labor ultimately separates the human subject from an integral sense of being "human." Estranged labor transforms the human "*species being*, both nature and his spiritual species property, into a being alien to him, into a *means* to his *individual existence*. It estranges man's own body from him, as it does external nature and his spiritual essence, his *human* being."[48]

What does Marx mean by this curious term, the *species being*?[49] Within a discussion on political economy, we as readers are suddenly confronted with explicit references to biology and natural history. Marx mentions the estrangement of the species being only after his explanation of the estrangement from the products and process of labor: "Man is a species being, not only because in practice and in theory he adopts the species as his object . . . but also because he treats himself as the actual, living species; because he treats himself as a *universal* and therefore a free being."[50] Again, such statements make sense only if we begin from the premise that nonalienated production and productive activity are in some ways the core of what it means to be "human." For something to become estranged, it must first be integral and integrated. The estrangement from the species being is, then, the most wide-reaching form of estranged labor and, possibly, the most fundamental. However, in the context of political economy and biology, this conclusion raises more questions than it answers. What exactly is this supposedly nonestranged, nonalienated form of production? How exactly does the very biology of the human being play into the production and productive activity of the "species being"?

Marx's concept of the species being is actually a cluster of related concepts. As noted, the term *species being* refers simply and directly to the capacity for productive activity that is specific to human beings. On a basic level, then, a consideration of production in the context of political economy also necessitates some consideration of production as a biological activity. As Marx notes in volume 1 of *Capital*:

> Labor is, first of all, a process between man and nature, a process by which man, through his own actions, mediates, regulates and controls the metabolism between himself and nature. He confronts the materials of nature as a force of nature. He sets in motion the natural forces which belong to his own body, his arms, legs, head and hands, in order to appropriate the materials of nature in a form adapted to his own needs. Through this movement he acts upon external nature and changes it, and in this way he simultaneously changes his own nature.[51]

However, production in general is not specific to human beings. Marx does note that animals also produce, but what defines the species being for Marx is the way in which human production and productive activity differ from production by animals. For instance, whereas animals produce for the short term and for immediate need, human beings also produce for the long term, and they can produce in absence of immediate need. Further, whereas animals display complex modes of producing (Marx often cites "social insects" such as ants and bees), human beings produce intellectually and physically, and both forms of production constitute a material production than can exist ideally before it exists actually.[52]

Perhaps, for Marx, the key differentiator between human and animal production lies in the concept of "life activity." As previously intimated, production and productive activity are the crux of the species being. Marx notes:

> For, in the first place labor, *life-activity, productive life* itself, appears to man merely as a means of satisfying a need—the need to maintain the physical existence. Yet the productive life is the life of the species. It is life-engendering life. The whole character of a species—its species character—is contained in the character of its life-activity; and free, conscious activity is man's species character. . . . Conscious life-activity directly distinguishes man from animal life-activity. It is just because of this that he is a species being.[53]

The particular life activity of the human species is not only the ability to work conceptually or to work in the long-term or even to work outside of immediate necessity. The particular life activity of the human is that its being is in its doing, and the character of its doing is that it always has the whole species being as the goal. Conscious, reflective life activity directed toward the species being as a whole is thus the aim of the species being. This goal is most emphasized when life-activity is expropriated in a capitalist system:

The object of labor is, therefore, the *objectification of man's species life*: for he duplicates himself not only, as in consciousness, intellectually, but also actively, in reality, and therefore he contemplates himself in a world that he has created. In tearing away from man the object of his production, therefore, estranged labor tears from him his *species life*, his real species objectivity, and transforms his advantage over animals into the disadvantage that his inorganic body, nature, is taken from him.[54]

Life activity can thus operate at several levels, depending on the context. The context provides what Marx calls the "means of life," of production. Life activity can certainly be an immediate "means of life" in the sense of providing basic necessities for survival and continued maintenance. Life activity can also be a more general "means of life" in the sense that production and productive activity are understood to be an integral, nonestranged activity that defines what it means to be "human."[55] Finally, life activity is also the object of expropriation in economic systems such as industrial capitalism, in which biological life activity (or labor power) is rendered into a commodity to be purchased and sold. In this case, life activity becomes an estranged "means of life," a life activity that is forced to produce that which it does not use or to carry out productive activities that bear no direct relation to the human subject or the species being.

Our conceptual cluster is growing: species being, life activity, means of life. One final, important concept follows from these others, and it has to do with how the species being's life activity (productive activity) relates to its environment. Although this is not the place to discuss Marx's complicated attitudes toward "nature," what I can note is that the species being is constituted as much by its internal biology as by its being embedded in an environment. Marx refers to this integral relation between species being and environment as an "inorganic body":

The universality of man is in practice manifested precisely in the universality which makes all nature his inorganic body—both inasmuch as nature is (1) his direct means of life, and (2) the material, the object, and the instrument of his life-activity. Nature is man's *inorganic body*—nature, that is, insofar as it is not itself the human body. Man lives on nature—means that nature is his *body*, with which he must remain in continuous intercourse if he is not to die. That man's physical and spiritual life is linked to nature means simply that nature is linked to itself, for man is a part of nature.[56]

The environmental milieu of the species being is as defining a characteristic as its anatomy, physiology, or biology. The "inorganic body" Marx speaks of is also a nonhuman body, a collective body that is, despite its collective character, highly differentiated. However, to keep such statements from descending into the naive ecological idealism of the 1960s (or indeed of German romanticism prior to Marx), it is important to emphasize the biological and biotechnological lens through which Marx is being considered here. Marx's notion of the inorganic body takes on greater significance in light of modern debates concerning epigenetics ("nature versus nurture"; genetic effects of environmental factors), deep ecology (borrowing from systems theory approaches), and the ongoing research into the polygenetic factors of disease (studies in gene expression networks and biopathways). In effect, Marx's inorganic body already provides a link between a more complex systems view of the organism and the manifold ways in which the organism is "put to work" in various biotech industries.

These views can be summarized by noting that in bringing together biology and political economy, Marx's notion of the species being may serve as a means of emphasizing the biological dimensions of production, *without assuming biological essentialism*. This last point is crucial, for it serves to indicate one of the critical claims Marx makes: *the species being is more than merely biological*. Marx outlines several elements that distinguish the human species being from other forms (e.g., the emphasis on conscious, reflective action), but, even more important, he also refuses any "state of nature" narrative that would presuppose a primordial species being prior to estranged labor in society. This is as evident in the early writings as in the later writings. As Marx notes in his critique of classical political economy, "Do not let us go back to a ficticious primordial condition as the political economist does, when he tries to explain. Such a primordial condition explains nothing. He merely pushes the question away into a gray nebulous distance."[57] "Production by an isolated individual outside society . . . is as much of an absurdity as is the development of language without individuals living together and talking to each other."[58] It is thus important to read Marx's critique of political economy alongside the particular biologism of the concepts of the species being, life activity, and so on.

For Marx, the species being is constituted by several factors. It is constituted by production, productive activity, or the "life activity" of the species being. This means that productive activity is what it means to be "human," but the "human" is understood here to be fully social and historical. That is, productive activity is specifically conscious, reflective production and not simply "production in general." There is no life activity outside of society; to posit an "outside" of society is as impossible as positing an "outside" of the species being. The species being not only is life activity, but is divided among various "means of life," which can be immediate or long-term activity, productive or unproductive activity, necessary or excess activity. Finally, what qualifies this species being is that it functions in some integrative, systemic way with its surroundings or its environment. It is this material, bodily "space" between the species being and its surroundings or environment that constitutes an "inorganic body."

Biomaterial Labor I: Estranged Biologies

Between Foucault's notion of biopolitics and Marx's constellation of concepts surrounding the species being, what can we derive about the particular logic of the biotech industry?

As we have seen, Foucault's work on the biopolitics of the population during the eighteenth century raises a number of issues that are still pertinent in today's global biotech industry. Of particular note are new biological theories and protoinformatic techniques, which Foucault locates in the emergence of vital statistics, demographics, and state concerns over public-health management. Again, one of the key transitions he highlights is the shift in the object of political control, from a geopolitical "territory" to a biopolitical "population," the latter described as a "global mass that is affected by overall processes characteristic of birth, death, production, illness, and so on."[59] This shift is informed by science, but a science contextualized by concerns over the health and welfare of the population as a political entity. We can say that, for Foucault, biopolitics is the moment in which "population" folds into "nation," the event in which public health folds into national security, the process through which biology folds into politics. From this vantage point, what is required are not draconian prohibitions, but rather practices that intervene, modulate, regulate, manage, and oversee the ongoing dynamics of the population's health. The political force of biopolitics is, then, not command, but rather a mode of control. The knowledges of medicine, natural history, and an emerging biological science play central roles in mediating between citizen and species, nation and population, politics and biology.

This emphasis on the population—as a biological and a political entity—also means that the citizen who is part of a nation is also an organism that is part of a species. As Foucault notes, biopolitics, as a set of measured interventions and regulations, is "applied not to man-as-body but to the living man, to man-as-living-being; ultimately, if you like, to man-as-species."[60] It is worth pausing for a while on this formulation, in which the category of the "human" is defined both biologically as a species and politically as a part of a citizenry. We might ask: What exactly does it mean to consider the population as both citizenry and species? What, in fact, is the relation between citizen and species in the political contexts of public health, national security, political economy, and so forth?

> Within this set of problems, the "body"—the body of individuals and the body of populations—appears as the bearer of new variables, not merely as between the scarce and the numerous, the submissive and the restive, rich and poor, healthy and sick, strong and weak, but also between the more or less utilizable, more or less amenable to profitable investment, those with greater or lesser prospects for survival, death and illness, and with more or less capacity for being usefully trained.[61]

It is this connection—species, production, citizen—that Marx explored in his early writings on political economy. Whereas Foucault suggests that biopoltics configures the population as a biological species, Marx had already noted the unique "species being" of human labor and the way in which this species being was in the process of being transformed by industrial capitalism. Paolo Virno has recently highlighted this intersection of life and labor, noting that in post-Fordist capitalism "the living body of the worker is the substratum of that labor-power which, in itself, has no independent existence."[62] Life becomes a sort of medium for the biological capacity for labor. But this is also a political instance as well. For Marx, the "imperative of health" is also, in many ways, an imperative of wealth. In this sense, production and labor form a core part of biopolitical practices, for it is production and labor that form the essential processes of the continued health of the population and the nation: the material production of goods and the biological reproduction of the population; the economic generation of value and the biological management of health. What is the relation between "species" and "citizen" in biopolitics? One answer is that species and citizen are mediated by processes of labor and production, in which economics has biological effects and in which biology serves as the guarantee for political and economic security.

Between Foucault's notion of a biopolitical population ("human beings insofar as they are a species") and Marx's notion of species being ("labor, life activity, productive life itself"), we can deduce several elements that will lead us back to a consideration of contemporary biotechnologies. First, in Marx as in Foucault, the relation between human and nature is complex. *There is no essential, extratechnological relation between human and nature.* In fact, Marx's notion of human beings' "inorganic body" frustrates any attempt to isolate a state of nature separate from society.[63] The very approach of political economy—in Foucault as in Marx—is this consideration of biological processes that are thoroughly embedded within political and economic systems. The condition of the human is not that it is first a natural state and then gets transformed into the political state; rather, what the biopolitical species being shows us is that there is already in place a certain political economy, a certain system through which human beings biologically and politically organize and manage their life activity.

Yet—and this is the second point—the particular kind of labor of which both Marx and Foucault speak is not simply the labor of machines or robots in a factory. *The labor of which the biopolitical species being speaks is a thoroughly biological labor*: birth, death, reproduction, disease, growth, sanitation, and, as Marx notes, a whole "social metabolism." Through Marx's species being and Foucault's biopolitical population, we can see how labor is configured not just as a technical, artificial means of production, but as a process that is isomorphic with biological processes such as metabolism, reproduction, and biological growth and decay. In the biopolitical species being, we have not only the body *in* the factory, but the body *as* a factory. In biotech fields such as transgenics—in which animals are genetically engineered to produce compounds needed for human medical therapies—we have a biological system that produces only biology, genes that express only certain proteins, cells that metabolize only certain compounds, milk from an animal that contains human insulin. Biology produces biology, through completely "biological" means, but in totally novel conditions, and this is where the *tech* in *biotech* comes into play.

From this, we can deduce a third point, which is that *the biopolitical species being labors, even when it is not at "work."* That is, the particular kind of biological labor performed by the species being, by the population, is unique in that it is a nonhuman type of labor. Even when we are asleep, our bodies are performing biological labor in our circulatory system, respiratory system, even at the cellular and molecular levels of metabolism and gene expression. This is an obvious, even ridiculous point, but it does bear reflection on in light of the biotech industry's focus on genetic engineering, GMOs, and transgenics. A body that never stops laboring is also a biology defined by production, a species being defined by its own particular type of labor.

It is this peculiar type of biological labor that forms the domain of contemporary biopolitical control. Certainly, it is not the only kind of labor we find in the biotech industry—innumerable laboring bodies of all kinds are needed to provide the right conditions for this kind of biological labor, and they are framed by larger organizations such as university labs, pharmaceutical corporations, and hospitals or health-care centers. But, arguably, all these individuals—lab technicians, CEOs, factory workers, graduate students, and so on—are constellated around one or more biological processes that are recontextualized in novel ways (new genetic drugs, novel diagnostic systems, techniques for genetic analysis, manufacture of model organisms for laboratory study). At the heart of this constellation is this strange biology that is also a technology, a biological labor that does not work, genetic "information" that is materialized in a range of contexts (databases, test tubes, DNA chips, bacterial plasmids).

For Marx, the transformation of the species being's productive capacity into a mere commodity means that the vital labor of the species being (ideally integrated with the human being's larger "inorganic body") was turned into "dead labor," labor that goes nowhere except into exchange and surplus value. As is well known, Marx's critical point is that although this transformation is beneficial for industry as a whole and for those who own the means of production, it has disastrous effects on the working class—and on the larger species being. Industrial capitalism has the effect, then, of totally instrumentalizing the biological labor of the species being, thereby liquidating the integrated "inorganic body" of the living world in favor of a total objectification of labor in commodities.

Our own views today may be less reductive and more complicated, given the ways in which emerging technologies and new modes of production have transformed our views of the relationship between human and nature, biology and technology. Critics of globalization such as Hardt and Negri suggest that Marx's notion of dead labor has been replaced, in the era of the Internet and information technologies, by an "immaterial labor" that encompasses cultural production, services, and the general trend toward an informatic economy. Although this replacement may have taken place, we can also suggest that a novel kind of material labor has evolved out of the biotech industry, at least since the 1970s and the development of the first genetic engineering techniques. This kind of labor, as already noted, is fully biological and yet is inseparable from political and economic effects. It is strangely nonhuman, at work in the molecular spaces of cells, enzymes, and DNA, and it never stops working, whether or not we as individuals pay attention. It is fully informatic, constituted by genetic codes, protein-folding motifs, and biopathways, but it never completely relinquishes its affiliation with the material, messy "stuff" of biology. It is something that we can perhaps call *biomaterial labor*, or even *living dead labor*.[64] Biomaterial labor is the kind of biological-yet-technological mode of production that is the foundation of the biotech industry: a technology that is fully biological. Biomaterial labor is the condition of Foucault's biopolitical population and Marx's species being, within the context of the practices of fields such as genetic engineering, transgenics, GMOs, genetic therapies, and regenerative medicine.

Biomaterial Labor II: "I Walked with a Zombie"

My elaboration of Foucault's notion of biopolitics and Marx's concept of the species being was necessary to make the following proposition concerning biotechnology: *the biotech industry appropriates the species being at the molecular, genetic, and informatic levels, and, in doing so, it refashions human biological life activity or labor power as a form of nonhuman production*. Paradoxically, in the biotech industry, what is valued is not human labor, but the specific labor power or life activity of cells, molecules, and genes. The upshot of this approach—an approach not that different from the industrial capitalism Marx describes—is that Marx's species being is transformed into a "molecular species being," a species being in which labor power is cellular, enzymatic, and genetic.

We can further explain this phenomenon by noting a few examples of biomaterial labor. Consider the following applications of biotechnology:

RLC/Mo-cell Line The Mo-cell line is an example of "immortalized cells" or cell lines maintained in laboratory conditions for research. Cell lines enable multiple copies of a cell to exist in a cell culture for study, and they also enable the cells to exist in a state that is close to their original, biological state in the organism. Although the practice of deriving cell lines has been in existence since the 1950s (the HLA tumor cell line), it was not until the late 1970s that cell lines became a center of controversy. In 1976, businessman John Moore became ill with a rare form of leukemia. His physicians at UCLA's medical center diagnosed him and took a number of tissue samples from him in subsequent visits. In addition to treating Moore, the UCLA physicians had the idea of isolating Moore's T-lymphocytes in a cell line and selling that line to interested pharmaceutical corporations. In 1984, they were granted a patent on Moore's derived T-lymphocytes (U.S. PTO no. 4438032) and had negotiated some $15 million with Sandoz Pharmaceuticals for the development of Moore's cells. Upon learning of this, Moore filed suit, claiming that he was not informed of the marketing of his cells. An initial court ruling in 1984 rejected Moore's claim that an individual has property rights over his or her own discarded tissues. This decision was reversed in 1988 by the Court of Appeals, which accepted that Moore did have property rights over his own biological materials. The final decision in 1990 by the Supreme Court was a compromise: individuals cannot sell their biological materials, but Moore was entitled to remuneration because he was not informed of the sale. The ambiguity—which still exists to this day—is whether property relations over one's own body include the ability to market and sell those materials. Today the Mo-cell line is held at the American Type Culture Collection, and is estimated to be worth more than $3 billion.[65] Researchers can purchase cells from the Mo-cell line as ATCC catalog number CRL-8066.

OncoMouse The famous OncoMouse is an example of genetic engineering used for bioscience research. Mice have long been used as "model organisms" for the study of human disease, and, indeed, the recent efforts to map the genomes of various organisms has found a large number of similarities or disease homologs between mouse and human genomes. A novel kind of laboratory mouse was introduced in the late 1980s. This particular type of mouse was genetically engineered with genes (oncogenes) that would reliably produce breast cancer tumors. In 1988, the U.S. PTO granted a patent for the OncoMouse (U.S. PTO no. 4736866), as it was called, the first such patent for a living animal. Although the mouse was dubbed the "Harvard Mouse" because it was developed by researchers at Harvard University, the actual patent on the mouse was assigned to DuPont Chemicals, who financed the research. The OncoMouse could then be marketed to laboratories for the study of breast cancer, in effect initiating a new kind of "biological supply" business. (Similar such model organisms have been developed; in 1987, for instance, an NIH team engineered mice with AIDS in their germ line. Researchers such as Robert Gallo warned of the high mutation rate in HIV and the possibility of the AIDS mouse unleashing a "super AIDS.") The OncoMouse is a mass-produced mouse whose sole purpose is to develop cancer; in this sense, it functions for researchers as a kind of universal biological system for the molecular study of cancer.[66]

t-PA Clotbuster Protein The so-called clotbuster protein is an example of the use of transgenic technology in creating "mammalian bioreactors." In the 1980s, the techniques of genetic engineering had enabled the recombination of genetic material from different species. Although genetic material from humans and other organisms had been inserted into animals and plants for the purposes of livestock breeding and agriculture, the same techniques were also used to transform animals into vehicles for the production of desired compounds for human medical use. These mammalian bioreactors, as they were called, were most often cows, pigs, sheep, and goats, who produced a particular compound (such as human insulin) in their milk as a result of an inserted human gene. The desired compound could then be extracted from the milk and then processed and purified, ready for drug manufacturing. One such example was a goat engineered to produce a blood-clotting protein (tissue plasminogen factor, t-PA) in its milk, a compound useful for treating blood clots in cardiac patients. The t-PA goat was developed by researchers at Genzyme Corporation, along with Tufts University. Similar mammalian bioreactors have been developed, including a sheep that produces the protein AAT in its milk (for the treatment of emphysema), developed by Pharmaceutical Proteins Ltd. and the Scottish Agriculture and Food Research Council, as well as a failed attempt by the U.S. Department of Agriculture (USDA) to develop a pig that can produce human growth hormone (HGH). To date, a wide range of medically useful compounds have been produced in this manner, including human interleukin-2, human lactoferrin, antitrypsin, human protein C, collagen, and fibrinogen.[67]

These are only the most obvious examples of biomaterial labor. In these examples—immortal cell lines, model organisms, mammalian bioreactors—

it is not difficult to see a novel type of technology developing in which the production process and product are thoroughly biological, a labor carried out by cells, enzymes, and DNA. However, although each of these cases is different scientifically, there are a number of important consistencies between them. To begin with, biotechnology by definition implies an instrumental approach to the biological domain. This, in itself, is not necessarily problematic, for it can also serve as the basis for alternative practices such as sustainable agriculture and complimentary or alternative medicine. But with technologies such as industrial-driven genetic engineering or transgenics, we see a biologicomedical model that is inseparable from an economic-business model. Indeed, nearly every account of the biotech industry notes the simultaneity of scientific and economic innovation: recombinant DNA and the formation of Genentech, polymerase chair reaction (PCR) and the Cetus Corporation, transgenics and DuPont, and a wide range of examples related to the pharmaceutical industry. However, we should be cautious of any attempt to limit such views to the commercial sector, for there also exists a "governmentality" surrounding biotech that is in many ways inseparable from the economic and business models: alliances between the corporate and public sectors patenting disputes between developed and underdeveloped countries, and policies that allow universities and nonprofit organizations to apply for patents based on federally funded research.

Shall we, then, demand the separation of biology and economics, science and commerce? This would be, to say the least, a naive position, but also a nearly impossible one at a time when governments deregulate and apply for patents, when "Big Pharma" increasingly globalizes its operations, and when the hegemony of U.S. and European science research largely dictates the terms of global health programs. Perhaps, instead, the question is, How can we identify and assess those moments at which the imperatives of biology and economics are at odds with each other? It can be argued that the majority of high-profile ethical disputes within the biotech industry have to do with a conflict between interests, a conflict between medical benefit and economic benefit.

Understanding such differences requires an understanding of the assumptions that inform biotechnology as a hybrid of science and industry, biology and economics. I have already noted one central aspect of the biotech industry—its general approach toward the biological domain as a source of material production and the production of economic value. But what is also unique about the three examples given earlier is that the productive activity or "life activity" specific to them is strangely nonhuman. We see cells that are "immortalized" in the lab, genetically engineered animals that automatically produce compounds for medicines, and genetic sequences moving from an organism to a computer database and then into a genetically designed drug. We can ask: Who or what does the "work" in the biotech industry? As an increasingly global industry, biotechnology certainly employs human labor at all levels, from business and marketing to basic research, to the development of products and services, to public relations and media campaigns. But it can be easy to overlook another, equally significant type of labor in biotechnology: the uncanny, nonhuman "work" carried out by the biological components of cells, proteins, and DNA.

This is, indeed one of the lessons of the biotech industry: production is not limited to human subjects or groups, but can be extended to the nonconscious, "molecular" level of cells, enzymes, and DNA. Early biotech companies such as Genentech and Cetus were built on the capacity of their biological, molecular labor power—that certain enzymes can splice DNA at particular locations (recombinant DNA), that a set of genes together produce certain proteins under the right conditions (human insulin), and that DNA's base-pairing processes can be harnessed in a machine (PCR). Without the core of this biomaterial labor, the company arguably does not exist. Thus, not only does the biotech industry adopt an instrumental approach to biology, but in doing so it also locates the productive life activity of biology at the level of cellular metabolism, gene expression, protein synthesis, and so on. The life activity of the species being is not just biological, but molecular. It is biological and economic, a biomaterial labor power.

This emphasis on the biomaterial labor of biotechnology is not in any way meant to deny the presence of the labor of human subjects in, for example, the pharmaceutical industry or the high-tech field of genomics. In the same way that labor power appears as something strangely nonhuman in biotech, so we can say the same for the estrangement or alienation of biology, but it is an estrangement quite different from the way Marx used the term. In the biotech industry, the estrangement of biology occurs as a separation of process from context, biology from milieu. For instance, the culturing of cell lines requires a separation of the "natural" processes of cell replication from the milieu of the organism and from the subsequent recontextualizing of cell replication in the laboratory (Petri dishes, nutrient, incubator). Similarly, medical transgenics involves the separation of the protein synthesis process and its recontextualization into another host organism (genetic engineering, cloning, extraction of compounds from animal milk or meat). In short, in industrial biotech, biological life-activity can be technically estranged to such an extent that cells, enzymes, and genes can biologically function outside of the body.[68]

To summarize, in the biotech industry a number of assumptions constitute its approach to the biological domain. First, the biotech industry approaches biology in an instrumental fashion. Biotech sidesteps the traditional biology-technology division, understanding biology as a technology, as a productive technology in its own right. This means that the core labor power of the biotech industry is the life activity of biological components and processes, which are constellated by R&D, product pipelines, marketing, and corporate management. But further, the biotech industry also specifies biological labor as the labor of cells, enzymes, and DNA. The labor power of the biotech industry is not just the manifold labor power of human beings, but the molecular labor power of cells, enzymes, and DNA. In these circumstances, biology produces both biological and economic value: genes synthesizing protein products, enzymes catalyzing cellular reactions, and cells metabolizing useful compounds for medical and economic ends (human insulin, HGH). This type of labor power is biomaterial labor, a form of labor power that is at once biological and economic. *In the biotech industry, labor power is cellular, enzymatic, and genetic.*

Given these assumptions of the biotech industry, we might be led to ask a deceptively simple question: Who sells the labor power in the biotech industry? On one level, the answer is quite straightforward—the productive capacity of cells, molecules, and genes is "sold" most often through licenses on patents, "informed consent" agreements, clinical trials for new drugs, selective government funding of research, and a host of related practices that are simultaneously medical and economic in nature. Again, not all of these practices are problematic in themselves, simply by their mere existence. Rather, the question is, In what contexts, under what conditions, and in whose interests do we see the conflict between medical and economic benefit? But the question of who sells the labor power of biotechnology also has another, less visible layer to it. Recall that in the traditional scenario elaborated by Marx and other political economists, the capacity for labor or life activity was sold by the individual human subject, most often because that subject had nothing except his or her capacity for work to sell. The "buying and selling of labor power" thus implies at a basic level the conscious, voluntary (if coercive), and pragmatic act of transforming the life activity of the species being into a commodity.

Yet consider the examples cited earlier—genetic engineering, transgenics, cell lines, mammalian bioreactors, and so on. Certain entities—government-funded labs, biotech companies, pharmaceutical corporations—are involved in the exchange and circulation of biological materials, but who sells or offers up labor power? The answers that come up are almost comical: it is not the individual human beings from which genetic samples are taken, for the molecules in those samples are made to perform quite different tasks within laboratory bacterial plasmids or animal cells. It certainly is not the cells, molecules, or gene themselves, and it certainly is not the range of sheep, mice, and pigs used as industrial bioreactors for drug production. Thus, one of the central contradictions in the political economy of the biotech industry is: biology is ceaselessly made to work, and biological life activity is constantly producing, yet the seller of labor power is absent from the production formula.

This curious absence points to a theme that runs throughout this book. On the one hand, the biotech industry configures biology as "natural" and thus something that is morally (if not legally) outside of the economic sphere of commodity production. Biotech deals with "life itself," life that cannot be bought or sold, life that is—we are told—served by the particular technologies that the biotech industry develops. Biotech is safe, healthy, and beneficial, largely because the "technology " it develops is seen to be "natural." On the other hand, the biotech industry also explicitly configures biology as a technology, as a commodity, as economic property. Biology is that which can be "worked up" and worked on, thereby transforming biology into property. Selected gene sequences can be isolated using novel techniques; genetic sequences can be derived into mutated forms, into complementary DNA (cDNA) sequences, into messenger RNA (mRNA) transcripts; and genetic engineering can produce novel hybrid organisms, from oil-eating bacteria to human insulin–producing pigs. All these developments are biological, but also technological and thus subject to current patent laws.

This transformation of biology into property is in many ways the core activity of biotechnology. In fact, we can propose that in the political economy of biotechnology there is no "buying" and "selling" of labor power in the conventional sense. Or, at the very least, there are more levels of valuation, production, and exchange in biotechnology than the direct buying and selling of genes, cell lines, or transgenic organisms. *In the biotech industry, we see the continual transformation of biological value and medical value into economic value, a continual refashioning of the species being as at once biological and economic, as a form of "biomaterial labor."*

Genomics Is Globalization

I began this chapter by noting two related trends in biotechnology: an increasing tendency toward the globalizing of the biotech industry and a tendency toward the integration between biology and informatics, genetic and computer "codes." These two developments are clearly related, for the global scope of the biotech industry has greatly benefited from advancements in information and network technologies. Likewise, the development of global genome projects, of public-private alliances, and of expanding markets for the genetic drug industry provide the conditions for online bioinformatics software and databases. So, on one level, the global organization of the biotech industry is consonant with the technical integration of bioscience and computer science.

But it should also be apparent that the biotech industry is unlike the information technology or communications industries. Although there may be a general trend toward the "informatization" of production, toward the immateriality of products and services, this trend has only meant a reinvention of material production for the biotech industry. This is the key to understanding the unique properties of the biotech industry: its simultaneous embracing of the information technology revolution and its equal interest in biological materiality, the "stuff" of biology. This position is at once ontological and economic. The biotech industry vigorously incorporates information technologies with the same fervor that it enables the synthesis of DNA, lab-grown tissues, and genetic drugs. This would appear to be a contradiction, but, as a range of examples show, it is not. In fact, it can even be said that the "industry" part of biotechnology is predicated on the operative, functional nature of this apparent contradiction: biology is information, *and* biology is materiality. This is, as I have suggested, only an apparent contradiction, for the tension between biology and informatics is only a tension if they are mutually exclusive: if biology cannot be immaterial, and if informatics cannot be material.

What, then, is the "global genome"? The global genome suggests, first, that genomics is globalization and vice-versa. The act of literally stretching a human genome across selected parts of the globe creates a distribution network for that biological data. Despite its international billing, the IHGSC network structure still led to the universal body, *the* human genome. But alongside the IHGSC's activities are a panoply of genomics programs dedicated to the individualizing aspects between one genome and another: minute mutations such as SNPs, or the effort to map the genomes of ethnic populations. *The human genome thus gains a kind of biological universality at the same time that it partitions and fragments biological types.* The global genome creates the conditions for new forms of standardization, be they in terms of technical standards (e.g., GenBank "flat file" format), economic standards (WIPs), or cultural standards (the pop cultural "icon" of DNA and genetic essentialism). The field of genomics and the example of the IHGSC form a kind of paradigm for the globalizing tendencies in the biotech industry.

But beneath this literal distribution of a human genome via the Internet lies an important argument concerning the "nature" of biology and biological "life itself." The concept of a genetic "code"—the product of the postwar discourses of genetics and cybernetics—has culminated in a set of new attitudes toward biology and technology, nature and artifice. In particular, there is no extratechnological, preinformatic biology. All biology is informatic from the beginning. Marx's species being is always already bioinformatics, a natural organism whose naturalness consists in the fact of its informatic constitution. This, however, does not mean that biology has ceased to be "natural." In fact, it is a primary strategy of the biotech industry to insist on this dual aspect of "life itself": biology is at once the "stuff of life" and essentially informatic. *Not only is biology accounted for via informatics, but informatics is always qualified by biological materiality.* For instance, in the practices of genetic screening, DNA fingerprinting, or medical surveillance networks for epidemics, "information" comes to supplement or even to stand in for the subject. In a related vein, all information derives from and culminates in some form of biological materiality, such as lab-grown tissues, cell cultures, or genetically engineered plasmids.

This complex relationship between biology and informatics means that the biotech industry can always be identified by the novel, chimeric artifacts it produces. DNA chips, online genomes, oligonucleotide synthesizers, RNA "antisense" drugs, recombinant viral vectors, and artificial chromosomes are just some of the biotechnologies to be found within the biotech industry. But such chimeras are not random mixtures of biology and technology. They are the direct products of the biotech industry: the tools and technologies that surround the biomaterial labor of cells, enzymes, and DNA; the many different types of biological databases; and the systems that generate and assign value to the artifacts of biotechnology. *Technical artifacts, storehouses of knowledge, and banks of value.*

2

Bioinformatic Bodies and the Problem of "Life Itself"

The Two Faces of DNA

In popular culture, it is becoming increasingly difficult to separate DNA from its computer-generated representation. Whereas earlier science fiction films such as *Bride of Frankenstein* could reference only in language the existence of the molecular level, in the late twentieth and early twenty-first centuries DNA can be represented in all its informatic and three-dimensional splendor. Whether the DNA in question be that of human-insect hybrids (*The Fly*; *Spider-Man*), mutants (*The Hulk*, the *X-Men* films), alien invaders (*The Thing*, *Virus*), or plain old genetic surveillance (*Gattaca*), wherever one finds biology in popular culture, one also finds that which generates biology: computer graphics and imaging technologies. But beyond this, popular culture also posits a further proposition: not only do computers represent DNA, but DNA is in some strange way translatable with information itself: consider the "rage virus" of *28 Days Later*, the biotechnological hybrid of virus and media violence, or the viral logic of the media virus in *Ringu*.

But in a sense, none of this is new and was already presaged in the life sciences. After all, in the 1940s didn't physicist Erwin Schrödinger already hypothesize that the genetic material in all living beings was a "hereditary code-script"?[1] And didn't Watson and Crick's 1953 papers on the structure of DNA simply affirm what earlier biologists had guessed, that DNA is a code?[2] When François Jacob and Jacques Monod published their research on genetic regulatory mechanisms, wasn't their formulation of the "cybernetic enzymatics" a further postulation that DNA is a computer?[3] And when Heinrich Matthaei and Marshall Nirenberg announced that they had "cracked the genetic code" in 1962, didn't this once and for all establish the fact of DNA as a code and biology as information?[4] This history has been well documented by historians of science such as Lily Kay and Hans-Jörg Rheinberger, especially in the exchanges between molecular biology, on the one hand, and cybernetics and information theory, on the other.[5] In addition, cyberneticians, information theorists, and computer scientists were similarly paying attention to DNA as well. Claude Shannon's information theory grew out of his dissertation dealing with the algorithmic and combinatoric properties of a genetic code, and computer scientist John von Neumann, in speaking of the relationship between the brain and the computer, noted how DNA forms, in the biological organism, as a kind of memory system.[6]

Thus, historically speaking, there is plenty of precedent within the hard sciences for this intersection between genetics and informatics, biology and technology. Yet, in a way, the motto "biology is information" is not spoken by anyone, but rather demonstrated by the surreal artifacts that populate the biotech lab: online genome databases, DNA chips, combinatorial chemistry, three-dimensional molecular modeling, gene-targeting software, and "wet-dry cycles" in drug development. In these and other hybrid artifacts, we see not only the integration of the biological and the technological, but also the integration of a material and an immaterial understanding of biological life. The field of bioinformatics is a good example in this regard, for it directly brings together bioscience and computer science, genetic and computer "codes." Simply defined as the application of computer and networking technologies to the research problems of molecular biology, bioinformatics has quickly grown into a discipline of its own and into a significant industry as well. High-profile projects such as the sequencing of the human genome have been largely bioinformatics-driven projects, and bioinformatics research on the Internet is increasingly becoming an indispensable tool for molecular biology labs.

In the current context of bioinformatics, we have a set of questions to consider. How does bioinformatics reconfigure the relationship between biology (as natural) and technology (as artificial)? In the negotiation between biology and technology, how does bioinformatics balance the inherent tension between the material and immaterial? Contemporary discourses surrounding cyberspace, virtual bodies, and cyborgs endlessly play out the informatization of the biological, material body. These discourses not only assume a pretechnological body, but also foster a vision of the body that is highly textualized and semiotic ("material-semiotic nodes" and so forth). Biotechnology, however, defined in part by its connections to biology, is preoccupied with the materiality of the biological body and not with the immaterial domain of virtual bodies and avatars. And yet the body's materiality is also always contextualized by a discourse of materiality.[7] So what is happening in a field such as bioinformatics? Is bioinformatics simply the "computerization" of biology? No doubt bioinformatics and biotechnology generally take us far beyond the well-worn tropes of the cyborg and its affiliates. Any technique or technology used in biotech research demonstrates this: PCR, gene therapy, microarrays, even tissue engineering. In the myriad of techniques and technologies that can be observed in the biotechnology lab, a common ontological and political-economical viewpoint stretches across them all. A "natural" biological process (such as DNA base-pair complementarity) is repurposed and serves as the "technology" for each process, but in a different context (such as the precise heating and cooling cycles of PCR). This technology is purely biological—a technology *because* it is purely biological.

This point assumes that bioinformatics—and by extension biotechnology—must work in either the material domain of biology or the immaterial domain of informatics. Bioinformatics refuses this split in its practices and products, however. For instance, the online genome database was at one time the living cells of human volunteers and subsequently cDNA libraries in the lab. The same can be said for the U.S. PTO database as well. Similarly, all the software and databases count for nothing if they do not in some minimal way reconnect and "touch" the biology of living cells and living patients via therapies, tests, or drugs. Thus, we can rephrase our question another way: How does bioinformatics—as an increasingly foundational approach in biotechnology research—strategically bring together the informatic and the biological in such a way that it can accommodate both the material and the immaterial, both medical benefit and economic value?

In this chapter, I address this tension-filled zone between the biological and the informatic, the material and the immaterial. The key concept is the concept of "life itself": the notion of an unmediated, self-present core of biological life, and yet a notion that is inseparable from all philosophical, scientific, and economic mediations. Thus, my opening questions lead us to a consideration of "life itself" in both its philosophical-historical dimensions (via Canguilhem's work on the history of biology) and its political-economic dimensions (via Marx's analysis of capital). The question I ask throughout is, How, in fields such as bioinformatics and pharmacogenomics, does "life itself" constitute at once a foundation *and* a problem for biotechnology?

Hacking the Genome

As previously stated, the discipline of bioinformatics encapsulates many of the tendencies in biotech research to bring together genetic and computer codes.[8] Although molecular biology and the life sciences generally have always relied to some extent on a set of techniques, tools, and technologies for research, the recent entrance of modern computers and networking technologies into the biotech lab has meant that the molecular biologist must also become in some sense a computer programmer. The emerging field of bioinformatics is a good example in this regard. It not only brings together genetic and computer codes, but in doing so it also brings together the biotech and computer industries, and combines the disciplines of molecular biology and computer science.

In fact, *bioinformatics* can simply be defined as the application (or integration) of computer science to molecular biology.[9] There are degree-granting bioinformatics programs at a number of U.S. universities, and there are even "for dummies" books on bioinformatics as well as more specialized books on computational biology. As a scientific field, bioinformatics became known to the general public owing to its key role in the various human genome projects. Both the publicly funded consortium (IHGSC) and Celera have noted how the sequencing of the human genome would not have been possible were it not for bioinformatics. In general, bioinformatics has specialized in three areas of biotechnology research: sequence and structure analysis (genomics, proteomics), data management (large-scale sequence repositories such as GenBank), and the development of integrated systems for "in silico biology" (simulation systems for the testing of drug compounds).

Early genetic databases, influenced in part by the innovations in genetic engineering during the late 1970s, were primarily information-management databases.[10] Their primary goal was to organize large amounts of data efficiently in a standardized database structure operating in a computer system. Most of these systems were locally run and operated; that is, they were laboratory-specific databases with limited "time-sharing" capacities over networks.[11] Thus, their main advantage was to function like a library's card catalog system—local use in the accommodation of data management. An early example was the PC/GENE software package produced by Intelli-Genetics, which not only performed database-management tasks, but could also automatically translate a given gene sequence into its corresponding protein or amino acid code, based on the prior knowledge of gene-protein correspondences.[12] Other software applications are working on either providing relationships not previously known or functioning on the level of statistical probability and prediction analysis. Instead of manually (one is tempted to say "physically") cross-checking genome databases and protein databases, intelligent software agents and data-mining programs can be set to target a particular gene-protein relationship or gene pattern. When the first threads of data began to come in from the government-funded HGP in the early 1990s, most of that data was archived in a computer database called GenBank, now managed by the NCBI (figure 2.1). Because the rate of sequencing was

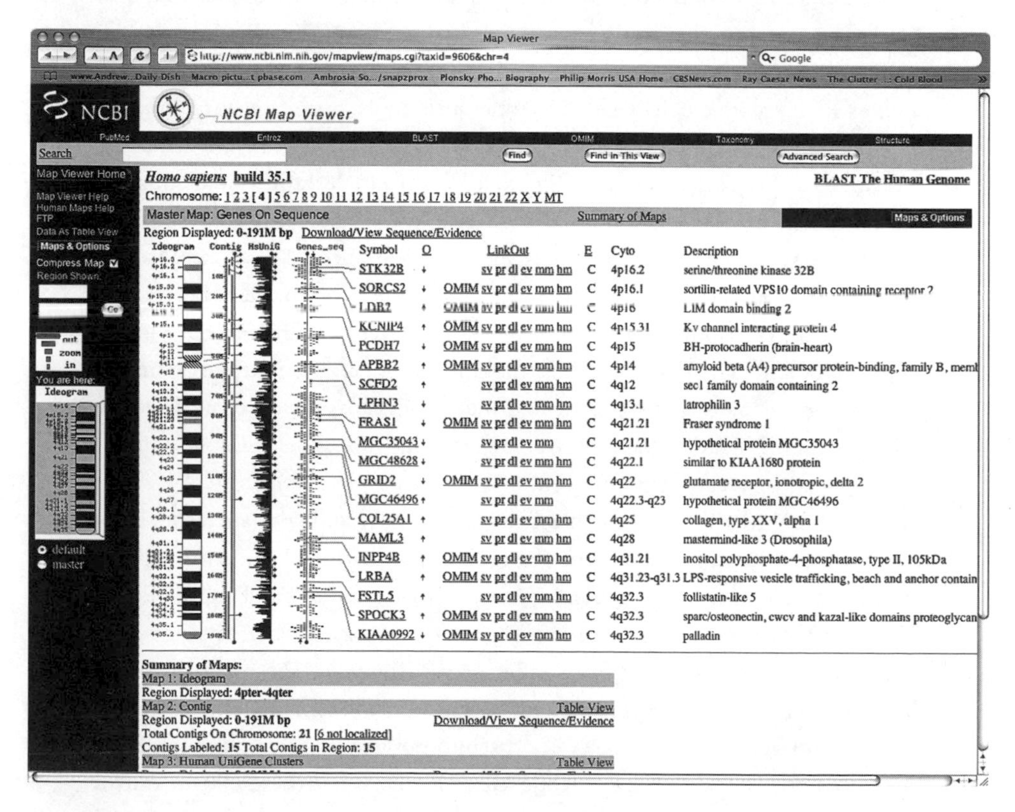

Figure 2.1 The ENTREZ online genome browser, hosted by the NCBI.

slow, and because only a handful of labs were working on the genome, the management of the GenBank database was something that could be handled by a small staff of biologists and computer scientists. As sequence data came in, they would be entered into the database, where researchers accessing the network (this is pre-Web, recall) could access the most recent sequencing results.[13] Many such tools can run off of a laptop or are made to operate exclusively online (as with the online Basic Local Alignment Search Tool [BLAST]).[14] The gradual networking of universities and research institutions and the tendril-like expansion of the Internet provided technical means of consolidating research results into centralized information nodes. The introduction of desktop computers and the Web, as well as educational spending on information technologies, helped to advance molecular biology into the information age.

Several factors contributed to the emergence of a distinct discipline of "bioinformatics" as the HGP progressed during the 1990s. Within the lab itself, several improvements were made in genetic sequencing and replication, such as PCR, more efficient gel electrophoresis tools, and, finally, fully automated gene sequencing computers.[15] In addition, although the HGP remained a U.S.-based project, it increasingly expanded itself, forming more substantial partnerships with the United Kingdom's Wellcome Trust and establishing outposts in Germany, Japan, China, and France. These two factors combined to produce an unprecedented rise in the genomic data being produced—a far cry from a small group of labs manually funneling data into a single database. The deluge of information has in recent years necessitated an efficient, accurate, and sophisticated means of managing all of it. The new field of bioinformatics has thus come to play an indispensable role in genetics and biotechnology research.

The bioinformatics response has been to diversify and specialize the existing technology, while also supporting a cross-platform compatibility in product development and implementation. For example, the PROSITE and SWISS-PROT databases are specifically protein databases, whereas other databases focus on the genomes from other organisms (e.g., the mouse genome database, pioneered by Celera), RNA sequences, SNP or genetic polymorphisms, and gene pools of ethnically distinct populations (e.g., the Icelandic Health Sector Database from deCode Genetics).[16] What these efforts helped to promote were the standardizing potential of networked computer-based labs and the notion of a distributed information reservoir.[17] With the introduction of the Web and more user-friendly software applications into universities and research institutions, the possibility of the online lab has become a reality. Biological information is thus not a matter of local reference, but of global distribution and access (or access limitations), operating according to protocols of uploading and downloading. In addition, as an intersection of the fields of molecular biology and computer science, bioinformatics has become a notable topic within the scientific community; the number of articles in science publications dealing with bioinformatics saw a sharp rise during the 1990s, and online hubs such as Biospace.com regularly post numerous job opportunities for bioinformatics specialists.

As a business, bioinformatics has shown steady growth and has promoted an optimistic future. A July 2000 report described bioinformatics as a $300 million industry, which was preceded by a March 2000 report that predicted the bioinformatics industry would be worth $2 billion by 2005.[18] Bioinformatics companies tend to work in one of three areas: so-called pick-and-shovel companies, which are low-risk ventures that provide the tools and hardware necessary for laboratory research; software and service companies, which perform data management and software-based analysis; and the pharmacogenomics-based companies, which utilize the products of the first two areas in the development of drugs, gene-based therapies, or genetic and diagnostic tests.

As an industry, bioinformatics has many overlaps with the pharmaceutical and health-care industries. This has meant, among other things, that bioinformatics has been shaped by economic as well as scientific and medical imperatives. According to a 2000 report by the Biotechnology Industry Organization (BIO), more than 90 FDA-approved, biotech drugs and therapies were for sale in the United States, with an additional 370 pending FDA approval.[19] Since 1995 (concurrent with the dotcom boom and the rise of bioinformatics as a field), this boom has meant that an average of 16–24 drugs or therapies per year have made it to the market.[20] This trend is noteworthy in light of the fact that the total time for most drug-development programs, from concept to FDA approval, is often 15 years or more.

Bioinformatics has not only helped to speed up this process, but it has also helped to attract further investment: according to a report by Pharmaceutical Research and Manufacturers of America (PhRMA), U.S.-based pharmaceutical companies invested an estimated $26 billion in R&D in 2000, an incredible jump from the estimated $8 billion spent just a decade earlier.[21] Part of this increase is owing to the nature of the drug-development process. Many pharmaceutical companies require 15 years to take a drug from research to market, with an average cost upward of $500 million per drug candidate. In addition, of the 5–10,000 compounds initially screened, only one actually receives U.S. FDA approval.[22] Even when a drug is granted approval and on the market, there is a significant rate of adverse drug reactions (ADRs) and drug nonresponsiveness; in some cases, drugs can be withdrawn from the market owing to severe or underconsidered ADR factors.[23]

With all this investment and risk, how are profits made? One primary area is in patents, especially in the United States but also in the European Union. A recent study by GeneWatch UK, a policy research and watchdog group, listed the top-ten genetic patent holders from both the public and private sectors. What is noteworthy is that this study was carried out by accessing the GENSEQ database, a commercial database that contains all genetic sequences patented worldwide. The report's top-ten list includes Genset (applications on more than 36,000 gene sequences), Genzyme (patent claims on more than 8,500 sequences), and the U.S. NIH (patent claims on just less than 3,000 sequences).[24]

Bioinformatics, according to many reports, promises to make the drug-development process more accurate and efficient. In a standard drug-development process, which includes a discovery phase, a preclinical phase, and three types of clinical phases, bioinformatics plays a key role in the initial discovery and preclinical phases—the most time-consuming and costly phases of the process. In the discovery phase, researchers identify potential "drug targets," or compounds upon which a drug might act. Through experiment, database research, and the mining of scientific literature, thousands of targets are identified that are connected to the particular disease being studied. These compounds are then tested using "high-throughput screening" (HTS) techniques, in which a large number of compounds can be simultaneously tested against the target for any responses. The "hits" from these tests constitute a "lead series," or set of candidate compounds whose selectivity and potency serve as the starting point for a drug therapy. In the preclinical phase that follows, the lead series candidates are tested in the lab and on animals in order to understand better the mechanism of the compounds (pharmacokinetics), possible side effects (toxicology), and drug-design approaches (computational chemistry). Once these studies have been completed, an Investigational New Drug (IND) application may be filed with the U.S. FDA, which, if accepted, will allow the candidate compounds to begin human clinical trials. In addition, bioinformatics also comes into play at a later stage of the process, in the testing of individual patients for potential ADRs based on their genetic makeup. The "input" data of the drug-discovery process are therefore directly linked to the "output" data of drug screening and prescription for patients.

In summary, the boost given to the pharmaceutical industry through bioinformatics has been in three primary areas: the identification of "drug targets" (molecular compounds upon which to act); "lead discovery," or the creation of novel drug compounds; and the production of genomics-based tests for already-existing drugs. The latter is broadly called *pharmacogenomics*, which, according to one definition, is "the study of the entire spectrum of genes that determine drug response, including the assessment of the diversity of the human genome sequence and its clinical consequences."[25] Along the way, computer and informatics-based technologies play a central role, from the mining of genome databases to the study of metabolic pathways in the cell, to the computational study of common drug candidate classes, such as proteases, ion channels, and G-protein coupled receptors (GPCRs, a type of membrane protein). Computer technologies are promising to play a key role in drug development, but the process is obviously still very much tied to the material and biological conditions of the clinical trial phases. Thus, although a novel drug candidate may be driven by bioinformatics and pharmacogenomics technologies, at some point the in silico world of drug discovery must directly and materially interface with the in vivo world of the human body (mediated by the in vitro experiments in the discovery and preclinical phases). This set of interactions is of interest here in both ontological and political-economic terms. The manifold passages from a file in a genome database to a patent file, to the pharmacogenomics study of a lead series, to a genetics-based test for patient ADRs are the primary interests of this chapter.

Thus, bioinformatics is not simply a subset of computer science, nor is it simply the newest tool for molecular biology research. It is a set of novel informatically driven practices and knowledge that have taken shape alongside bioinformatics as a discipline, a business, and an industry. Two main things are interesting about bioinformatics. First, bioinformatics is as much an ontological endeavor as it is a biological one. This endeavor goes beyond the rhetorical exchanges of postwar molecular biology (the metaphor of the genetic "code" and so on), without ever absolutely denying the metaphorical role that informatics plays in molecular genetics. In a sense, bioinformatics takes all the terminology, concepts, and metaphors of informatics at face value; it moves with great ease between the metaphor of DNA as information and the construction of online genome databases. As we shall see, assumptions concerning the division between the material and immaterial, or the biological and the informatic, are redefined and reshaped in various ways along the way. Bioinformatics is, first and foremost, an ontological practice, demonstrating the ways in which DNA is information and the ways in which information materializes, and, above all, participating in the reconfiguration of dominant ways of understanding the relation between the living and nonliving, the biological and the technological.

But—and this is the second interesting point—bioinformatics is not just in the business of philosophy; as an emerging field, it has outlined for itself a set of specific aims and areas of application (e.g., pattern matching in sequence alignment, database management in genomics, three-dimensional graphics in molecular modeling). Bioinformatics does not exist in a vacuum, in which the niceties of the formal relationships between DNA and information are endlessly examined. In many cases, the same genetic code in a genome database such as GenBank is also found in the U.S. PTO database in a patent file for a derived gene or gene product. Indeed, a number of commercially available databases are dedicated to the correlation of genetic and patent data for researchers, academic departments, and biotech companies. A single gene sequence can potentially generate a great deal of economic value, from the genomics-based development of drugs to the forging of temporary alliances between biotech start-ups and large pharmaceutical corporations, to the creation and marketing of genetic tests, to patent licenses for the development of other laboratory techniques and technologies. Bioinformatics is not just an informatic view of biology; it is also at the same time a biological view of economic value. In other words, the ontological question "What is life?" is always folded into a set of political-economic questions, such as, "Can biology be turned into a technology?" or "Can life be property?"

This twofold character of bioinformatics—at once ontological and political—can be called *the politics of "life itself."*[26] One of the defining features of this politics of "life itself" today is the unique view of the biological body that has been instantiated by molecular biology. As Nikolas Rose notes, this "molecularization . . . was a reorganization of the gaze of the life sciences, their institutions, procedures, instruments, spaces of operation and forms of capitalization."[27] The phrase "life itself" is in scare quotes—and will remain so—

for a number of reasons. For one, it denotes a term used again and again by molecular biologists during the 1950s and 1960s in popular-science books on the genetic code—the idea that there was a master code that coded for the very biological foundations of life or "life itself." This twofold aspect, at once biological and yet informational, signals a reformulation of what counts as "life itself." Sarah Franklin summarizes this process: "nature becomes biology becomes genetics, through which life itself becomes reprogrammable information."[28] But the phrase "life itself" also denotes the slipperiness of any claim to have discovered an essence—mechanistic or vitalist—of biological life. As the philosopher and historian of science Georges Canguilhem notes, this emphasis on "life itself" drives much biological thinking in the West, from Aristotle's animating "Soul" to Darwin's selection mechanisms, to the emergence of molecular biology during the postwar era. The elusive nature of "life itself" seems to be both the basis of biology and the point that always stands outside of biology.

Molecular genetics and biotechnology are the latest phases in this ongoing interest. The politics of "life itself" occurs alongside the discourses of cybernetics and information theory as much as it occurs alongside biology and biochemistry. Biotechnology as an industry adds another twist: that the pursuit of "life itself" is more than a purely intellectual inquiry; it has always been about the ability to instrumentalize whatever comes under the phrase "life itself." There is another, underconsidered history of biotechnology (including modern forms of breeding, agriculture, and fermentation processes) in which "life itself" is synonymous with the "uses of life."

SB-462795 (Not THX-1138)

Because much of bioinformatics can be highly abstract, it is helpful to consider a case study. The aim is twofold: first, to provide a concrete example through which to consider the ontological and economic aspects of bioinformatics, and second, simply to illustrate the increasingly formalized and even routine character of the pharmacogenomics process in drug discovery. The example is not necessarily a high-profile or controversial drug, but neither is it an insignificant or failed drug candidate. In fact, the example is actually an unfinished account because, as of 2004, the drug itself is still in human clinical trials. The example is the drug candidate SB-462795, a drug for osteoporosis, a degenerative disease that destroys the bone marrow cells. It is estimated that osteoporosis affects some 10 million people in the United States alone, with an additional 18 million affected by low bone mass; a majority of those affected are women in middle to late age.[29] SB-462795 was developed by the biotech company Human Genome Sciences in conjunction with the pharmaceutical corporation GlaxoSmithKline; in 2002, Human Genome Sciences and GlaxoSmithKline entered SB-462795 into Phase I human clinical trials.[30]

According to Human Genome Sciences, SB-462795 is the first drug candidate to have been derived specifically from genomics and bioinformatics technologies.[31] This claim has been countered by Celera Genomics, which, along with the pharmaceutical corporation Merck, also claims to have an osteoporosis drug candidate entering clinical trials.[32] Whatever the case may be, what is clear is that SB-462795 represents what may be a new generation of drugs produced largely—if not exclusively—via bioinformatics. Whereas conventional drug development predominantly "black-boxed" the genetic mechanisms of disease, bioinformatics and pharmacogenomics aim to mine genome data as the starting point for the drug-discovery process. PhRMA, the primary voice in the United States for the pharmaceutical industry, notes that "while yesterday's scientists relied on a combination of persistence and serendipity to find compounds that might work against diseases, today's scientists are supplementing the old approaches by using new technologies that increase efficiency and enable researchers to create new medicines."[33] Human Genome Sciences has led in the development of other, similar drug candidates along these lines, including an enzyme inhibitor used for the treatment of cardiovascular disease (Lp-PLA2), a protein that speeds the repair of damaged tissues (KGF-2), and a drug candidate that displays revascularization properties in the heart (VEGF-2). However, SB-462795 is an interesting example because the story of its evolution brings together the scientific, technical, institutional, and economic aspects of bioinformatics.

SB-462795 is also referred to as a "cathepsin-K inhibitor." In the normal functioning of the body, an enzyme called cathepsin-K (figure 2.2) breaks down bone marrow cells (osteoclasts) as new ones (osteoblasts) are generated. This process is usually regulated, such that enough cells are continuously regenerated in order to maintain the structural integrity of bones. In some types of osteoporosis, the cathepsin-K enzyme is overexpressed in osteoclast cells; that is, the genes that code for the synthesis of the cathepsin-K enzyme are overactive, producing excessive quantities of the enzyme, which in turn

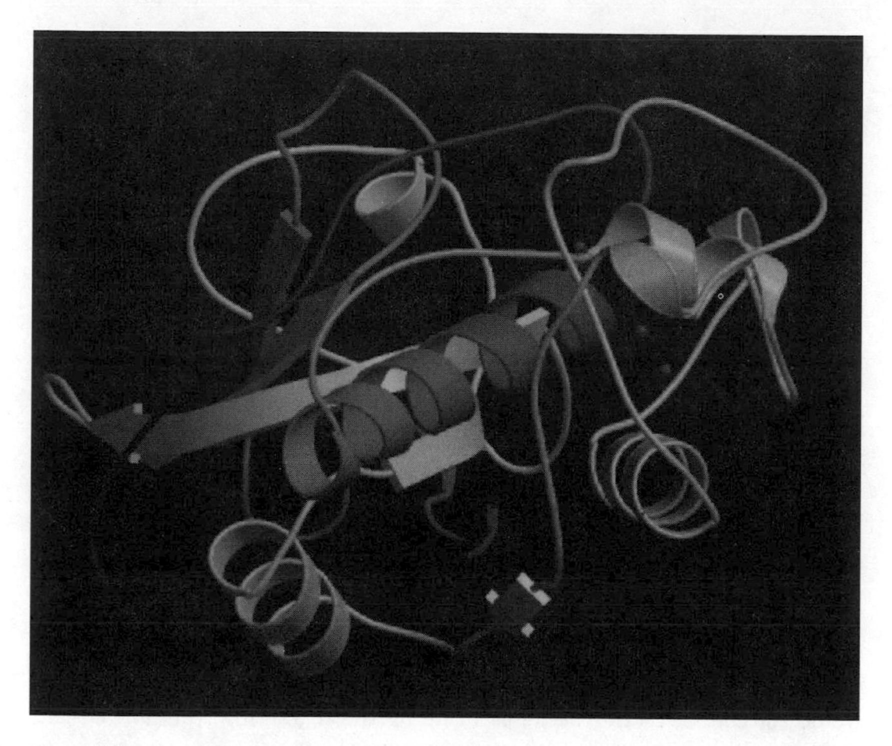

Figure 2.2 A simplified molecular model of a cathepsin-K molecule (Protein Data Bank ID: 1ATK), showing motifs and domains. From the Protein Data Bank. See B. Zhao, et al., "Crystal Structure of Human Osteoclast Cathepsin K Complex with E-64," *Nature Structural Biology* 109 (1997).

results in cathepsin-K enzymes breaking down healthy or even new bone marrow cells at a greater rate than new cells are generated. The result is that the bone marrow literally becomes shot through with tiny "holes" owing to the degradation of bone marrow cells, making the bones themselves weaker and brittle. The SB-462795 drug candidate is based on research that has shown how the inhibition of the cathepsin-K genes improves bone resorption in mice. SB-462795 thus aims to decrease or lower the overexpression of cathepsin-K genes (and thus of the cathepsin-K enzymes). It is for this reason that SB-462795 is also referred to in a less-cumbersome way as a cathepsin-K inhibitor.

In 1993, GlaxoSmithKline (then SmithKline Beecham), at the time working on a drug for osteoporosis, asked Human Genome Sciences to analyze an osteoclast sample. Although GlaxoSmithKline had long been in the pharmaceutical industry, the then-emerging technologies of genome sequencing and analysis attracted it to Human Genome Sciences. Human Genome Sciences took the cell samples and within three weeks had produced cDNA libraries and fully analyzed the genome of the osteoclast cells. Company researchers found that a certain class of the cathepsin enzyme was overexpressed in the osteoclast cells and that a relationship existed between this overexpression and the degradation of bone marrow cells in osteoporosis. Based on this research, Human Genome Sciences received two patents: one in 1996 for Human Osteoclast Derived Cathepsin and another in 1999 for a derived cathepsin-K gene.[34] At the same time, in 1997, GlaxoSmithKline and Incyte formed diaDexus, a company whose to aim was to develop and market tests for osteoporosis and other diseases.

The industry side of SB-462795 can be seen from the original 1993 agreement established between Human Genome Sciences and GlaxoSmithKline. Under this agreement, Human Genome Sciences would receive clinical development "milestone payments" for the drug candidate as well as royalties for any subsequent compounds discovered by GlaxoSmithKline through the use of Human Genome Sciences' techniques and technology. GlaxoSmithKline would, of course, retain the gains made by the marketing, promotion, and sales of the drug once it reached the market, though some of this profit might be shared with Human Genome Sciences. From this agreement, both Human Genome Sciences and GlaxoSmithKline formed the Human Gene Therapeutic Consortium, a group that included Takeda Chemicals Ltd., Schering-Plough, Merck KGaA, and Sanofi-Synthelabo as a way of facilitating alliances and potential product development centering around a bioinformatics- and genomics-based technology platform. Although the initial term of the consortium expired in 2001, it identified more than 400 research programs along the lines of SB-462795, and it also serves as an indicator of the increasingly common trend of situational, temporary alliances between the different sectors of the biotech industry.

The development of the cathepsin-K drug SB-462795 can be seen as a paradigm for how bioinformatics and pharmacogenomics research proceeds. Whether the aim is to develop a novel drug or to study the gene expression patterns for a particular biopathway, the components are often the same: a combination of more traditional "wet-lab" technologies such as PCR or BAC libraries and informatics-based tools such as computers, databases, software for imaging and analysis, and so forth. Indeed, a technology such as the automated genome sequencing computer is a literal hybrid of both types of technologies: a wet-lab component of biological samples and an in silico component of scanning and analysis technologies. In many molecular biology labs, the practices of bioinformatics are becoming more and more routine. At every step of the way, both the computer itself and the concept of a genetic "code" is central to how the research is done. How was Human Genome Sciences able to analyze and develop the cathepsin-K drug candidate so rapidly? We can use this example to outline what has now become a somewhat standard protocol in bioinformatics-based drug development:

- Acquisition of biological sample and sample preparation;
- Isolation and purification of DNA or RNA from cell sample;
- Sequencing of DNA or RNA using automated genome sequencing computers (including creation of cDNA libraries);
- Translation of DNA into an amino acid sequence using software;
- Analysis of both DNA and amino acid sequences against the human genome databases (including use of EST and/or sequence-tagged site [STS] information);
- Possible application for patents on novel techniques or derived gene sequences;
- Analysis of both DNA and amino acid sequence against other model organism genome databases (e.g., the mouse genome, the fruit fly genome, etc.);
- Analysis of both DNA and amino acid sequence against structural properties and homologs of protein databases (from primary to tertiary protein structure);
- Design of novel compound (drug or therapy) based on analysis and homology studies;
- Selection of model organisms and in vitro systems for preclinical experiments;
- Clinical trials in humans (phases I–III);
- New Drug Application (NDA) to U.S. FDA;
- Pending approval, ongoing "postmarket" analysis;
- Development and marketing of genomic tests specific to drug[35]

Although this list is incomplete and brief, what should be evident is the degree to which computer and information technologies play a central role in molecular biology and the drug-development process—*without ever implying the wholesale transformation of biology into an immaterial, "virtual" practice.* This dual assertion—informatic and material—is the key to understanding the promises, potentials, and contradictions in the foundational assumptions that bioinformatics and pharmacogenomics set forth.

In a way, then, we can invert Timothy Leary's famous maxim that "computers are the drugs of the 1990s" and say that for the biotech industry drugs are the computers of the twenty-first century. Even within the nonspecialist discourse, the idea of a genetic "code" has become second nature, having found its way even into science education, science journalism, and popular culture. It seems that both in science research and in culture generally, the distance separating biology and informatics has begun to collapse, but without biology simply becoming virtual and without informatics simply becoming an extension of the biological. What has happened, then, to the notion of "life itself" in this process? Addressing this question requires us to consider bioinformatics and pharmacogenomics from both an ontological and an economic perspective.

Canguilhem and "Life Itself"

From the most general perspective, thinking about biological life is as old as thinking about thinking. Canguilhem notes that "the theory of the concept and the theory of life have the same age and the same author."[36] He is speaking of Aristotle, "the logician of the concept and the systematic philosopher of living things."[37] The links between philosophy and life, ontology and biology, are, at first vague and abstract. In any definition of biological life, be it mechanist or vitalistic, there is always something that escapes, something that exists beyond the ability of conceptualization. And yet nothing is more self-evident and less in need of proof than the fact of life. What could be more impractical—and, more important, impossible—than the age-old, unanswerable question "What is life?" For Canguilhem, the seemingly vague question concerning "life" is both ancient and contemporary. Consider, as an example, the titles of books published by molecular biologists during the postwar era, books about the then-emerging field of molecular genetics and aimed at nonspecialist, popular audiences: Francis Crick's *Life Itself*, François Jacob's *The Logic of Life*, J. B. S. Haldane's *What Is Life?* Andre Lowff's *The Biological Order*, Henry Quastler's *The Emergence of Biological Organization*, and George Beadle and Muriel Beadle's *The Language of Life*. Such books owe a great deal to Erwin

Schrödinger's lectures entitled "What Is Life?" in which the existence of a "hereditary code script" was hypothesized in decidedly informatic terms. More recently, diverse books such as Robert Sinsheimer's *What Is Life?* Lynn Margulis and Dorion Sagan's *What Is Life?* and bioinformatician Pierre Baldi's *The Shattered Self* continue to pose the questions first elaborated in a specifically genetic manner by molecular biologists of the postwar era.[38]

If the question concerning "life" is at once very old and yet very contemporary, how can we understand its most recent permutation in the age of the HGP, bioinformatics, and pharmacogenomics? We can begin by taking as our primary object of study the term "*life itself*." As previously stated, this seemingly innocuous phrase remains terminally in scare quotes in this volume in order to evoke constantly a reflexivity concerning the relationships between biology, philosophy, and political economy. More accurately, we can refer to the concept of "life itself," a concept that has changed a great deal over time and at any given moment is constituted by more than one approach. The concept of "life itself" refers not only to scientific knowledge concerning biological life (today, something articulated at the molecular-genetic level), but also to the ways in which that knowledge is never just scientific. As we have already seen in the example of cathepsin-K (a case study to which I return later), the concept of "life itself" today brings together digital DNA, engineered in vitro proteins, intellectual property, alliances between companies, and genetic tests for use in medicine. Understanding how the concept of "life itself" is transforming and transformed by the current integration of biology and computers requires a further consideration not just of the concept of "life itself" but also of the *concept* of life itself. It is to this topic that I now turn.

In 1966, Canguilhem gave two lectures in Brussels, both entitled "The Concept and Life."[39] He gave the lectures just four years after Marshall Nirenberg and Heinrich Matthaei had succeeded in "cracking the genetic code" and some thirteen years following the research of James Watson, Francis Crick, Maurice Wilkins, and Rosalind Franklin on the structure of DNA. Though Canguilhem had long been interested in the history of biology and in particular in the relationship between philosophy and biology, this new stage in the development of biological science signaled a significant new twist in the long-standing quest after "life itself."

The title of Canguilhem's lectures is instructive. As he notes at the outset, to have a concept of life is, in a sense, to be life itself: "For Aristotle, soul was not only the nature but also the form of the living thing. Soul was at once life's reality (*ousia*) and definition (*logos*). Thus, the concept of the living thing was, in the end, the living thing itself."[40] In tracing the interest in the relation between concepts and life back to Aristotle, Canguilhem is also suggesting that, despite the reliance on quantitative analysis and the techniques borrowed from information theory, the then-emerging field of genetics still displayed an Aristotelian interest in life-giving "form" or "Soul." As Canguilhem notes, "when we say that biological heredity is the communication of a certain kind of information, we hark back in a way to the Aristotelian philosophy with which we began. . . . If we are to understand life, its message must be decoded before it can be read."[41] Yet, at the same time, genetics was not simply Aristotelian in the same way that seventeenth-century natural history or eighteenth-century vitalism were; its unique appropriation of information theory and cybernetics made it at once the most modern of biological sciences and yet a science in which Aristotle's life-giving "form" reappears as genetic sequence or molecular structure. "Messages, information, programs, code, instructions, decoding: these are the new concepts of the life sciences."[42]

The value of Canguilhem's contribution to the philosophy and history of biology is in the way in which he understands biochemistry and molecular genetics as endeavors that are at once philosophical and technical, ontological and "economic" (in the broadest sense of the term). In developing a more critical understanding of contemporary bioinformatics and pharmacogenomics, I would like to spend some time elaborating the problem of "life itself" in Canguilhem's writings. In one sense, for Canguilhem the history of biological thought can be seen as a footnote to Aristotle. In Aristotle can be found the seeds of what would later become natural history, physiology, even the modern emphasis on self-regulation and organization.[43]

Canguilhem highlights three elements from Aristotle's biological writings. First, Aristotle's notion of the life-giving "form" or "Soul" in *De anima* makes a distinction between matter and form in the organism, the latter providing the invisible yet immanent principle of organization for the former.[44] In the passage from an acorn to a fully grown tree, the form of the organism exists in potential and is eventually actualized through the biological processes of growth and development.[45] It is this basic notion—an organizing principle distinct from, yet immanent to the material of the organism—that Canguilhem detects even in modern biochemistry and genetics. Although the passage from Aristotelian form to genetic information is not a straight, progressive line, Canguilhem nevertheless points to the concepts of "biochemical specificity" and cellular "errors in metabolism" as examples of a new type of genetic Aristotelianism.

Second, Canguilhem highlights the role of "organization" in the Aristotelian teleology. Although he does not claim that the history of biology is a history of variations on the teleological argument (form is function), he does give great significance to the various terms derived from *organ* (*organon*) in Aristotle:

> it was Aristotle who coined the term "organized body." A body is organized if it provides the soul with instruments or organs indispensable to the exercise of its powers. . . . The concept of organism developed in the eighteenth century, as naturalists, physicians and philosophers sought semantic substitutes or equivalents for the word "soul" in order to explain how systems composed of distinct components nevertheless work in a unified manner to perform a function.[46]

Terms that occupy a central place in biological thought—*organ, organic, organization*—are, for Canguilhem, versions of Aristotle's emphasis on the organization of the living. Although the emphasis on teleology changes (especially with Darwinian evolution), the emphasis on formal causes does not. Aristotle's writings on biology reiterate this point: if you want to know what it is, look at what it does. Aristotle's famous example of the eye is exemplary in this regard: if the organism is the eye, then the life-giving "form" or "Soul" is sight. Canguilhem's interest in Aristotle's concept of *organon* is directly linked to the modern interest in biological organization—biocomplexity, autocatalysis, and self-organization in biological systems. Indeed, Canguilhem traces the biological concept of self-organization back to eighteenth-century vitalism and natural philosophy, examples of what he calls "the autocracy of nature."[47]

Finally, along with the Aristotelian concepts of "form" and "organ," Canguilhem highlights the principle of motion or animation, the principle that stood out for Aristotle among all others as the defining character of life itself. Encapsulating Aristotle's biological philosophy, Canguilhem notes that the "fundamental concepts in Aristotle's definition of life are those of soul and organ. A living body is an animate and organized body. It is animate because it is organized. Its soul is in fact act, form, and end."[48] Furthermore, "life, identified with animation, thus differs from matter; the life-soul is the form, or act, of which the living natural body is the content: such was Aristotle's conception of life."[49] If it moves, it is alive. Motion or animation is thus directly related to organization and form:

> A remarkable and interesting fact from the epistemological standpoint is the proliferation of terms containing the prefix auto-, used today by biologists to describe the functions and behavior of organized systems: auto-organization, auto-reproduction, auto-regulation, auto-immunization, and so on. . . . Living systems [in these sciences] are open, non-equilibrium systems that maintain their organization both because they are open to the external world and in spite of being open to the external world.[50]

This emphasis on animation implicitly brings in a dynamic, temporal quality to the organism, but it also emphasizes the automobility of the organism as well. The organism moves itself, and, in doing so, it provides for a range of resistances to the processes of decay, degradation, and disorder or entropy. In this regard, Canguilhem often cites the Xavier Bichat maxim: "Life is the collection of functions that resist death."[51] But this resistive quality of life is also a resistance of order against entropy, of organization against randomness, of information against noise. It is this more modern formulation of Bichat's maxim that is taken up by biochemistry and genetics. Thus, at the molecular level, it is not so much that something is alive because it moves, but rather that it is informed that it moves and thus is alive.[52]

The relationship between Aristotelian "Soul" and genetic "information" is part of a broader, more complex historical transformation that Canguilhem briefly outlines in a 1973 encyclopedia article simply entitled "Vie" (Life).[53] Canguilhem identifies four, largely overlapping threads in the history of biological thinking on "life itself." Each thread is dominated by a principle understood to be nontranscendent and yet distinct from the particular materiality of the organism: *animation* (which is the most elaborated in Aristotle), *mechanism* (articulated in philosophy and biology by Descartes, and adumbrated by Harvey and Malpighi), *organization* (again stemming from Aristotle but taking fuller shape in Comte, cell theory, and Bernardian physiology), and, finally, *information* (resulting from the intersections between cybernetics/information theory and molecular biology).[54]

For Canguilhem, the point in outlining these threads is precisely to emphasize the nonlinear, overlapping, discontinuous character of biological thinking. In the case of molecular genetics, the influence of cybernetic "feedback" and the concepts of information theory are decisive in reformulating the question concerning "life itself." Canguilhem points specifically to information theory in this regard: "Claude Shannon's work on communications and information theory and its relation to thermodynamics (1948) appeared to offer a partial answer to an age-old question about life. . . . Is organization order amidst disorder? Is it the maintenance of a quantity of information proportional to the complexity of the structure?"[55] Whereas the second law of thermodynamics describes how an object will tend toward a more disordered state (a state of greater entropy), the example of living organisms seems to provide a counterexample, where growth, development, regeneration, and adaptation resist the tendency toward entropy.

Keeping in mind the three Aristotelian elements that Canguilhem points to—form, organization, animation—we can further elaborate on how genetics and bioinformatics transform the question of "life itself." Modern genetics, in the significance given to the notion of "genetic information" or a "genetic code," implicitly refashions Aristotle's notion of form, organ, and animation. Instead of a teleologically driven form (recall Aristotle's example of the eye), genetics gives us a technically driven information (genes are not determined or determining, but they do express, mutate, and replicate via protein intermediaries). Likewise, instead of an organ's serving as the principle of order for the organism, genetics and biochemistry emphasize the notion of organization, even self-organization (e.g., protein folding, bonding specificity between enzymes, metabolic networks in the cell). Finally, instead of the more physiological principle of animation or motion, genetics offers the model of political economy: the production, distribution, and consumption of molecular compounds within the cell (transcription/translation, membrane signaling and transport, cellular metabolism).

However, in this transformation of the concept of "life itself," a number of important questions arise. Canguilhem points to one: "Where does biological information originate?"[56] That is, the model of genetics presupposes the existence of genetic information already as part and parcel of the living system. Is there a stable quantity of information in the living cell? Is new information created or lost in the lifespan of a cell? For that matter, we can ask a deceptively simple question: What counts as biological information? Is it the number of genes, the number of proteins, the number of interactions between them, or some statistical measure of gene expression levels or metabolic activity? Canguilhem notes the quasi-evolutionary solution given by Henri Atlan, in which information in the cell arises through the selection from a background of "noise" (which mimes Norbert Wiener's definition). We might also mention biocomplexity researcher Stuart Kauffman, who has proposed a similar solution, but one based on the self-propagating, self-regulating principles of simple RNA-type molecules ("collective autocatalytic sets").[57] Yet, despite these interesting approaches, the problem of "life itself" still hinges on the priority of some nonmaterial yet immanent principle of organization. Canguilhem asks, "[M]ight the meaning of organization lie in the ability to make use of disorganization?"[58]

I can summarize Canguilhem's approach to genetics by deriving three propositions from his work:

- The first proposition is that molecular genetics is an attenuated Aristotelianism; the concept of a "genetic code" is, in the framework established by Canguilhem, a more technical, operative version of Aristotle's "form." Genetics, however, is not completely Aristotelian because it is, as a biological science, predicated on the materiality of the organism, on the "stuff" of life.
- A second proposition is that genetics concerns itself less with the constitution or morphology of the organism and more with the ways in which an organism is "informed." The emphasis on codes, sequences, and other data types makes genetics—and bioinformatics—very distant from the kind of visible analogies established by natural historians. Indeed, Canguilhem wonders what the discourse of nature's anomalies, its monsters and teratologies, might be in genetics: "According to Aristotle, a monster is an error of nature which was mistaken about matter. If in contemporary molecular pathology, error generates formal flaws, hereditary biochemical errors are always considered as a microanomaly, a micromonstrosity."[59]
- This relationship between information and error leads to a third proposition: in the medical context, genetics facilitates, or at least inculcates, a shift from the organism to the sequence as the basic unit of "life itself." At issue here is much more than mere scientific reductionism, for a number of genomic and so-called systems biology approaches have been able to incorporate molecular genetics into a systemswide view that includes technologies such as DNA microarrays and computer databases.[60] Yet the shift away from organism to sequence, with all its connotations of "code" and "information," also reveals a significant fissure in the concept of "life itself."

Consider the ways in which information theorists such as Claude Shannon and molecular biologists such as Francis Crick define "information." For Shannon and his colleague Warren Weaver, information is explicitly defined as a quantitative unit, a statistical measure of a message moving from point A to its destination at point B.[61] As they state, "two messages, one of which is heavily loaded with meaning and the other of which is pure nonsense, can be exactly equivalent, from the present viewpoint, as regards information."[62] Thus, in the context of information theory, *information* is a quantity, a pattern, or a form that exists irrespective of, from the perspective of information transfer, content or meaning.

As Evelyn Fox Keller notes, precisely the opposite occurs in molecular biology's appropriation of terms such as *information*, *code*, and *sequence* from information theory and cybernetics.[63] In genetics, a single base pair mutation—from an A (adenine) to a C (cytosine), for instance—can result in a drastic change in phenotypic expression. Diseases such as sickle cell and cystic fibrosis do in fact result from, respectively, single base pair and single gene mutations. And, as noted, new fields such as pharmacogenomics are predicated on the search for the minute genetic differences (SNPs) that make some patients more susceptible to side effects than others. Thus, when Watson and Crick talk about "genetical information," they define *information* in exactly the opposite way in which the term operates in the information and computer sciences.[64] This tension—between an informatic definition of information and a genetic definition of information—has become even more complicated today, in which biological research and computer research are increasingly integrated into genome databases, gene-finding algorithms, and molecular-modeling software. In short, fields such as bioinformatics contain the residues of the Aristotelian tension between "form" and "information," a tension that is, paradoxically, materialized in artifacts such as in vitro plasmid libraries and online genome databases.

"In a Nature Twofold They Shine"

We can now outline several key characteristics of the concept of "life itself" in contemporary fields such as bioinformatics and pharmacogenomcis.[65] Let us return to our example of the cathepsin-K drug candidate SB-462795. Recall that this cathepsin-K drug developed by Human Genome Sciences is currently in clinical trials and is among the first genetic pharmaceuticals to have been developed from the use of genomics and bioinformatics (as opposed to more traditional methods grounded in wet-lab research). What are the characteristics of "life itself" in such bioinformatic contexts?

To begin with, the concept of "life itself," as Canguilhem shows, always involves a reliance on Aristotelian notions of "form," or some organizing principle that coordinates discrete parts (e.g., gene sequences) into functional wholes (e.g., gene expression pathways). For bioinformatics this principle is actually twofold: sequence and structure. Indeed, the major gap within bioinformatics is the so-called sequence-structure gap, for it is in this intermediary zone that the one-dimensional feature of sequence is transformed into a three-dimensional molecular structure, from gene to protein. In the case of the cathepsin-K drug candidate, the drug-discovery process began from informatics, from the sequencing and analysis of osteoclast cells; that is, initial analysis began in silico. What this implies is that although the other properties of osteoclasts are not without import, the sequence and the patterns of that sequence served as the starting points for the discovery process. What is the life-giving form to osteoclast cells? In this case, it is implicitly the principle of organization within the genetic sequence—a sequence derived and analyzed within the computer.

However, "life itself" is not merely concerned with the Aristotelian form, or the life-giving principle of a cell or protein, but also with the biological processes through which that form is expressed and made manifest. In the case of the cathepsin-K compound, the identification and isolation of the cathepsin-K genes was only a first step; the real results would come only from understanding the role these genes played in the phenotypic overexpression of the cathepsin-K protein in the living cell. Form is thus incomplete without the efficient cause of some biological process (in this case, gene expression). For this, researchers at Human Genome Sciences used comparative genomics to analyze the cathepsin-K genes against their homologs in other species such as mice (a common model organism for humans). Again, the recent progress in both the human genome and the mouse genome has made such comparative analyses possible using the Internet to access these databases. Using not only genomics databases, but also enzyme classification, EST, and biopathway databases, the drug-discovery process can generate a great deal of data without ever entering the wet lab (though this is never done without wet-lab research). These first two characteristics—form and process—suggest that in the era of bioinformatics-driven drug discovery, "life itself" is a dynamic principle of organization that can be analyzed irrespective of specific material instantiation.

In spite of what would seem to be a tendency to make the biological fully "virtual," neither bioinformatics nor the drug-discovery process takes place completely in the computer or on the Internet. This point cannot be stressed enough: *bioinformatics is as material as it is immaterial, all the while constituted by an informatic approach to the biological domain.* What this means is that, for bioinformatics, Aristotelian form and process is not enough; in a way, bioinformatics seems to be a dissatisfied Aristotelianism. Computer-assisted sequencing and analysis can go so far, but the drug-discovery process demands, at some point, a return to the wet lab. In the case of the cathepsin-K drug, this return happened after the in silico drug-target identification and validation processes (identifying the compound to act upon and matching that compound with the manifestation of disease). In this case, animal studies using the cathepsin-K therapy were the first loop from the digital to the biological in the "wet-dry cycle" of the drug-discovery process. This tension is crucial to understanding fields such as bioinformatics. Though, as Canguilhem notes, modern genetics and biochemistry may smuggle in Aristotelian "form," they also remain resolutely materialist, if for no other reason than that their endpoint or application must be something material (a gene therapy, a drug) or must in some way materially touch the body of the medical patient.

Although these three elements of form, process, and matter significantly inform the role of "life itself" in bioinformatics, another element is equally important: the role that various discursive exchanges play in the transformations of the concept of "life itself" within these scientific fields. I have already noted the significant roles that the sciences of cybernetics, information theory, and early computer science played in the formation of molecular genetics during the postwar era; this influence can be seen today in the range of hybrid artifacts that populate many biotechnology labs, from genome databases to DNA microarrays. The intersection between genetic and computer "codes" is in many ways the dominant theme of the recent history of molecular biology and genetics, and it will arguably take new forms in the leading-edge fields of bioinformatics and "in silico drug discovery." However, as also noted, such conceptual exchanges are at best incomplete exchanges and approximate appropriations. Such approximation, in some instances, can affect the way in which scientific knowledge is determined. For instance, the fields of cybernetics and information theory, largely contextualized by World War II military research, configured "command and control" within communications and weaponry systems in a particular way; agency was human based and largely centralized, though augmented by an array of mechanical or informational prosthetics or both. In importing the concepts of "information," "code," "noise," and so forth, molecular geneticists such as Francis Crick implicitly imported the notions of command and control that were part of those concepts. Thus, the emphasis on the "central dogma" of genetics during the 1960s ("DNA makes RNA makes proteins, and proteins make us") configured DNA as a kind of command-and-control system that deployed instructions outward to the synthesis and organization of proteins in the cell. Though one still finds versions of the central dogma in textbooks, it has taken some time for this largely reductive maxim to be complemented by more complex approaches emphasizing gene regulation, biopathways, and protein-protein interactions.[66]

The concept of "life itself" becomes even more complicated if we consider not just the discursive exchanges between disciplines, but also the end product or application for fields such as bioinformatics and pharmacogenomics. A large number of biotechnology fields are directly contextualized by their potential application in medicine and health care. This is, on the surface, the rationale for the need for patents as well as publication, for incentives for application as well as incentives for knowledge production. And, certainly, the very definition of *biotechnology* implies such an instrumental approach: biotechnology is a technology of biological life, a technology in which biological "life itself" is at once the processes of production, distribution, and consumption. In biotechnology, "life itself" is both the technology and that which the technology aims to benefit. For instance, the cathepsin-K drug candidate is both a biological compound and a technological mode of intervention into an organism. It is at once fully "informatic," being based on sequence analysis and database research, and fully "biological," being engineered to interact in a biological manner in the cells of the organism or patient. There is, we should not forget, another important layer, and that is the economic layer. The cathepsin-K compound may perform biologically in the cell, but that performance will directly affect the economics of companies such as Human Genome Sciences or GlaxoSmithKline, and it will also indirectly affect the business of diagnostics for osteoporosis based on this knowledge. In this sense, the cathepskin-K drug functions biologically as well as economically; indeed, it functions economically to the degree that it functions biologically. Because the business of drug development is costly and time-consuming (as noted, it often takes 15–20 years from basic R&D to FDA approval), the concept of "life itself" is, in cases such as these, always more than biological. It is a biological "life itself" that is in many ways not separate from a technical and economically functional concept of "life itself."

So far I have offered only a partial listing of the characteristics of "life itself" in the domain of bioinformatics and drug development. What should be apparent is how the ontological concerns of Aristotelian "form" are closely tied to the political-economic concerns of application and instrumentality. Clearly, there is nothing wrong with the development of successful genetic therapies and drugs in general; but, all the same, it is important to acknowledge and pay attention to the complicated, manifold interests in fields such as bioinformatics. Biotechnology is both a science and an industry, and in examples such as the cathepsin-K drug candidate it is not difficult to see how the medical and economic values are not always in sync. In some instances, a drug candidate may perform biologically in a way that runs counter to the economic interests of its development (e.g., the HER2 test for cancer). In other cases, potential economic value is facilitated by new diagnostic technologies (e.g., the HLA-B test for the AIDS drug Abacavir). Depending on the direction in which the balance between medical and economic value tips, we may see in the future the combination of population genomics, genetic screening, and pharmacogenomics result in the "medicalization of lifestyle," or what one report calls "pills for the healthy ill."[67]

In the example of cathespin-K, the concept of "life itself" is at once ontological and economic, not only in the sense that a particular molecular compound performs both biologically and economically, but also in the broader sense of *economy*, meaning efficiency, productivity, minimum energy expenditure. We can return to Canguilhem's work in the history of biology on this point. In discussing the "thematic conservation" of the theme of normality in the biological sciences, Canguilhem notes how in the seventeenth century the metaphor of "animal economy" was first put forth to account for the ways in which the organism regulated its own biological functioning. If Descartes's mechanism could not account for the self-regulating and self-moving properties of animals (as opposed to machines), then perhaps the notion of an animal economy could account for it: "Like the domestic economy, the animal economy required wise government of a complex entity in order to promote the general welfare. In the history of physiology the idea of 'animal economy' was responsible for a gradual shift from the notion of animal machine to the notion of organism over the course of the eighteenth century."[68] The concept of a biological or animal economy is significant historically given the concurrent rise of natural history's systems of classification and the development of modern notions of political sovereignty in Thomas Hobbes's *Leviathan*. It may also be seen to function as a precursor to the later development of classical political economy in Ricardo, Smith, and the biologically inspired work of Malthus. For Canguilhem, the notion of animal economy (instead of animal machine) served as a hinge to the prevalence of concepts derived from Aristotle's term *organon*: order, organization, organic, and so forth, "the idea that a certain *order* obtains in the relations of the parts of a mechanism to the whole."[69] The economic model would persist through the nineteenth century, albeit in a different form, in the guise of industrialism's factory model in Bernardian physiology. Here, the specialization of the organism's parts is analogized as a biological division of labor. As Canguilhem notes, "since the organism was conceived as a sort of workshop or factory, it was only logical to measure the perfection of living beings in terms of the increasing structural differentiation and functional specialization of their parts, and thus in terms of relative complexity."[70]

Economy in this sense is broader than any specific mode of exchange or production, though, obviously, not isolated from it. In this sense, an animal economy or biological economy would be what Aristotle meant by *entelechy*, or the ways in which the principle of living organization resulted in a specific, efficient end or aim. Yet the end of the organism, like the end of "life itself," is, in another sense, death. The organism ceases at the moment of its end, and yet its end is consistently to "resist death," as Bichat put it. It is perhaps the notion of animal economy that serves as a bridge here, for it articulates biological function as a correlative of the economizing activity of the organism, of what is costly and of what comes cheaply. If, as Canguilhem suggests, the economic model plays a significant role in the concept of "life itself" (though in historically conditioned, different ways), then perhaps we can elaborate on the impact of this connection between biology and economy in bioinformatics.

The Economy of DNA, or Marx as Bioinformatician

Consider again our case study of Human Genome Science's cathepsin-K compound. We can ask, What are the different types of value in the development of such a drug candidate? From one perspective, it is the drug itself that is of the most value, for the measure of the success of the drug is directly related to the potential economic gains for Human Genome Sciences and Glaxo-SmithKline. Thus, the measure of economic value is predicated on the measure of biological or medical value; or, put another way, animal economy determines, in part, financial economy. The task of the drug-development process from this vantage point is therefore to turn information effectively into a product, to turn an abstract, immaterial entity (e.g., genetic sequence on a database) into a concrete, material entity (an FDA-approved prescription drug). The exclusive reliance on information in itself goes only so far. In the business of health care and pharmaceuticals, the endpoint is always the material, biological body of the patient, and it is this biological baseline that must be confronted.[71] Although information may be of value in itself, it is only part of the picture. Thus, *connecting information to the biological body is the primary challenge of "life itself" in the age of bioinformatics*. The measure of the two types of value—medical and economic—are based on effectively transforming something immaterial that is exchanged into something material that is consumed as its endpoint.

Yet, from another perspective, the biological, material body of the patient is not the endpoint of the drug-development process, despite the newfound role bioinformatics promises to play in the process. The average drug developed using biotechnology arguably benefits only partially from direct drug sales. The booming industry of diagnostic tests—many genetics based—and the linking of such tests to computer databases are also significant parts of the drug-discovery process and the pharmaceutical industry. These two aspects—diagnostics and databases—are more services than products and, as such, are dependent on information technologies to deliver results accurately. In the case of diagnostics, pharmacogenomics techniques, including the use of DNA microarrays, are promising to be able to match the genetic makeup of individual patients to particular drugs in order to minimize ADRs. This possibility, combined with the increasing diversity of biological databases (including SNP databases, population-specific genome databases, and traditional patient data in clinics), means that both medical and economic value are predicated on the ability of information to perform computational and analytical work. Here, economic value is predicated on medical value, but in a way that is different from the case of drug discovery. From a purely economic standpoint, there is less risk involved in the offering of these types of services, for a diagnostic test can be medically and economically beneficial, whether or not an ADR is discovered in a drug that is tested. Indeed, a common critique leveled at the idea of pharmacogenomics is that it offers a way for pharmaceutical companies to sell better the drugs already on the market, instead of transforming or even questioning the logic of drug development. Instead of a traditional "one size fits all" approach, pharmacogenomics simply shifts the scale of its operations into a "many sizes fit all" approach. As a number of more critical reports suggest, the basic principles of drug effectiveness, regulation, and testing are still left largely unquestioned.

In the first approach, we have a product-based economy, which takes the biological and the material as its endpoint. In this model, the challenge for bioinformatics-based drug discovery is to transform information effectively into a material product or therapy; information must minimally benefit the biological body of the patient so that any minimal economic benefit may be possible. This seems to be common sense, and it fits many of our experiences in the health-care sector: a visit to the doctor leads to one or more tests, and the results from those tests (combined with physician consultation) lead to one or more prescriptions, which then takes the patient to the pharmacy. Yet any number of factors limit this approach from the economic standpoint: for instance, the patent on a drug may run out, and the competition from the generic drug market may force a company to develop a "new" drug that takes its place.

But if we consider this same process from the larger picture of a bioinformatics industry, there is more than this simple information-into-drugs equation. As can be witnessed in the development of many drugs, including the cathepsin-K candidate, the process includes not only an actual pill, but also patents, licenses on patents, tests, and databases, not to mention direct-to-consumer marketing for new drugs. From this perspective, the challenge for a bioinformatics-based drug-development process is different. The challenge is now to maintain the recirculation of products (pills, testing technologies) back into information (databases, test results, marketing and media campaigns). If drugs currently exist in a cycle of technological obsolescence similar to the information technology industry, then the main challenge put forth to the pharmaceutical industry is not how to develop sustainable and effective treatments, but rather *how to transform temporary material products continually into the long-term generation of information*. What generates economic value, from this standpoint, is an infrastructure for the production of information, yet without ever completely severing the link to the patient's biological body.

These two perspectives on the medical-economic valuation of the cathepsin-K drug candidate correspond to what Karl Marx has famously called the "general formula of capital." Speaking about nineteenth-century industrial capitalism, he notes that the nature of capital can be considered through two perspectives on exchange. The first is the perspective of the individual worker and consumer. In this perspective, the individual works a certain number of hours manufacturing a commodity, for which a wage is given in the form of money. That person then takes the money and purchases other commodities produced by other individuals—food, clothing, or housing. Or, in the context of health care, an individual who works at an information technology job receives wages (part of which may go toward health insurance), and, if the person is ill, that money is exchanged for a prescription drug, which is then consumed.[72] This admittedly simplified context includes two subprocesses: the transformation of a commodity into money and the transformation of money into another commodity. Another way of stating this is that the product of the individual's labor power is exchanged for money, and then that money is in turn exchanged for the product of the another individual's labor power.[73] In a sense, even in this basic relationship, capitalist exchange already coordinates living, biological labor power via immaterial, money intermediaries. Marx was fond of using biological metaphors for the process of exchange. As he notes, "in so far as the process of exchange transfers commodities from hands in which they are non-use-values to hands in which they are use-values, it is a process of social metabolism." Thus, the exchange of money for commodities and commodities for money constituted the core "metabolic interaction of social labor." Although in a modern context the emergence of service industries, flexible accumulation, and information-based or immaterial labor has made this basic relationship more complicated, what we can retain from Marx's analysis is this basic "metamorphosis of commodities through which the social metabolism is mediated."[74]

This perspective—that of the individual worker-consumer—is what Marx calls "selling in order to buy."[75] Commodities as material objects and use values are mediated by the immaterial exchange values represented by money. The endpoint is the use or consumption of the commodity, be it food, clothing, or housing. Marx abbreviates this relationship by the formula C-M-C, in which the capitalist exchange process is seen to be the dual transformation of commodities (C) into money (M) and then back into commodities, which are then used or consumed. In describing this process as a "metamorphosis," Marx means quite literally that one type of thing is transformed into another type through a dynamic, morphological process mediated by the equivalence of quantitative values. Thus, "[f]rom the mere look of a piece of money, we cannot tell what breed of commodity has been transformed into it. In their money-form all commodities look alike."[76] Thus, in the formula C-M-C, each moment in the metamorphosis of commodities necessitates the intermediary of money; in other words, each moment in the transformation of material objects necessitates their mediation by immaterial means of equivalence.

However, as Marx notes, this description represents the individual worker-consumer's perspective. A different picture is created when we consider not only the discrete individual, but the dynamics of the system as a whole. From this perspective, the formula C-M-C is inverted, and the endpoint of the system is precisely that it has no endpoint. From this systemwide perspective, the C-M-C circuit is only a kind of subroutine within a larger set of dynamic processes. This larger process is the inverse of the preceding formula: M-C-M′, in which money is exchanged for money via the intermediary of commodities and in which the money that returns is more than the money originally invested. Keeping with our simplified example of health care, a pharmaceutical company may pay a licensing fee for a patent to produce a prescription drug, but that drug will not be the endpoint, for the drug serves the function of boosting investments in the company as well as generating sales from spinoff products such as diagnostic tests. Marx calls this "buying in order to sell." The real goal of capital, then, is not merely to mediate between material commodities or use values, but rather to use commodities as intermediaries in the ongoing, expansive generation of capital. "In the circulation of C-M-C, the money is in the end converted into a commodity which serves as a use-value; it has therefore been spent once and for all. In the inverted form M-C-M, on the contrary, the buyer lays out money in order that, as a seller, he may recover money."[77]

Two consequences result from this perspective. The first is that in the mediation of money through commodities, quantitative value becomes the ground for exchange. "The process M-C-M does not therefore owe its content to any qualitative difference between its extremes, for they are both money, but solely to quantitative changes."[78] Though Marx is careful to note that this quantitative exchange process is itself conditioned by a range of qualitative factors (such as the social creation of demand and the process of "valorization"), the process of capital is predicated on the transformation of M into M′, or of value into greater value, of investment into profit (and back into investment). The second consequence is therefore that "the circulation of money as capital is an end in itself, for the valorization of value takes place only within this constantly renewed movement. The movement of capital is therefore limitless."[79] Again, Marx does note that commodities are consumed as use values in this process, but that the use value or pattern of consumption is often conditioned by the larger aim of exchange value.

Thus, in the example of the cathepsin-K drug candidate produced by Human Genome Sciences and GlaxoSmithKline, we can see a process similar to the one Marx describes as the "metamorphosis of commodities" and the general formula of capital (M-C-M′). Certainly, from one perspective, the drug-discovery process results in a consumable "thing," But from the perspective of the movement of the economic process as a whole, that moment of material production and consumption exists within a larger framework that emphasizes the further generation of value, a kind of biological valorization process. The drug-discovery process for SB-462795 resulted not only in an actual drug for testing in clinical trials, but also in a number of patents (one for human osteoclast-derived cathepsin, another for a derived version of the cathepsin-K gene), an osteoporosis test sold to Quest Diagnostics, and the formation of a gene therapy consortium of pharmaceutical and biotech companies. What at one level appears to be a relatively straightforward economic relation—wages into insurance, insurance into drugs, drugs into the body—is at another level a dynamical, unending process of valorization—patents into drugs, drugs into tests, tests into patents.

However, the example of bioinformatics and cathepsin-K is not fully accounted for by the model Marx proposes, for fields such as bioinformatics and pharmacogenomics must, as we have seen, constantly negotiate the space separating the material and the immaterial, the biological and the informatic, the complex of Aristotelian form, organ, and animation. Indeed, Marx does note that the process of capital is forced to deal temporarily with material commodities, but that the ultimate aim of capital is to obtain a kind of immaterial, free-floating state. For Marx, the "material variety of the commodities is the material driving force behind their exchange, and it makes buyers and sellers mutually dependent, because none of them possesses the object of his own need, and each holds in his hand the object of another's need." Yet the contradiction in the general process of capital is that if the process is taken purely in terms of the exchange of equivalencies, there can be no surplus value and no profit: "If commodities, or commodities and money, of equal exchange-value, and consequently equivalents, are exchanged, it is plain that no one abstracts more value from circulation that he throws into it. The formation of surplus-value does not take place."[80]

Thus, on the one hand, capital is defined as the process in which money is turned into more money via commodities, and yet this exchange of equivalencies denies the creation of surplus value, the very thing it is supposed to make possible. A pharmaceutical company invests a certain amount of money in the development of a drug with the intention of extracting some surplus value or profit from this drug (either in direct sales or in licenses or spinoff services). In the case of cathepsin-K, the C in the M-C-M′ formula is thus the material pill, the drug itself. Yet the potential profits gained do not arise from pricing alone, for pricing does not address the more fundamental issue of how a differential can emerge from an equivalency. In this case, even the pricing mechanisms of the drug industry, although they calculate potential gains, cannot account for the conditions that make possible the derivation of a differential from equivalencies.[81]

If this is the case, from where might the profit differential arise? The pricing of the cathepsin-K drug must take into account the investments made by GlaxoSmithKline, but this inclusion only establishes equivalency. Where does the difference of profit arise? Perhaps it is in the ambivalent, yet necessary descent into the material domain that the differential of profit can be found. In a sense, the C in the M-C-M′ formula is not just the drug itself, but the drug functioning (or not functioning or malfunctioning) inside the body of the patient, the drug in a biological context. Indeed, in drug development the delicate balance between making a profit, suffering a loss, and breaking even is predicated on the biological functioning of the drug-body relationship. It is for this reason that Marx turns his attention to the labor process and the valuation of labor power. In speaking of industrial capitalism, he notes that innovations on the level of production serve as the profit differential, the difference that arises from the equivalency of exchange. In the nineteenth-century context, this meant, among other things, the technological development in "large-scale manufacturing," which not only maximized the output per unit of time, but also redefined labor power according to specialization and the division of labor.

When we turn to the case of bioinformatics and pharmacogenomics, we can similarly notice two major transformations in the drug-development process that, the companies hope, will create the profit differential. The first is that the commodities in the biotech industry, like in many other industries, are increasingly becoming immaterial and informatic. The medical application of biotechnology is directed primarily toward prescription drugs, but a panoply of service-based industries satellite the actual development of drugs: database management, data analysis, software design, "infomedicine," and, of course, diagnostics. But this can be detected in a range of other industries as well (most notably the information technology and entertainment sectors). What is unique about biotechnology generally and about the bioinformatics and pharmacogenomics industries specifically is that *it is not only commodities that are becoming more immaterial, but also the bodies that consume those commodities*. To be sure, patients' biological bodies are not being "uploaded" in some kind of science fiction scenario. But the hegemony of molecular genetics, when combined with new information and computer technologies, has created a context in which the biological body—"life itself"—is increasingly understood and analyzed in informatic ways. The push toward pharmacogenomics and, in some cases, toward preemptive genetic testing has as its aim the development of a totally in silico monitoring and diagnostic system for biological "life itself."

One possible consequence of this tendency—a tendency supported by a great deal of investment capital in pharmacogenomics technologies—is that genetic medicine will "touch" the body only to the degree that the body and "life itself" are understood in informatic ways. If we accept Marx's description of the general process of circulation of capital and the ambivalent necessity for capital to descend temporarily into the material domain, then, with the body reconfigured as an informatic entity, capital can potentially bypass to a greater degree the vulnerabilities of the C intermediary in the M-C-M′ formula. By supporting the creation of genome databases, patient-specific genetic tests, and integrated medical and pharmacy data networks, the biotech industry facilitates minimal contact with the biological body. Or, to be more specific, the general tendency toward a bioinformatics-based approach to drug development makes contact with the biological body determined by the degree to which that body and "life itself" are understood as informatic. *Thus, the passage from Aristotelian form to genetic information is mediated by capital.* We can even extend Aristotle's metaphor into Marx's analysis. In the same way that, for Aristotle, the true function of the eye is sight, the true biological function of DNA becomes the generation of capital (as information): the "form" of DNA is information.

As previously stated numerous times, this tendency toward an informatic understanding of "life itself" in no way implies the total liquidation of the material, biological body. The point is that the material biological body—"life itself"—is also the informatic body. The body materially counts only inasmuch as it is understood as information and as genetic information. From Marx's political-economic perspective, fields such as bioinformatics and pharmacogenomics can be seen as an attempt to create a new type of commodity, one that is not based on services, affective entertainment, or immaterial goods. This new type of commodity is just material enough to permit the M-C-M′ cycle to continue; indeed, another name for Marx's general formula is what the pharmaceutical industry calls "wet-dry cycles." This minimally material commodity is just material enough to permit the necessity of further genetic tests and the ongoing consumption of new drugs in the cycle of obsolescence. But as a form of "life itself" that is informatic and bioinformatic, this commodity creates a more seamless transition from M to M′; the area of greatest vulnerability and risk—the living body—is thus placed at one or several removes from its context in novel artifacts such as genome databases and SNP-based diagnostic tests. If, as the hopeful forecasts for pharmacogenomics state, drug development is to take place almost entirely in silico in the future, and if, in such a context, the body is only minimally material, then this mode of economic valuation would seem to be the most beneficial for the biotech and pharmaceutical industries.[82]

Yet, despite this attenuated materialism, the pharmaceutical industry is not always able to derive a profit differential from information alone. In a sense, "life itself," being ambivalently material and immaterial, also resists the M-C-M′ process. We can see this resistance most explicitly in cases where a drug is inaccurately prescribed, where the regulation process fails, or where severe ADRs are reported for drugs on the market. The human clinical trial is still the true testing ground of a drug. But even a human clinical trial is framed by a range of factors, including the population chosen for the trial and the difficulty of maintaining standards of drug effectiveness. In many cases, the real clinical trial for new drugs happens after FDA approval; drugs are really tested when they are already on the market.

Milieu and Life

At once informatic and material, dealing with the "code" of life and "life itself," fields such as bioinformatics and pharmacogenomics are much more than the mere disembodiment of bioscience and medicine through information technologies. Although the concept of a genetic or protein "code" is, to be sure, widespread in the research, the deployment of this concept in a range of practices—from sequence analysis to genetic testing for ADRs—reminds us that biology has, if anything, become more material, not less.

Yet in all the discussions concerning "pairwise sequence alignment," "homology modeling" in proteins, and "knockout experiments" for selected biopathways, one factor has curiously been sidelined: environment. The relation between organism and environment raises a great many issues, including causality, determinism, individuation, and adaptation. In the history of biological thought, the relation between organism and environment has been a constant concern, and, in a sense, biological theories ranging from mechanism to vitalism to evolution can be seen as different attempts to bridge the gap between the two. Have bioinformatics and pharmacogenomics, through their use of new technologies, been able to integrate the organism-environment division successfully? Or has that division simply been reconfigured as a division between "lead" and "target" molecules in pharmacogenomics?

Canguilhem, writing about the paradigm shift taking place with biochemistry and molecular genetics, notes that the medical concept of pathology may be replaced with that of informational "error." He traces the emergence of this paradigm to early-twentieth-century biochemisty and in geneticist Archibald Garrod's phrase, "inborn errors of metabolism." In biochemistry, Canguilhem finds the twofold emphasis on biological processes that function "in error" and that are also inherited from one generation to the next. What exactly is inherited in such cases? Not the molecules themselves, but some mode of functioning or malfunctioning, a pattern of relationships between molecules (overexpression of certain genes, inability to metabolize certain compounds in the cell, etc.). Everything "works" biologically in that the basic processes of protein synthesis and enzymatic reactions take place, except that the rate or relations through which they take place result in a pathological state for the organism: "Insofar as the fundamental concepts of the biochemistry of amino acids and macromolecules are concepts borrowed from information theory, such as code or message; and insofar as the structures of the matter of life are linear structures, the negative of order is inversion, the negative of sequence is confusion, and the substitution of one arrangement for another is error. Health is genetic and enzymatic correction."[83] For Canguilhem, two effects of this transformation are the defusion of the role of individual responsibility in illness and, in some of the more extreme cases, the determination of the individual organism's biological state by the effects of chance, probability, and noise. Such a view of the organism alters the notion of "adaptation" in its Darwinian usage (and results, in part, in the "new synthesis" of evolution and genetics in the early twentieth-century). Adaptation is no longer a principle of the organism's flexibility, but rather a kind of informational processing of the "factual data of the environment."[84] In this case, it is not only the organism that is understood in informatic terms, but the environment, context, or situation in which the organism is living. Thus, the question, If the organism is essentially informatic, is the environment of the organism also essentially informatic?

Canguilhem addressed this question in a lecture dedicated to the concept of "milieu." In the eighteenth and nineteenth centuries, the concept of "milieu" was positioned against the organism (despite the differences between Lamarckian and Darwinian notions of adaptation) and had inherited dual physical and geographical notions.[85] However, the organism-milieu distinction was further complicated by the growing research in biology and ethology, which showed how different organisms respond to a single environment in radically different ways. Influenced by both nineteenth-century biology and the then-emerging fields of biochemistry and genetics, Jacob von Uexküll's research in biosemiotics began to distinguish further between milieu and environment. The former—milieu—was constructed by the organism through its particular sensory apparatus, whereas the latter—environment—was simply the external, natural world. For Uexküll, the milieu (or *Umwelt*) was therefore quite distinct from the environment (or *Umgebung*) as such.[86] An organism may have any number of external forces acting upon it, but its sensory apparatus enables it to take note only of certain ones; in this way, the organism actively engages the environment through the establishment of a milieu. As Canguilhem notes, for the organism, "the specific behavioral milieu (*Umwelt*) is a set of stimuli that have the value and significance of signals. To act on a living thing, it is not enough that physical stimuli be produced; they must also be noticed. . . . *Umwelt* is therefore a voluntary sample drawn from the

Umgebung, the geographical environment."[87] Thus, we have one environment and many milieus corresponding to many organisms. In addition, the milieus are stratified: a milieu for the organism may be the nucleus, the cell, or the external environment. "From the biological point of view, one must understand that between organism and environment there is the same relationship that exists between the parts and the whole within the organism itself. . . . The biological relationship between the being and its milieu is a functional one, and as a result it changes as the variables successively change roles."[88]

This general distinction between milieu and environment has been elaborated more recently by biologist Richard Lewontin, who proposes a new view of genetics as a "triple helix" of genes, organism, and environment. As Lewontin sharply notes, "the organism does not compute itself from the information in its genes nor even from the information in the genes and the sequence of environments."[89] This sort of determinism is the point of critique for Lewontin, for it fails to take adequately into account the "chance" properties of random processes such as "developmental noise." Lewontin proposes a view of the organism-environment divide based on interrelationality: "Taken together, the relations of genes, organisms, and environments are reciprocal relations in which all three elements are both causes and effects." But, beyond this, Lewontin concurs with Canguilhem regarding the emphasis on the active, creative processes of the organism. As Lewontin states, "organisms not only determine what aspects of the outside world are relevant to them by peculiarities of their shape and metabolism, but they actively construct, in the literal sense of the word, a world around themselves."[90]

For Lewontin, the point of critique against the reductionism of much genomics-based research is that it often excludes both milieu and environment in its hunt for "disease-related genes." Or, when a notion of environment is allowed, too often it implodes the milieu into the environment, effectively polarizing the organism against a sharply demarcated, external environment. Lewontin's triple helix serves to show how the concept of milieu operates at many levels: gene, organism, and environment. As Canguilhem notes, these relationships are not simply a matter of information processing, but of an informatic-based understanding of biological life that is inseparable from the material, meaning-making processes of the organism: "Biology must therefore first consider the living as a meaningful being. . . . To live is to spread out; it is to organize a milieu starting from a central reference point that cannot itself be referred to without losing its original meaning."[91]

The question raised by the concept of the milieu is whether or not there can be an informatics-based understanding of the relations between organism, milieu, and environment. *Can there be a milieu of bioinformatics?* In other words, is it possible to account adequately for significant, extragenetic factors in the biology of the organism without having to renounce wholesale the informatic paradigm of molecular genetics? Biologists such as Lewontin and Lynn Margulis seem to suggest that it is possible. Indeed, such a view is implicit in the emphasis on networks in biochemistry and in the study of gene expression networks and biopathways. However, the question is what changes would be required in the process, especially in cases where—as we have seen—the informatic paradigm of genetics smoothly dovetails into the economic paradigm of patenting and the production of diagnostic data as ends in themselves.

3

A Political Economy of the Genomic Body

The Burdens of Data

During the summer of 2000, both the "publicly funded" IHGSC and Celera Genomics announced the completion of the sequencing of the human genome.[1] The milestone event was given much press, including a White House press conference, in which President Clinton described the genome map as "the most important, most wondrous map ever produced by humankind." However, the joint announcement was preceded by a long string of tensions and disputes between the two projects, primarily over the issues of access to information, decentralized operations, and joint publication. Whereas Celera was concerned over its ability to develop proprietary database subscription services, the IHGSC was concerned over simply keeping up with the pace at which Celera's sequencing machines were churning out genome data. The familiar public-private tensions continue to this day, but it is also important to note that despite the organizational and infrastructural differences between the two projects (Celera's more centralized operations in Maryland versus the IHGSC's international outposts), both genome projects were unified in their use of new genome sequencing technologies: automated sequencing computers, genome analysis software, and visualization tools.

In early 2001, when both teams published their reports, one newspaper headline wondered, "What Do We Do with All That 'Junk DNA'?" The article was referring to both projects' finding that approximately 97 percent of the total human genome actually consists of "noncoding regions" of DNA.

This means that only 3 percent of the genome contains information for the production of proteins, leaving vast stretches of sequence as "junk DNA." In truth, the article's deduction is inaccurate because it is thought that noncoding regions may play a significant regulatory, repair, or maintenance role in the genome.[2] The term *junk DNA*—used more in the popular science discourse than in research—is worth paying attention to, however, especially in the larger context of managing the so-called deluge of data the genome projects have generated. "Junk," of course, is not garbage. Garbage is waste that is disposed of or that can be disposed of, whereas junk is waste that may possibly be of use—someday. And, thus, junk sits around, waiting to fulfill its potentiality at some undefined, later date. The idea of junk DNA is in this sense accurate because that 97 percent of the human genome has increasingly become a source of anxiety for scientists in the attempt to communicate to a nonspecialist public the findings of the genome projects. Not only were there far fewer actual genes than expected (from the ambitious estimates of 100,000 to the more modest number of 30,000), and not only were the smallest differences found between the genome of me and you (an average of 0.1 percent differences between one individual's genome and another), but, on top of these findings, only a small fraction of the total human genome was said to contain any meaningful "information." It seems that the average human genome is almost nothing but junk; or, put another way, the human genome seems to be determined largely by genetic excess—the problem of too much information.

The strange "problem" of junk DNA in genomics is contrasted by the enormous level of investment in genomics as a significant part of the biotech industry, including the various *-omic* fields made possible by genomics: proteomics; structural genomics; pharmacogenomics; genomics dedicated to other organisms (the mouse, the fruit fly, viruses); and the development of projects dedicated to the transcriptome, spliceosome, interactomes, and so forth. For instance, in 2000, Celera Genomics reportedly received $900 million in financing (much of which went to their various genome efforts), a sum that dwarfs the U.S. NIH's HGP budget of $250 million.[3] Celera's genome databases, as of 2000, could be accessed through a subscription of between $5 and $15 million (for corporate subscribers; academic labs would pay some $8,000).[4] Other companies, such as Millennium Pharmaceuticals, are taking a different route, concentrating on partnerships with pharmaceutical companies for the development of genetic tests. Millennium has established partnerships with "Big Pharma" corporations such as Bayer, Eli Lily, and Pfizer for the development of drugs and diagnostics. Yet another business model is presented by Human Genome Sciences, which has been focusing on the use of genome information to revolutionize the drug-development process. Aside from its more than 7,000 patent files on genes or gene-related compounds (many of them pending approval), Human Genome Sciences has three of the industry's first genome-based drugs currently in human clinical trials.

The potential benefits of genomics, both medical and economic, are not just the stuff of media hype. It has also been given substantial support by those working in the field. For instance, in a 2003 article in *Science* on the promises of "large-scale biology," Francis Collins, Michael Morgan, and Ari Patrinos suggest that the "organization of large, international collaborative projects" such as the IHGSC can help decentralize bioscience research and thereby help it to move more effectively from basic research to medical application. As the authors note, "genomics holds the promise of 'individualized medicine,' tailoring prescribing practices and management of patients to each person's genetic profile."[5] The benefits, the authors propose, not only will come to those in the developed world, but will have "an enormous impact on health in the developing world," as well as promising to help environmental scientists to "create better methods of cleaning up toxic material, production experts [to] streamline industrial processes, and energy researchers [to] work toward sustainable, nonpolluting energy sources."[6] Likewise, in an article published just after the 2000 genome announcement, Samuel Broder, G. Subramanian, and Craig Venter suggest that one of the primary medical benefits of the human genome findings is in the vision of a personalized medicine, a model of prescription drug health care tailored to the genomic makeup of the individual patient.[7] The authors specifically point to what they call the genomic differences in "host-pathogen interactions," or the ways in which gene expression and other factors affect the pathogenic potential of a virus or bacterium.[8] As they state,

> The genome will provide practical benefit when there is integration of genomic information (genetic loci) with the phenotypes of clinical disease. We are in the midst of a major paradigm shift in biology and medicine: the process of looking at genes in isolation has now shifted towards exploring "networks" of genes involved in cellular processes and disease, identifying molecular "portraits" of disease based on tissue or organ involvement, and ultimately defining the biochemical readouts that are specific

to clinical conditions. The era of "personalized medicine" will evolve as a parallel process, wherein DNA variations recorded in human populations will be integrated into the above paradigm, to guide a new generation of diagnostic, prognostic and therapeutic modalities designed to improve patient care.[9]

Although numerous differences are apparent in the approaches and specific positions of the IHGSC and companies such as Celera, both public and private efforts support this broad vision of a more specific, more efficient, personalized medicine stemming from genomics and its related fields. If we take into account the factors mentioned here—an excess of genetic information (junk DNA, repeat elements, etc.), a growing genome industry, and an effort to streamline the linkages between genome research and medical application—then it is not difficult to understand genomics as the medical, technological, and economic management of information—that is, as a political economy.

In this chapter, my aim is to consider the human genome and the field of genomics as a technoscientific effort to produce and manage huge amounts of information. The perspective of political economy can be useful in such an endeavor, for fields such as genomics, proteomics, and bioinformatics can in many ways be understood as sets of techniques for producing, distributing, exchanging, and making use of biological information. That biological information may be in a material form (e.g., plasmid libraries, cell cultures) or in an immaterial form (e.g., online databases), and it may be used, applied, or consumed in a range of contexts (e.g., drug development, genetic tests, DNA fingerprinting, etc.). What makes fields such as genomics interesting in this light is that they combine two forms of value together: the medical and scientific value of "discovery science," and the institutional and commercial value of products, services, and property rights. Although the aims may ultimately weigh in on the side of the medical and scientific benefits of such research, motive (individual or institutional) is not the issue here. Rather, what is at stake is the development of an understanding of how, in some cases, the dual investment in medical-scientific value and commercial value may not always work together in a smooth way.[10]

This chapter covers a great deal of ground, moving from a consideration of genome databases to a consideration of the textual and informatic tropes in genetics, to a consideration of various hardware systems used in genomics. The thread connecting these topics is the view of genomics as a political economy of the genetic body. However, the overall aim is to show how genomics performs two intertwined functions: *the production of an excess of genetic information and the development of new technologies for managing that excess*. This basic process is seen to be consonant with the economic interests of the biotech industry, which endeavors to link basic research in genomics to its application in the pharmaceutical industry (particularly in drug development and diagnostics). In particular, the tension between an excess of data and the management of that excess is resolved through a novel form of "biomaterial labor," in which the nonhuman labor power of cells, enzymes, and genes is redefined as both material (that is, as living labor) and as informatic (that is, as a biotechnology). The concept of political economy, then, is taken as both a philosophical concept and a critical one. It is for this reason that the critiques of political economy found in Karl Marx and Georges Bataille are helpful in elucidating how the practices of genomics manage all that "junk DNA." First, however, more should be said about the concept of political economy itself.

Critique I: DNA as Source Code

Although the term is not often used today, *political economy* brings together its eighteenth-century definition—the production, distribution, and consumption of wealth—with its more modern connotations regarding the political impact of economic practices and policy legislation. In the eighteenth century, the European mercantilist approach to political economy emphasized a form of state-controlled production and trade, especially regarding monopolies and barriers to free trade. The classical political economists who responded to mercantilism—Adam Smith, David Ricardo, John Stuart Mill—emphasized just the opposite. Their more liberal view toward free trade and laissez-faire economics, unfettered by state intervention, can still be witnessed in contemporary debates surrounding globalization. For Smith, the "mode of subsistence" in the Western world of the eighteenth century was a capitalism based on the division of labor, an economic system that made "liberty" dependent on and consonant with commerce. Freedom meant the freedom to buy, sell, and exchange goods on the level playing field called the market. Smith's famous—or infamous—description of the "invisible hand" of capitalist market forces enabled a coexistence between the freedom to pursue selfish interests in the competitive market and at the same time the responsibility to benefit the community as a whole. Thus, the central question for classical political economists was poised between the benefit of the people who constituted a nation and the need of that nation to develop wealth to support its populace. The challenge was thus, in Smith's words, "providing a plentiful revenue of subsistence for the people" at the same time as supplying "the state or commonwealth with a revenue sufficient for the public service. It proposes to enrich both the people and the sovereign."[11]

However, the concept of political economy elaborated in this chapter is a different version, a *critique* of the political economy elaborated by Marx, primarily in his 1857–58 notebooks, which have been given the title *Grundrisse*. In the introduction to these notebooks and manuscripts, Marx outlines a number of criticisms of classical political economy. The first of these criticisms is his questioning of the ways in which classical political economists rendered categories such as labor, wage, rent, and profit as universal, naturalized categories. Marx's criticisms of authors such as Smith and Ricardo are that they not only assumed such categories as universal, but they projected the universality of the categories of political economy into the past, thereby making it seem as if eighteenth-century capitalism were the inevitable fulfillment of a tendency that had always been nascent in economic relations.[12] In particular, Marx points to how political economists such as Smith and Ricardo began their analyses from the assumption of the property-owning and property-seeking individual subject, rather than accounting for how such a subject had been formed by historical forces. Classical political economy thus took the economic subject "not as a historical result but as history's point of departure," in effect "an ideal, whose existence they project[ed] into the past."[13] Marx's strategy, in both early and later writings, is to emphasize constantly the historical character of economic relations: "Only in the eighteenth century, in 'civil society,' do the various forms of social connectedness confront the individual as a mere means towards his private purposes, as external necessity."[14] This critique will be helpful for us in emphasizing the historicity of the concepts of "information" and "code" in molecular genetics and genomics, especially when seen as technologies that produce large amounts of information.

Genetic information enters here in a redoubled manner. First, it is produced through the methods, techniques, and procedures for physical and sequence mapping, often in literalized carbon-silicon hybrids (e.g., the DNA chip). Second, it is reprocessed on an informatic level, rerouting the specificity of individual genetic polymorphism into a mathematical result of quantitative data across a designated population field.[15] What results is a kind of bioinformatic montage of social normalization across three lines: that of genetic reductionism (ideologies of causality and biological determinism), that of genetic homogenization (either pathologization or marginalization of polymorphisms or point mutations), and that of a fragmented statistical averaging assembled to form a genetic model corresponding to hegemonic notions of normativity and health.

As we have seen, the informatic tropes in molecular genetics during the postwar era were owing in part to Francis Crick's efforts to pursue the genetic code ("the book of life," "the code of codes," and so forth).[16] In the late 1940s, physicist Erwin Schrödinger suggested in a lecture entitled "What Is Life?" that the development of genetic science had made possible a new understanding of biological life in terms of information mechanisms and quantum mechanics. As Schrödinger stated (prior, of course, to the elucidation of DNA's structure), the chromosomes of every cell "contain in some kind of code-script the entire pattern of the individual's future development and of its functioning in the mature state."[17] He would go further, calling the hereditary units in the chromosomes "law-code and executive power," endowing genetic material not only with information-storage capacity, but with instrumentality as well.[18] This reference to information was no accident; the complex roles that mainframe computers played in the war and thereafter the increasing dissemination of computers in business and industry made for a fertile discourse concerning computerization and society.

It was during this time that Norbert Wiener and Claude Shannon separately began to develop theories of information.[19] Along with colleague Warren Weaver, Shannon developed a simple model for studying information transmission, one that envisioned a linear process moving from the sender to encoder, through a medium, to decoder, and then to receiver. Though Wiener and Shannon's theories of information differ in important ways, both were led toward a consideration of information as a process independent of content.[20] What this consideration amounted to was that information theory was, for Shannon, concerned exclusively with the transmission of data, irrespective of what the "message" actually said—it was a thoroughly quantitative and statistical issue. Thus, essential aspects of information theory were not specific to the contents of the message to be transmitted, but were concerned primarily with the efficiency and accuracy of the encoding/decoding mechanisms, as well as with the medium of transmission. The implications of this classical version of information theory is that, as Katherine Hayles points out, "information lost its body."[21] As Hayles suggests, when information is equated with disembodiment, a critical intervention is needed in order to point out the embodied contingencies of information and the subjects immersed in information.[22]

This historically and socially rooted equation of information with disembodiedness is one of the major assumptions behind the gradual "encoding" of genetics during the same period in which Wiener and Shannon were doing their research. Well aware of the then-current developments in cybernetics and information theory, Watson and Crick explicitly adopted a language in their articles referencing *information*, *messages*, and *codes*.[23] The influence of this general terminology persists to this day, especially in light of the advances made in molecular engineering, biotech, and genetic engineering—not to mention in popular science books, media reports, and classroom textbooks. One of the greatest challenges put to genomics has been how this universal informatic code operates with such diversity and variety within and across species.

Modern genetics has for many years supposed that this process occurred in a linear fashion, from the "master molecule" of DNA to the end product of a particular protein.[24] This notion of DNA as a controlling mechanism—that is, as a "steering" mechanism, as in classical cybernetics—implies both a causality and the positing of a point of origin in the biological regulation of the body. Influenced in part by contemporary developments in cybernetics, nonlinear dynamics, and complexity, many contemporary approaches are radically reconfiguring the role DNA plays in the organism.[25] Such perspectives place a great deal less emphasis on DNA as of central concern; rather, what becomes important are patterns and networks, the multitude of heterogeneous elements that collectively form an operational matrix. The telos of such operational collectivities is, certainly, the production of key molecules such as amino acids, but it is also the reproduction of the system itself. Current genetics research suggests not only that DNA is much less responsible for the production of proteins, but that one of its primary functions seems to be the reproduction of its own state. As Susan Oyama and others make clear, DNA is, by current standards, relatively inert, controlling nothing and containing only "lists" or "recipes" that are referenced by certain other molecules.[26] For Oyama, concerned with deconstructing the assumptions regarding the genetic basis of evolutionary and developmental change, DNA serves the function of a kind of guarantee of agency in the body's smallest regions: "The discovery of DNA and its confirmation of a gene theory that had long been in search of its material agent offered an enormously attractive apparent solution to the puzzle of the origin and perpetuation of living form. . . . [I]n fact, genetic information, by virtue of the meanings of in-formation as 'shaping' and as 'animating,' promised to supply just the cognitive and causal functions needed to make a heap of chemicals into a being."[27] According to such emerging perspectives, the genome map (that the essence of the genetic body is found in the DNA) is only a small part of the picture, not taking into account the actual pathways, "subsidiary" molecules, intracellular environment, and a host of other parameters. Despite these advances in deconstructing the centrality of DNA, genomic mapping—largely for political and economic reasons—still assumes the primacy and indeed causality of DNA as the instigator if not the reservoir of the totality of genetic meaning.

We can again extend our analysis thus far and suggest that one of the primary manifestations of the close ties between genetic and computer information has been the formation of an epistemology known as "genetic reductionism." As a critique, genetic reductionism suggests that genetics relies on and is dependent on the metaphors that historically have been appropriated from a technical field.[28] In its worst-case scenarios, this view of genetics comes with the consequences of eliding the differences between information (as disembodied, as quantitative) and the organism (as embodied, as qualitative). More important, it becomes especially problematic when ways of understanding biological life as informatic are naturalized and their conditions of emergence effaced. Reductionism in the biotech industry is, in a sense, nothing new, nor is the well-known relationship between molecular biology and information theory, genetics and cybernetics. Much of this history has been exhaustively covered by historians such as Lily Kay, as well as by those writing from the perspective of science studies (Evelyn Fox Keller, Richard Doyle, Donna Haraway, and many others). These and other studies often note the following key points:

- The scientific language of codes, information, sequences, texts, books, and maps includes historically derived tropes and figures that significantly influence the theory and practice of molecular biology and genetics. Far from being a transparent and self-evident "code" or "map"—as in popular accounts, such as Matt Ridley's *Genome*—these scientific tropes and figures play a central role in the articulation and construction of objects of knowledge. In the life sciences, the hegemony of the "genetic code" is one such object.

▪ More specifically, it is in the discursive exchanges between, on the one hand, biochemistry and molecular biology and, on the other, information theory and cybernetics that we can see the gradual, uneven, and polyvalent shaping of the language of genetic codes, genetic "transcription" and "translation," and the metaphors of the "book of life." Molecular biologists such as Francis Crick, James Watson, François Jacob, Jacques Monod, George Gamow, Max Delbrück, Martynas Ycas, Henry Quastler, and others worked together and separately on what came to be known in the 1950s as "the coding problem": how a mere four base pairs (A, T, C, G) generated 20 amino acids, which then went on to form complex proteins of different specializations in the cell. In attacking this Cold War "enemy code," biologists were influenced by Claude Shannon and Warren Weaver's work in information theory, Norbert Wiener's work in cybernetics, and John von Neumann's work in cellular automata.

▪ However, this appropriation of information theory, cybernetics, and early computer science was not a direct one; in the process, the definition of *information* changed several times because the pairing between, say, Wiener's cybernetic feedback model and gene expression was not always a one-to-one match. In some instances, conceptual and metaphorical mutations occurred (as in Watson and Crick's misappropriation of Shannon's definition of *information*); in other cases, the informatic paradigm came to shape biological and genetic knowledge directly (as in Jacob and Monod's configuration of gene regulation as akin to informational feedback in a computer).

▪ In these studies of the relationship between genetics and informatics, not only are the information sciences seen to have significantly shaped postwar molecular biology research, but the particular way in which it was shaped tended to emphasize the abstract, immaterial pattern, the form or information, of molecular biology. It is not difficult to see how a genetic reductionism followed from a perspective that placed undue emphasis on DNA and genes, on what Crick and Watson called "the master molecule." In our contemporary context, the headline-making news of the discovery of "genes for X" as well as the health-care and pharmaceutical industry's emphasis on gene-based drugs and diagnostics can also been seen as outgrowths of this discursive exchange. Indeed, many of the current debates over genetic patenting and the sharing of information also take place within this broad historical shift.

This is, briefly, the crux of much of the science studies research on molecular biology and genetics. Although this research has been extremely helpful in elucidating the processes through which scientific paradigms arise and the artifacts they construct, we might go further and extend these studies to make a few more points of clarification, given our context of genomics, bioinformatics, and pharmacogenomics.

First, the intersection of molecular biology and informatics does not mean that the former is dematerialized by the latter. Information theory and cybernetics do emphasize abstract pattern over concrete materiality, but it is precisely in their incorporation into molecular biology and genetics that we see a new concept of biological materiality emerge. The mere existence of tropes denoting codes and sequences does not imply the "informatization" or "virtualization" of biology. If anything, the introduction of information theory and cybernetics into molecular biology and genetics has meant the opposite: fields such as pharmacogenomics utilize information technologies to enable new practical modes of materialization—for example, in the genomically designed drugs and diagnostics that molecularly touch patients' bodies. Similarly, laboratory technologies such as oligonucleotide synthesizers, DNA microarrays, and genome sequencing computers demonstrate the extent to which informatics is a material science.

Second, despite this newfound materiality made possible via informatics, there is also a strong emphasis on the valuation of genes or DNA. Genetic reductionism implies a certain mode of valuation, and in a reductionist framework what is of greatest value is the code—even more so than the thing itself. This is at once a separation of molecule and code (in that a sequence can be studied without the cell from which it was extracted) and a new type of reintegration (in that code materializes molecule and molecule informationalizes code). The very idea of a genome database implies this twofold relation—at once nothing but the source code (minus the material or body) and yet somehow the core of "life itself." But the existence of actual computer databases is not a prerequisite for this type of thinking. As we saw earlier, the development of in vivo and in vitro "banks" such as plasmid libraries shows us the extent to which informatics had already infused biological laboratory techniques. A database can exist, in effect, without computers of any kind.

Today, this relationship between DNA, database, and value is made more concrete in commercial genome databases (such as those offered by Celera or Incyte), as well as in the U.S. PTO database categories containing patents on genes and gene-related compounds.

Third, the historical backdrop of information theory (Shannon's communications and code-breaking work), cybernetics (Wiener's research in cybernetic artillery systems), and computer science (von Neumann's consultancy on many wartime and postwar mainframe computer projects) has a direct effect on molecular biology and genetics. Not only were biologists aware of their postwar efforts to "crack the code of life," but the very introduction of code and code breaking into genetics reconfigured the study of life as the practice of command, control, and communication of "life itself." It is only later, with the waning of Cold War tensions and the boom of the biotech industry, that the HGP began to become a "big science" project in its own right. Yet the war-code-DNA relation can still be witnessed today, not only in the genetic "war" on cancer or AIDS, but in the pharmaceutical industry's emphasis on the discovery of "gene targets" and biomarkers, and in the complicated tension between health insurance and health security.

Finally, despite the manifold influence of informatics in molecular biology and genetics, something always escapes. Against the genetic-reductionist position, we can re-read the work of Jacob, Monod, Quastler, André Lwoff, Manfred Eigen, and Henri Atlan as implying a high degree of complexity in the process of gene expression and regulation. In a sense, the antireductionist position is already there in molecular biology research of the late 1950s and 1960s, with its emphasis on "organization," circuits, hypercycles, networks, and pathways. This alternate scientific history has been effectively marginalized in the attention given to DNA and genes, but, again, it is part of the research of the period, influenced as it was by Shannon's emphasis on networks, Wiener's emphasis on systems, and von Neumann's emphasis on combinatoric complexity.

Critique II: DNA, DBMS, and Biomaterial Labor

In the previous section, I discussed how Marx's first critique of political economy was directed at its universalist assumptions. A second critique Marx leveled at classical political economists was that writers such as Smith and Ricardo were unable to account for the development of the most fundamental concepts of political economy, concepts such as property, labor, and capital. Specifically, it was the tension-filled relationship between *labor* and *capital* that occupied much of Marx's attention in the *Grundrisse* and in later writings. The challenge for capital is that it must constantly develop techniques for the incorporation of what Marx calls "living labor" into its basic relationships of wage labor, commodity production, and the extraction of surplus value. In Marx's now-familiar analysis, the worker's living labor is separated from his or her means of production by the selling of his or her labor power to the capitalist, the owner of the modes of production. From this selling of labor power—a transformation of living labor into labor power—the capitalist extracts surplus value, which constitutes, for Marx, the primary mechanism of exploitation. For Marx, living labor is a capacity constitutive of the individual and social body: "Production is always a particular branch of production—e.g. agriculture, cattle-raising, manufactures etc. . . . [P]roduction also is not only a particular production. Rather, it is always a certain social body, a social subject, which is active in a greater or sparser totality of branches of production."[29] Thus, the relations of capital provide the conditions in which living labor becomes labor power and in which this labor power can be bought and sold as wage labor, as "the *objectivity* of living labor."[30]

Marx referred to labor as the "living, form-giving fire" of society and in the "fragment on technology" had contrasted living labor with the "dead labor" of fixed-capital machinery.[31] This means, among other things, that labor is not simply a passive force inexorably incorporated by capital. In the 1970s, Italian Marxists associated with the *autonomia* movement would redefine Marx's tension between labor and capital as a potential source for intervention. Reading Marx's concept of living labor in this light, Antonio Negri suggests that a certain antagonism exists between labor and capital:

> Labor can therefore be transformed into capital only if it assumes the form of exchange, the form of money. But that means that the relation is one of antagonism. . . . [I]t is this antagonism which is the specific difference of the exchange between capital and labor. . . . The opposition takes two *forms*: first, that of *exchange value against use value*—that the only use value of workers is the abstract and undifferentiated capacity to work—the opposition is also *objectified labor against subjective labor*."[32]

Labor is thus also active—or, put another way, living labor is defined precisely by the degree to which it cannot help but to escape, evade, or sidestep altogether the labor-capital relation. As Nick Dyer-Witheford notes, "labor is for capital always a problematic 'other' that must constantly be controlled and subdued."[33] Thus, despite Marx's rather dichotomous characterization of living versus dead labor, the identification of living labor looks forward to the biotech industry, in which genetically engineered mammals, microbes, and cells are literally put to work as biotechnologies. Here Marx's characterization of technology-driven production as a "living (active) machinery" and as a complex, "mighty organism" seems to shed new light on the development of genomics artifacts such as bacterial plasmid libraries and computer databases.[34] "The appropriation of living labor by objectified labor—of the power of activity which creates value by existing for-itself . . . is posited, in production resting on machinery, as the character of the production process itself."[35] Transported into the biotech industry, this view shows us a new kind of labor power being created within biotech fields such as genomics—a strange, uncanny, nonsubjective labor that is as far from the living human subject-worker as it is from the cold, mechanical factory machine. It is different from the so-called immaterial labor of the information, computer, and network industries. In effect, this form of biomaterial labor is at once fully technical and yet fully biological, both living and dead, a process-driven and yet fixed form of capital.

Previously, we saw how Marx's first critique of political economy's tendency to universalize its terms can thus be seen in contemporary biotech fields such as genomics, in which the figures of information theory and cybernetics contribute to the increasing naturalization of DNA as a "code" and to the definition of biology as in some essential way informatic. However, if the influence of information theory and cybernetics were limited to the conceptual level, the effect on the practices of genomics might not be as integral as it is today, for not only are informatic modes of understanding biological life attaining a certain hegemony, but, more important, they are materialized in a range of artifacts that reflect this informatic mode of defining and organizing "life itself." The database is one such artifact, a mode of organization that at once orders and produces that which it orders. In the "fragment on technology" (Notebook VII), Marx describes nineteenth-century innovations in automated machinery as redefining the relation between worker and commodity, thereby also changing the nature of labor power itself.[36] However, we also need to remind ourselves that Marx is writing specifically about industrial production in the nineteenth century. Many changes have occurred in the shift to the biotech industry of the late twentieth and early twenty-first centuries.

What are these changes? We can begin by taking a basic lesson from Marx's critique of political economy: that the categories of production, distribution, and consumption are determined in part by a scientific and technical infrastructure for defining what counts as labor, as a commodity, and as capital. (Recall Marx's proposition that in production the living labor of the worker is consumed, thereby making production a consumption.) Certainly, the biotech industry involves traditional modes of manufacturing, especially in the production of laboratory technologies (e.g., computers, thermal cyclers, bioreactors), as does the pharmaceutical industry (e.g., in the factory production of drugs). However, what distinguishes the biotech industry from other industries is the way in which it integrates biotechnologies and information technologies into a new synthesis. There is no better example to illustrate this point than the database. In computer science, databases have been used for decades in all different types of configurations. However, a database in biotech fields such as genomics can mean a number of things: databases can be standard computer databases (such as GenBank, the main genome database), but they can also be "wet" databases composed of microorganisms, racks of test tubes, or even genetically engineered artificial chromosomes. The way in which genomics and its related fields correlate these types of databases may very well structure (or restructure) the way in which "biological information" is defined in these fields and in drug development.

What is a database? From an archaeological perspective, modes of information storage are as old as civilization itself; the need to keep records of transactions, calendar systems, and valuable objects constitutes a given social formation. As Geoffrey Bowker and Susan Leigh Star point out, classification systems such as the computer database play an integral role in the articulation of knowledge about discrete objects, events, and situations.[37] Like its historical precedents—on the one hand, informational systems such as the dictionary, the encyclopedia, the field of statistics, and, on the other hand, representation systems such as cartography—the database has as its primary function the particular organization of a range of heterogeneous material. The particular manifestation of that organization depends on the technical specificity of the system; whereas the dictionary and encyclopedia utilize an alphabetic system, the anatomical atlas utilizes a diagrammatic logic, and statistics utilizes a series of tables, charts, and quantitative graphical elements. By comparison, the database can accommodate each of these types of organization, but can also integrate them depending on the "front-end" interface or search engine utilized to organize and find material in the database.

But it is specifically with technological modes of storing information that social formation and technical organization are increasingly separated (and thus allowed to transform each other). Friedrich Kittler provides an in-depth analysis of the "media revolution of the 1880s" in the mechanical-inscription technologies of the gramophone, film, and the typewriter.[38] Whereas these storage technologies tended toward the isolation, separation, and autonomization of the senses (sight in film, sound in phonograph, language in typewriter), Kittler sees the current digital media revolution as providing a synthesizing phase, in which all media operate through binary code. This standardization of the language of storage media means an effective reassemblage of previously incompatible media formats and a greater homogenization of media themselves.

According to contemporary textbooks and technical manuals, "a database is a collection of related data that contains information about an enterprise such as a university or an airline. Data includes facts and figures that can be represented as numbers, text strings, images, or voices stored in files on disk or other media. A database management system (DBMS) is a set of programs (a software package) that allows accessing and/or modification of the database."[39]

For computer databases, three properties characterize its design: first, a database is not just a randomly generated grouping of information, but a collection of logically related elements grouped into a single structure for a specific reason (e.g., the online genomic database contains data related to specific genes). Second, the database is functional; it is more than a knowledge base or an archive, but it is designed with a set of potential uses and applications in mind (e.g., the online genomic database is geared toward research in mapping and sequencing, drug development, and genetic medicine). Finally, the database always establishes, through its design and implementation, some direct connection to the "real world." This connection may occur simply through representation (e.g., molecular modeling) and bibliographic modes (e.g., a health network database of medical doctors; a medical library database), or it may occur through more ambiguous means (e.g., cell samples and high-throughput DNA replication).

The database contains, of course, the data files that provide its content; these data files ostensibly are sets of binary code, which translate into a given mode of presentation on the output side of the computer. In order to manage the large groupings of data files within the database structure, a database management system (DBMS) is required. The DBMS is essentially a software package that enables the design, creation, regulation, and accessing of a database; it is the production and reception gateway to the data files in the database. Within the DBMS, there are two main components: a database definition language (DDL), which contains descriptions and definitions of the database structure, the elements within it, and the kinds of transactions possible; and a database modification language (DML), a functional language that enables the updating, deletion, and modification of data files within the database.

A database and a DBMS have four main advantages over more conventional file-management systems (as in good old-fashioned file cabinets or the later file-based computer systems):

- The DBMS makes possible what is called "data abstraction," whereby data modified on the "back end" (producer) does not modify the representation of data at the "front end" (user). This means that, for the computer user, the database can be black-boxed, such that the data representation (say, a file on a particular gene) will not be modified if a specific part of the data file is modified (say, a change in the file describing its location in a folder or data compression in data transfers).
- In data-processing lingo, the DBMS can act as a "normalization" system, dynamically regulating the database as a whole so as to prevent unnecessary procedures (duplication and redundancy, deletion and insertion of errors, and other computing anomalies). The general goal of database normalization is to enable the database to be as efficient as possible, which translates to greater performance for database users and procedures.
- The DBMS separates the data (or information) from the program applications that run the database. Earlier file-management systems were composed of multiple systems, each of which managed its own share of data; thus, data were inextricably connected to structure and could encounter system-incompatibility problems. With the DBMS, several databases can act on a single set of data, resulting in greater efficiency and data consistency.
- The DBMS enables a greater degree of automation of database management, as can be seen today by the various Web-based services (including online

shopping, secure servers, automated listservs, "cookies" and CGI scripts, and so forth). This is especially convenient in terms of security and the automation of basic, repetitive procedures (from data entry to backups).

As a way of managing logically grouped sets of information, the database and the DBMS form an important link between the distribution and transmission of knowledge between computer users and storage systems. Many computer science and database-management textbooks present this relationship as an interface: between the hardware and software of the database as a computer system, on the one hand, and the sets of actions affected by human subjects, on the other, the database and DBMS serve as mediators, enabling or disabling access, addition, deletion, modification, transfer, and interactivity:

hardware/software ⟵⟶ DBMS ⟵⟶ actions/subject

Each of the properties described earlier are techniques through which the DBMS interfaces computer users with storage systems. In data abstraction, an "object" (in computer science terminology, a set that includes the "entity" or representation, the "attribute" or particular property, and the "relationship" or links between attributes) is transformed into a certain linear depth of data, between a source code (which can be modified) and a output modality (which is not affected); inscription can thus signify without producing any meaning. In normalization, data are streamlined, pared down, and optimized specifically for functional efficiency; data become aphoristically inscribed as data packets. In the separation of data from programs, information becomes a quantity that is not only universal (it can be used by several programs) but also polyvalent (it can be put to several uses); data become morphological, capable of being instituted in multiple ways. And with greater automation, the database becomes a stronger actant (to use Bruno Latour's term), or a stronger object-participant, in the mediation between the human computer user and computer storage media.

Of little mention is the fact that the HGP used two types of databases. One type was the more familiar system of networked computer databases such as GenBank; these databases were built and managed much like any computer database today, including database management software, secure networking protocols, and user accounts. But the other database used in the production of the human genome map was the BAC, or "bacterial artificial chromosome," a type of wet, biological "library" that has been used in one form or another since the 1970s. The BAC, as its name indicates, is an extrachromosomal, free-floating, circular loop of DNA called a "plasmid," which is found in bacteria such as *E. coli*. The free-floating DNA can be spliced with a gene sequence from any source (human or mouse), and, when cultured, the bacterial colonies will then contain the inserted gene. The "plasmid library" can be used to store in vitro samples of any desired gene, thereby forming a kind of living, biological database. In bacteria such as *E. coli*, the relatively short length of plasmids allows for an easily controlled storage of gene sequences of up to approximately 10 kbp (where one kilobase equals 1,000 bases), making plasmid libraries excellent storage systems for modest-size gene sequences.

The use of plasmid libraries (figure 3.1) was actually an off-shoot of the genetic engineering experiments of the early 1970s. Bacterial plasmids provided the ideal platform for experimenting with the splicing and editing features of certain "restriction endonucleases," enzymes that form the scissors and glue of DNA. In 1974, engineered plasmid vectors with obscure names such as pBR322 or pSC101 were used to demonstrate the effectiveness of recombinant DNA techniques—the foundation of the biotech industry. Researchers discovered that the use of engineered plasmids, when inserted into bacterial cells, also provided an efficient and stable means of replicating or cloning those genes through bacterial culturing (thus the name "plasmid cloning vectors"). In the late 1970s, these basic techniques were used to develop DNA libraries based on engineered plasmids. These "wet" databases consisted of two components: first, the creation of the database using the techniques of genetic engineering, which resulted in bacterial cultures containing plasmid libraries with foreign DNA, and, second, a means of "searching" these databases using a "DNA probe." The construction of plasmid libraries often makes use of cDNA, or complementary DNA, a double-stranded DNA produced from a single-stranded RNA (in effect reversing the normal process in the cell, in which genomic DNA produces RNA). This use enabled researchers to focus on the gene-coding segments of the genome, minus the junk DNA of introns. Once the cDNA is synthesized, it can be inserted into the bacterial plasmid and then cultured. Indeed, most uses of plasmid libraries contain more than one inserted gene sequence (just as a computer database's hard drive contains more than one file).

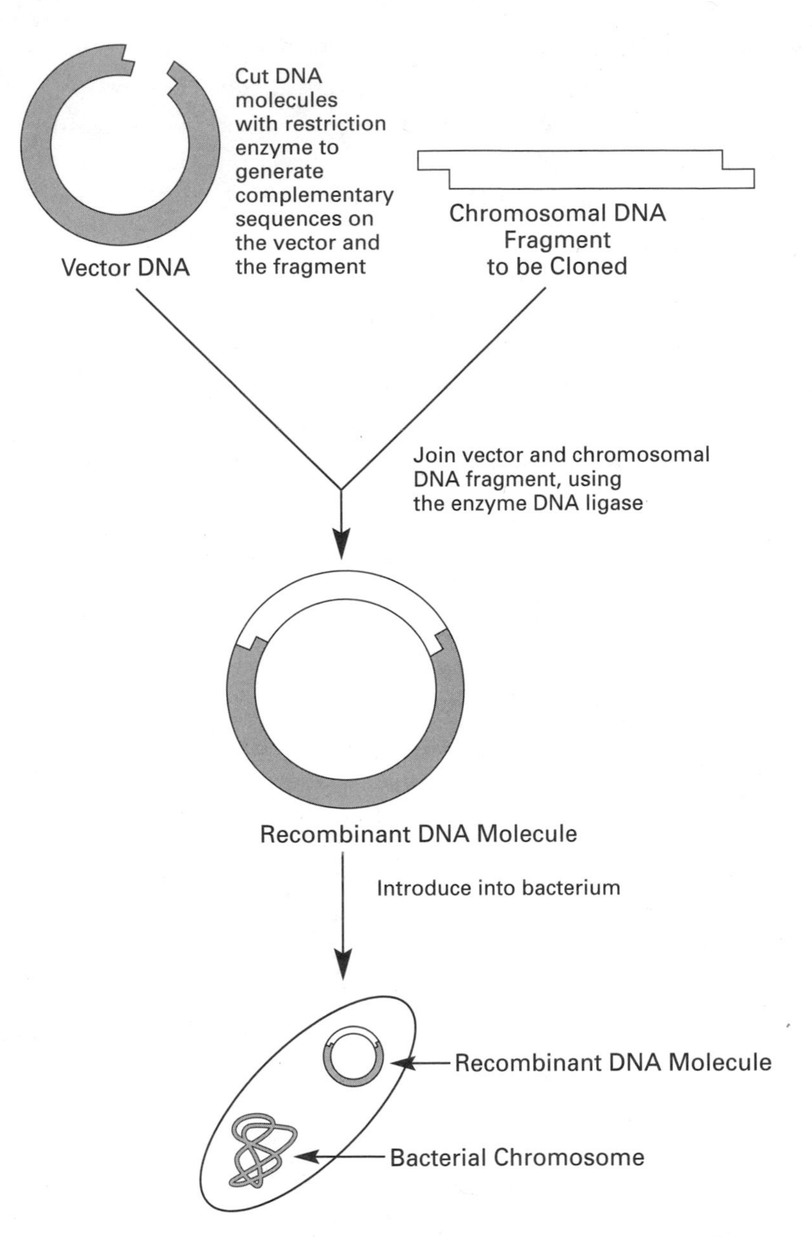

Figure 3.1 Constructing plasmid libraries for sequencing. Image courtesy of the U.S. DoE Genome Program, http://www.ornl.gov/hgmis.

However, when researchers need to access this information, a method of searching these "wet" databases is required. A technique often used is known as "colony hybridization." If, for instance, you have a plasmid library with two or more genes inserted into it, you can search for which cells contain the desired sequencing using a DNA probe. Let's say the gene you are after has the sequence ATTTGC (this is for the sake of simplicity; usually gene sequences will run anywhere from 1,500 kbp to 200,000 kbp or more). You can synthesize a single-stranded DNA molecule with the sequence TAAACG, which is the complement to the gene you are after. This DNA probe can then be applied to the cell cultures (after the cell membranes have been lysed and the DNA denatured). In addition, the DNA probe will be radioactively labeled, so that when it binds it can be detected.

Prior to and alongside the introduction of new computer technologies, the construction of plasmid libraries has been an essential tool for genomic studies. In the early years of the HGP, plasmid libraries already existed for bacteria such as *E. coli*, the yeast *S. cerevisiae*, the roundworm *C. elegans*, and the *Drosophilia* fruit fly. The time-consuming and tedious task of genome mapping at this time consisted in using plasmid libraries to construct overlapping clones of DNA fragments, which could then be painstakingly reconstructed using technologies such as gel electrophoresis. For the IHGSC and Celera genome projects, newer computer and robotics technologies that automated the previously manual process of sequencing enabled a faster, more efficient mode of gene mapping using plasmid libraries. The IHGSC, for instance, made extensive use of BAC plasmid libraries to store and clone the numerous fragments from the sample human genome. Many of the IHGSC's BAC libraries were derived from BAC repositories, such as the Cal Tech or RPCI "large-insert" genomewide libraries.

Although plasmid libraries are composed of the "wet," messy "stuff" of biology, they nevertheless employ a database logic—even prior to the development of current microchip-based computer memory. Certainly, the idea of the database can be traced as far back as the idea of a computing machine (storage of data was a problem in Babbage's nineteenth-century calculators and was not fully theorized until John von Neumann's draft for a computer architecture in the early 1950s). Yet what is interesting is that the creation and use of plasmid libraries show already the degree to which the informatic paradigm of molecular biology materializes its artifacts. In a sense, the plasmid library, complete with infrastructure (restriction enzymes, DNA complementarity) and a search engine (DNA probes), is a better example of how biotechnology integrates genetic and computer "codes" than is the computer databases of GenBank or the Protein Data Bank (PDB).

However, in the past 5 to 10 years (that is, during the period of the development of the Web), an equally significant mode of organization has combined with genetics research and mapping endeavors to produce a second, redoubled mode of organizing the genetic body—that of the online database.[40] When the government-funded HGP was launched in 1988–89, its organizers had already foreseen the problem of large-scale organization and archiving of vast amounts of genetic information.[41] At the time, the Internet was certainly not the ubiquitous medium it has since become, but the Internet's origins in both military-governmental and research-educational institutions meant that the possibility already existed for the effective computerization of an entire human genome. Significant changes in this field include the rapid developments in computer data storage capacities and processor speeds, along with the maturation of the Internet and Web as the much-hyped media of the future. Joan Fujimura and Michael Fortun have suggested that the introduction of computer databases has had the effect of transforming molecular biology into a mathematically based and theoretical science. Issues of classification and standardization impact not only the tools of biology, but the very concepts of nature produced through such artifacts. "We . . . have to question not simply how these sequence-based technologies will be used, but also what kind of nature we are constructing with these forms of knowledge."[42]

Numerous biotech corporations and research labs have moved from using the Web as an information network (e.g., Web sites of biotech corporations) to utilizing the Web as a constituent component of genetics and biotech research (in online bioinformatics search tools). This shift means that the computerization of research data can and in most cases does combine with networking technologies to produce a decentralized and disseminated data-transfer network that extends from one or more databases. Genomic databases such as GenBank or Ensembl offer examples of such organizational systems used on a regular basis by researchers at many remote locations and with many different institutional and industrial affiliations. Despite the changing perspectives within genetics research as to the actual role of DNA, such databases still reaffirm and reproduce modern genetics' narrative of the genome as the central "source code" and referent point for contemporary biological science and medicine.[43]

In his essay "Gramophone, Film, Typewriter," Kittler brings up the ways in which the media revolution of the late nineteenth century created a linkage between new technical modes, power, history, and the realm of the dead and the phantasmic. Transforming sound into acoustics, visual movement into moving images, and writing into text, these media effected, in their own way, the arresting and storing of sensory phenomena. They were able to do this through a process of standardization (an "alphabetism"), which made possible new modes of communication (new meanings for hearing, seeing, writing).[44] The archive gathers, filters, and organizes multiplicities—peoples, cultures, events, cartographies and so forth. The document is a representational tautology; it always authenticates and verifies only itself as a fact. These storage media make possible the separation and organization of sensory experience and, by extension, the world. When it becomes possible to technologically store something, the capacity to translate time, movement, and vision into information has been achieved. Once the senses can be translated into information, they can be effectively separated and "arrested." Once that information can be stored, it can be organized and made technical through considerations of data transmission, data access, and encoding-decoding procedures. In fact, storage media effect a transformation upon the interface between the body and technology on three levels: they incorporate earlier media as their content; they subjectify viewers and audiences through an interweaving of machine sense and the experience of the subject; and they objectify the sensory body through a capturing of objective properties as an index to sensory experience. As Kittler points out, "man becomes physiology on the one hand and information technology on the other."[45]

For Kittler, this capacity is made possible by these media's ability to arrest/capture, translate, and store/archive information. Within this capacity are not only encoding operations, but the potentiality of placing these body-technology relationships into an archive, or, to use a more current term, the database. In the same way that the mechanical storage media of the gramophone, film, and typewriter form "partially connected media systems," so do they also form the content for a future of *totally connected media*, based on computer code. "In computers everything becomes number: imageless, soundless, and wordless quantity . . . a total connection of all media on a digital base erases the notion of the medium itself."[46] In this way, the archive (the form of mechanical storage) becomes the database (the form of digital storage and computer-based memory). For Kittler, this process follows the broad transition from mechanical storage media to digital storage media and from partially connected systems to totally connected systems. Although the archive is often unable to be modified—as with printed records, gramophone discs, filmic and photographic plates, and typeset pages—the database is defined by its flexibility in the handling of information (and thus the supplementary need for "backup storage," data encryption, and security). Digital media's generative character means not only that computer media can encode themselves (automated processes, expert systems, and agents), but that through this recombinant logic information can come to be seen in new, morphological terms. The implication here is that information, as in the mechanical archive, is no longer simply a concrete materiality of fact; rather, with the database, information becomes the highly technical and automated reservoir for the proliferation and production of other media. The database is not only the foundation from which totally connected media systems emerge, but also the very interstices and linkages that constitute the possibility of connected media.

Such a situation applies directly to the various genome mapping endeavors and the digitization of the molecular and genetic domains. In cultures where storage media play a central role in the relationships between bodies and technologies, storage media fulfill a function of accountability. The human genome databases, SNP databases, population genome databases—all inscribe the subject in the contexts of health, medical normativity, genetic diagnostics. As a databasing endeavor, the HGP is just one isolated example of a much more pervasive technocultural trajectory. This trajectory is not the familiar transcendentalism of moving away from the materiality of the body toward the purity of disembodied cyberspace. Rather, it involves the concern with accounting for the materiality of the medical subject and of the molecular-genetic body. Understanding the broader meanings of genome databases (whether BACs or Oracle databases) requires a comprehension of the ways in which a unique type of genomic body is constructed in the extraction of information from the bodies of cells, proteins, and DNA. The database makes possible a series of potential extensions that exceed the mere recording and preservation of information. With the database, information becomes productive, generative, morphological. With the database, information also becomes organized, classified, and taxonomized according to a range of flexible uses.

According to standard computer science and database theory, the content of the database is of less importance than its structure and mode of organization and function. That is, the consideration of the content of the database (or the descriptions of the "data files") becomes important only in the process of creating a database, in which logically related data must be gathered and analyzed. After this design and gathering process, the content or data files are organized according to "attributes" (descriptive categories along the lines of natural histories) and "relationships" (differences between attributes); that is, after the design and gathering phase, the data files become functional nodes within the larger database structure. They do not lose their specificity as particular data files different from other data files, but they are nevertheless put through a process whereby they are partitioned and articulated according to the logics of a particular type of database (e.g., hierarchical, network, relational, object oriented, etc.).

Critique III: The Excesses of DNA

A brief review: in considering biotechnology fields such as genomics and bioinformatics, we can see how the "problem of too much information" is also a political-economical problem. Marx's critiques of political economy provide us with a way of understanding how biotechnology manages—and in manys produces—its excess. Marx's first critique is against the naturalization and universalism of the categories of political economy, and his second critique is poised on the particular form of the labor capital relation. Finally, a third critique against classical political economic thinking is his questioning of the division between the categories of production, distribution, and consumption. In his reading of classical political economy, Marx identifies production as the starting point (human labor upon nature), distribution as the economic allocation of goods to the people, and consumption as the use of those allocated goods. Whereas writers such as Smith, Ricardo, and Mill tend to treat these processes as isolated and linear (which nevertheless form a whole), Marx's emphasis is on the ways in which each process dialectically determines the other: production is consumption (in the use of laboring capacity), consumption is production (in the replenishing of labor power), and distribution determines both production and consumption (in that it is determined by social class hierarchies). Marx notes, "[t]he conclusion we reach is not that production, distribution, exchange and consumption are identical, but that they all form the members of a totality, distinctions within a unity."[47] What this really means, for Marx, is that the traditional categories of production, distribution, and consumption are superceded by a different kind of production, a production that is immanent to each category and that ties them all together.[48] Marx's critical point is that the system of liberal, market-driven capitalism produces its categories: "consumption produces production," just as "production creates the consumer"; and, in the same way, "distribution seems to precede production and to determine it," just as "distribution is itself a product of production."[49] In this strategic confusion of categories, Marx highlights the ways in which an economic system such as industrial capitalism continually produces its own conditions, its own categories of operation. Beyond the specific, terminal problematics within production, distribution, or consumption separately, Marx points to the way in which the "process always returns to production to begin anew," but a notion of production that is constitutive, both determining and determined. Negri refers to this method as "determinate abstraction," for it attempts to highlight the fundamental yet concrete forces behind specific instances. For our purposes, Marx's deconstruction of the categories of political economy is helpful in considering how the HGP and related technoscience projects conceive of political economy as the management of biological information and in this way actually produce the excess data that are subsequently managed via new technologies. Consider the following examples.

Statistical Genetics

The problem of genetic difference has been an issue with the HGP since its inception. As is known, genomics aims to derive a single genetic map by way of a statistical averaging.[50] Such tactics determine which "letter" or which gene will be entered into the database at a given locus. Given the claims by genetics researchers that each individual differs from others by approximately 0.1 percent of its DNA, such an averaging seems to be a sensible, practical way of working. However, given that there are at least some 30,000 genes and more than 3 billion base pairs ("letters") in the human genome, and that a number of known genetic disorders (such as Parkinson's or sickle cell anemia) are in part triggered by a single base-pair mutation, such homogenizing practices take on a different tone.

At first glance, this massive homogenization (and genetic simulation) of the human genome seemed to be corrected by the Human Genome Diversity Project (HGDP) when it was initiated in the early 1990s.[51] The HGDP's contribution to the human genome–mapping project would be to consider the ethnically based genetic differences from a range of "other" cultures, especially those cultures without dense histories of transethnic genetic migration. Its job would basically be to account for the excess genetic material not within the Eurocentric field of consideration of the human genome projects (that is, the genome projects aim above all for universality, but the HGDP would instead concentrate on the local). It has been only recently, though, that the methods and practices of the HGDP have come under fire, primarily owing to its ties with debates surrounding the patenting of genes and cell lines from indigenous cultures.[52] Certain HGDP practices thus not only constitute literal examples of genetic colonialism (many of the cell lines representing indigenous cultures were collected without proper consent and were intended to be archived—as a type of genetic museum—by the NIH and related government health organizations), but they also bring up questions about the ways in which excess is translated into a dominant system involving science, politics, and juridical-economic structures.

The HGDP epistemologically ends up reduplicating the universal, statistical assumptions of the human genome projects. Although it considers cultural and ethnic specificity, it is too tightly bound to a genetic determinism to be anything but a messy conflation of the discourse of nature and the biology of ethnicity. Here genetic difference does not simply account for cultural-ethnic specificity; it demands it. The always dangerous equation between the discourses of nature and culture formulates the excess of otherness through genetic reductionism. In producing a series of ethnic gene maps, projects such as the HGDP intimately forge a bond between genetic difference and ethnic-cultural difference. When one considers these ethnic gene-mapping projects alongside the human genome projects, one sees not only the development of a genetic technology of normativity, but also the situating of ethnic-cultural excess in relation to the HGP (that is, the mapping out of "otherness").

Technologies of Excess

Genomic mapping depends a great deal on a range of specifically designed laboratory and analysis technologies, most of which are geared toward the increasing automation, speed, and accuracy of genetic sequencing. Of particular concern here are two examples of such technologies, the DNA chip (or microarray) and the automated DNA sequencing machine. Both play key roles in making efficient genomic mapping possible, but whereas the DNA chip has been utilized primarily for the mapping out of minute differences in genetic sequences, the DNA sequencing machines are used for the mass analysis of DNA samples in general genome mapping endeavors.

In fields such as medical genetics and genetic drug design, when the need for quick and efficient modes of analyzing minute genetic differences arose, it was technologies such as the DNA chip that made such analysis possible. The DNA chip is indeed a hybrid of carbon and silicon—on a microcomputer-size silicon wafer thousands of cloned DNA samples are attached—but it is also a product of the biotech and genetics industry, intended to perform the analytical task of isolating and "reading" genetic difference. For example, one of the first uses of the DNA chip was in the field of breast and ovarian cancer research, where researchers were attempting to localize the gene or genes involved in the triggering and development of these types of cancer. Within a probable field of localization, researchers could analyze cancerous cells by using the DNA chip to pinpoint possible minute mutations in a genetic sequence. Such mutations, or SNPs, have been found in relation to a host of genetically related disorders. To achieve the results of this type of analysis, however, a large amount of genetic material must be generated; in other words, SNPs cannot be analyzed simply by using a single thread of DNA. An excess of genetic material is required for SNP mapping and analysis, but this excess does not precede the analytical process. Rather, the genetic material is synthesized or cloned, or both, on two fronts. First, the DNA chip contains thousands of identical strands of DNA, and these strands are precisely allocated sequences that correspond to the suspect gene. Thus, a multitude of DNA fragments must either be generated ("grown" or "assembled") or cloned in order that the DNA chip as a whole will have a greater degree of accuracy. Second, the genetic material to be analyzed (e.g., a sequence or strand of DNA from a cancerous cell) must be cloned en masse because given the current technology, a more accurate reading will result from a greater quantity of genetic sample.

In theory, such indulgence in assuring accuracy is unnecessary, but when considering the extreme importance of even a single base-pair mutation, and given the current technological limitations, a need for the production of a particular type of excess arises. This genetic excess is, above all, the product of a technique—a reading and inscription technique. It is an indication of the amount of genetic material (biological material alternately extracted, cloned, recombined, and regenerated) that must be both expended and produced as "useful excess" in order that the genetic body may be instantiated as a disembodied code or information.

The Accursed Share of Introns

According to current assessments among genetics researchers, the amount of "coding" material in the human DNA accounts for roughly 3 percent of the total DNA in a human being. As genetic science now tells us, this amount is primarily owing to the particular mechanism by which DNA relates to the production of a variety of different proteins in our bodies. Still considered as the master molecule, DNA is thought to initiate a process in which a given strand of a chromosome unwinds and makes itself available for a "transcription" onto a single-stranded RNA molecule. During this phase, large portions of DNA are simply excised or cut out, leaving only the coding portions of the DNA, which snap together, forming a more compact strand of RNA to be delivered to the ribosomes for "translation" into instructions for the building of a particular amino acid (which will later be assembled with other amino acid compounds to form a protein). The parts excised in the transcription process are called *introns*, and the coding portion retained on the RNA molecule are called *exons*. Genetics researchers have for years ambiguously referred to the introns (which, again, compose 97 percent of human genetic material) as "junk DNA," implying both something in excess but something one cannot do without (that is, "junk" as differentiated from "trash").

In addition, because modern genetics discourse and practice is founded on principles of an informatic teleology (that is, the genetic code as a text that exists to be decoded), there has been much speculative debate as to the introns' possible functions. Some evolutionary geneticists claim that the introns are genetic fossils, remnants of DNA that previously coded for biochemical compounds no longer needed by the species, a position termed the "introns-early" theory. By predictable contrast, the "introns-late" theory proposes that the introns are not in fact useless fragments or simply junk, but rather may be inserted into genes later and possibly involved in complex ways in the biochemical process by which DNA is related to the production of proteins and to its own reproduction as a molecule.[53]

What is important here is obviously not which position is correct, but that there is an explicit need among genetics researchers to reconfigure the introns as nonexcessive, productive elements in the larger narrative of a master code saturated with (mostly potential) signification. For this reason, there is debate among genome mapping projects as to whether to map out the human genome completely (as originally intended) or to concentrate exclusively on the coding portions, adding maps of the introns as supplements (as corporate biotech projects are doing). Whichever method is used, it is clear that the introns form a troubling source of anxiety for a discourse such as genetics and for practices such as genomic mapping. Whether the introns are included as part of the genomic map out of a concern for totality and universalism (informed by scientific narratives of totality and completeness), or whether the introns are appended as a secondary but not irrelevant addition (as Celera has proposed, mostly informed by economic imperative), the unthinkable for genomics in general is that the introns are a form of excess that can be understood through genetic science only by way of a recuperation and refiguration, as potentially useful. Entertaining such a notion would be the equivalent of stating that the majority of human genetic material in the body is "pure" or "empty" information, or simply "noise."

Gene Expression Markup Language

In 2001, Rosetta Inpharmatics, a bioinformatics company from the Pacific Northwest, announced that it had developed a cross-platform standard for computer-based molecular genetics and biotech research. Called Gene Expression Markup Language—or GEML—this programming language is designed to facilitate the current file-format and compatibility problems in computer-based biotech research. Now researchers in biotech can work, with the ease of cross-platform standardization, between widely different biological databases, from Celera's human genome sequence to the SWISS-PROT protein database to the Human Gene Mutation database. All online, all digital, and now all in a standardized backend programming language. GEML is based on Extensible Markup Language (XML) and operates independent of any particular database file-format schema. GEML manages two types of genetic data: genetic patterns (gene expression, or analyses of which sets of genes are switched on or off, and which may include biochemical pathway information or gene-protein relationships) and genetic profiles (digital scans of microarray chips, which are used to analyze genetic samples speedily and efficiently). GEML can keep meticulous track of a given piece of genetic data, noting what the original file format was, where the data were retrieved from, the type of database search method used, and the operations performed with a given data file—

presumably biotech research, like Photoshop, now includes several levels of "undo" as well.

All of which is to say that with biotech research the computer is no longer just a tool. It has become much more, extending its range of operations, bringing in computer science and programming, and transforming the "wet" biology lab into a networked computer lab. Unfortunately, researchers and companies still seem to see bioinformatics tools, such as the GEML language, as a transparent tool, something that will transparently aid in the advancement of science research without fundamentally altering research itself or the objects of study. When a GEML-based application accesses several genomic databases, is it also accessing genetic bodies? Or is this process simply just data acting on data, and if so, where exactly are the points of connection to material, biological bodies? No one is asking whether such bioinformatics techniques will fundamentally change our notion of what the body is or whether the level of complexity that bioinformatics can deal with will fundamentally challenge traditional bioscience and genetics research.

This places an "object" such as GEML in a very strange position. On the one hand, GEML, as a programming language, refers to or points to something in its tags, the same way that the tag ⟨IMG SRC="mycells.jpg"⟩ points to a digital image of my cells on a Web page. In this sense, GEML operates not only as hypertext markup language (HTML) does (with referring tags and attributes), but it also operates according to the traditional signifier-sign relationships that characterize modern linguistics. The "thing" the language points to, however, is another type of data, a genetic code, itself with its own set of rules and protocols for functioning. This also means that, as an XML-based language, GEML is developed from the ground up, so to speak, so that the types of tags and attributes used, as well as their interrelationships, will be dictated by the ways in which genetic data operate. Each use of a GEML implementation needs to be identified by a document type definition (DTD) file, which lists the types of attributes used. In the case of GEML, this DTD file will be based on the ways in which genetic code operates in the body—that is, according to sequences for genes, chromosomal positioning, gene-protein relationships, promoter-terminator regions, splice variants, gene polymorphisms, and so on. In other words, the DTD file for GEML is based on the current state of knowledge in biotech research—how reductive or complex that knowledge is, how rigid or flexible it is, and so on. What is being produced with GEML, then, is a kind of metacode for approaching the molecular code of the genome. In a sense, GEML does not add or modify anything in the genome; it is not a genetic engineering tool in the traditional sense of the term.

This interrelationship between molecular genetics and computer science means that bioinformatics will only be as complex, technically sophisticated, and potentially transformative as its DTD file—or the types of knowledge input into bioinformatics code. The conventional truism of molecular genetics—that in a causal, linear fashion "DNA makes RNA makes protein"—will produce a bioinformatics only to that level of complexity. However, as researchers and laboratories have been acknowledging, most diseases and phenotypic markers are the product of multiple genetic triggers and multiple biochemical pathways, not to mention networked interactions with context or environment. This is why the most interesting alternative approaches within biotech research, such as systems biology, have demanded that both molecular genetics and computer science transform the discourse of genetics and biotech, moving away from the overdeterminism of single-gene theories and toward more distributed and networked approaches.

"Curse the Skies While the Ground Slides Away"

Thus far, I have noted how contemporary genomics undertakes a particular organization of the bioinformatics body, through the incorporation of informatic tropes and in the design of "wet" and "dry" databases.[54] If such practices dictate what will be designed into a genome database, then we can ask how the political economy of genetic bodies in genomic mapping is also a set of organizational technologies based on the management of elements of excess. Georges Bataille's concept of general economy provides a helpful way of considering genomics not just as a political economy, but as a general economy in which *excess*, not scarcity, is the rule.

For Bataille, writing during the same period in which the structure of DNA was elucidated, traditional political economy represented a narrow mode of social analysis that often reduced the range of activity within the social field to utility, production, accumulation, and conservation.[55] Stemming from a long history of dissatisfaction with Marxism and Hegelianism, Bataille's thought outlines a trajectory that explores—in philosophy, in poetry, in political activism, in occultism, in pornography, in anthropology—the "negative" ways in which given social formations articulate themselves as productive, homogeneous structures.

Bataille's almost obsessive interest in the visceral, abject body, what he often termed "base materialism," has to do with the various means by which a tension-filled relationship was established between a potentially threatening, transgressive body (obscene, grotesque, unstable, dysfunctional) and the particular mode of social organization within which such bodies were found.[56] In other words, Bataille's base materialism is not about positing the transgressive body at a position of absolute exteriority to the social, but neither does its complex contingency simply predetermine it to a recuperation by discourse. The social formation does not simply repress or forbid the transgressive body; but neither does this body form the core of social hegemony. The social must—reluctantly, frustratedly, and even hesitantly—engage with this difficult, "dirty" body through a variety of means. It was central to Bataille's project, then, to consider the types of relationships between the productive dynamics of "homogeneous" society and the troubling excesses of "heterogeneous" bodies: "The very term heterogeneous indicates that it concerns elements that are impossible to assimilate; this impossibility, which has a fundamental impact on social assimilation, likewise has an impact on scientific assimilation . . . as a rule, science cannot know heterogeneous elements as such . . . the heterogeneous world includes everything resulting from unproductive expenditure."[57]

A key methodological move in this consideration was a reconfiguration of the field of political economy. Rather than considering the social from perspectives privileging utility, production, and conservation, Bataille attempts to analyze the social body as a kind of sociological metabolism, or as a waste-management system. That is, rather than consider political economy from the vantage point of utility, Bataille instead begins from the vantage point of uselessness and excess. For this reason, commodities do not form the center of modes of exchange for Bataille; instead, objects and instances form a unique mode of exchange based not on conservation or agglomeration, but on expenditure: jewels, poetry, festivals, war, eroticism, sacrifice, flowers. Abundance comes first, not scarcity. Only in the specific historical instance of capitalism are abundance and exuberance turned back on themselves. Noting this shift in perspective in Bataille's thought, Steven Shaviro suggests that "what needs to be explained is no longer the fact that spontaneous acts of expenditure, like sacrifices or gifts, also turn out to serve useful purposes for the people who perform them, but rather the fact that an economy entirely given over to utilitarian calculation, or to 'rational choice,' still continues to express . . . the delirious logic of unproductive expenditure."[58]

Such a theory must take into account the languages, structures, and sociohistorical contingencies of the organizational tactics of modes of social organization themselves. This perspective—that of a "general economy"—would work within but pass between the interstices of a "restricted economy," disallowing internal critical perspective.[59] Bataille's premise is at once biological, ontological, and economic: "The living organism, in a situation determined by the play of energy on the surface of the globe, ordinarily receives more energy than is necessary for maintaining life; the excess energy (wealth) can be used for the growth of the system (e.g., an organism); if the system can no longer grow, or if the excess cannot be completely absorbed in its growth, it must necessarily be lost without profit; it must be spent, willingly or not, gloriously or catastrophically." [60] If a restricted economy—the political economy of Mill, Ricardo, and Smith—is primarily concerned with the state's use of its resources, then a general economy would begin from the opposite position and consider social activity that implicitly challenges and goes outside of the state. This general economy thus asks two primary questions: How does a particular social formation interact with its "accursed share" (those elements that do not stand in a direct, productive relation to the social formation), and, To what degree are those elements of excess contingent upon their being situated within the social formation in which they are found (is there an element or dynamic to them that is not simply recuperated into production or affirmation)? The general trajectory of such an approach would be to question the ways in which both the inclusive and the exclusive are constitutive of the social formation.

What does Bataille's general economy mean for contemporary biotech fields such as genomics? If we approach genomics as an activity that rearticulates the body at the genetic and informatic levels, then it would seem that genomics is as much about the management of excess ("junk" or noncoding DNA) as it is about the organization of scarcity (coding DNA). In one sense, it goes without saying that the centrality of the mapping of the human genome constitutes one of the foundational fields within genetics and biotech research. Gene therapy, new reproductive technologies (NRTs), pharmacogenomics, medical technology, pathology, immunology, and a host of related fields are to varying degrees dependent on the promises, claims, and current implementations of the research performed in genomics. Part of this dependency requires that basic research produce as much data as possible to ensure accurate results (in the case of genome mapping) and to ensure application (in the case of drug development or preclinical trials). In a sense, the excess of genomic data—SNPs, introns, simple sequence repeats (SSRs), and so on—does not simply exist, but is actually *produced* by genomics technologies. If we look at current genomics, not in terms of official science (e.g., the progressive march toward the complete cataloging of the human genome), but rather in terms of Bataille's excess, then our examples demonstrate the degree to which the political economy of genomics is primarily concerned with the production and subsequent technical management of excess. But an excess of what exactly? If both the technology of genomics and its preliminary findings are considered, it is clear that the excess that is managed is an excess of biological information—*though, note, not necessarily an excess of biological "life itself."* As we have seen, examples of genomics technologies of excess include the statistical management of polymorphisms, or SNPs; the overproduction of DNA in tools such as oligonucleotide synthesizers; the study of junk DNA, or intron regions in the genome; and, finally, the challenges of establishing computer database ontology standards in XML. As a general economy—that is, as a political economy based on excess—genomics becomes defined according to a range of practices that involve the ongoing production and management of the inclusive *and* exclusive elements in the genomic body.

The Insubordination of Matter

This chapter has considered the ways in which genomics forms a flexible management of excess DNA (or information) from the perspective of Marx's critiques of political economy and from the perspective of Bataille's notion of a general economy. Genomics and its related fields perform this management not only through positive means (textualizing, universalizing, creating causality, constructing), but also through negative means (individualizing and grouping; managing excess) that define further data for genomic analysis.

Both Marx and Bataille pose a set of challenges to the conceptual underpinning of political economy. If we consider political economy in this way (that is, philosophically), then it is not difficult to understand how the particular modes of production, distribution, and consumption help to define what is or is not valuable or useful in biotechnology fields such as genomics. As Lev Manovich states with regard to the role of the database in new media,

> What we encounter here is an example of the general principle of new media: the projection of the ontology of a computer onto culture itself. If in physics the world is made of atoms and in genetics it is made of genes, computer programming encapsulates the world according to its own logic. The world is reduced to two kinds of software objects which are complementary to each other: data structures and algorithms. . . . [A]ny object in the world—be it the population of a city, or the weather over the course of a century, a chair, a human brain—is modeled as a data structure, i.e. data organized in a particular way for efficient search and retrieval.[61]

Fields such as genomics are unique, however, because they do not simply reiterate the abstraction of the material world into an immaterial data structure. In genomics, as noted, a database logic pervades both the wet-lab maintenance of BAC libraries and the more familiar construction of computer databases such as GenBank. In other words, an informatic mode of organization—an informatic mode of "banking"—spans both the material and immaterial aspects of biology.

This dual aspect of genomics, of being at once material and informatic, is key for an understanding of how the characteristics of a genetic reductionism defines the molecular-genetic understanding of "life itself." The artifacts of the database, the BAC library, the intron, and GEML contribute to this general notion that biological "life itself" is in some way identical to "information." And yet, as a number of authors have pointed out, this equivalence is always inexact, for the genetic "code" is not a language—it has no grammar, it contains no information, and it transfers no message separate from a medium. Something always seems to escape: on the one hand, there is the precision of sequences and structures, where a single base-pair polymorphism can result in a disease, yet, on the other hand, there is a high degree of "redundancy" in the sequence and a panoply of "error-correcting" mechanisms in the genetic code.

Whereas Marx, in the *Grundrisse*, points to the central tension between labor and capital as the motor of industrial capitalism, Bataille translates this tension into one between expenditure and accumulation. For Marx, it is "living labor" that always provides the challenge to capital and to which capital responds either through technological advances in production (e.g., the fixed capital of machinery) or through redefining the relationship between labor and machinery. Regarding this "appropriation of living labor by objectified labor,"[62] Marx points to the transformation of living labor into a technology of living labor, a transformation of nature into a technology of nature: "Nature builds no machines, no locomotives, railways, electric telegraphs, self-acting mules etc. These are products of human industry."[63] Yet—and this point is the key to understanding the uniqueness of biotechnology today—capitalism's revolutions of production are able to reconceive of biology and of nature itself as a technology: "In machinery, the appropriation of living labor by capital achieves a direct reality in this respect as well: It is, firstly, the analysis and application of mechanical and chemical laws, arising directly out of science, which enables the machine to perform the same labor as that previously performed by the worker."[64] Although Marx is speaking here of industrial manufacturing, it is not difficult to expand his comments to the biotech industry, in which we see the routine manufacture of molecules, enzymes, and cells. What has changed in the jump from industrial capitalism to the biotech industry is the role that modern notions of "information" play in relation to the biological domain. "Living labor" is no longer just the labor appropriated by workers, but the uncanny *biomaterial labor* of immortalized cell lines, cultured stem cells, DNA synthesizers, and even transgenic organisms. Marx had already begun to see this integration of living labor and the "dead labor" of machinery in biological terms ("the metabolism of capital"), and in the biotech industry this integration is pushed further still.

However, for Bataille, even Marx's critique of political economy had not gone far enough. Marx was able to critique the universalist presuppositions of political economy, but, for Bataille, political economy had erred in its understanding of the true nature of the processes of production, distribution, and consumption. Bataille's emphasis on a "general economy" based on excess, expenditure, and luxury is an attempt to invert further the categories of traditional political economic thought. From this perspective, Marx's tension between labor and capital is converted into a tension between expenditure and accumulation in capitalist societies. Instead of Marx's living labor in production, Bataille gives us the intensive production of a loss in expenditure; instead of Marx's general formula for capital, in which commodities are always transformed into money, Bataille gives us a general economy in which the drive for accumulation takes the form of a reinvestment in production with profit. The challenge for an economic system is thus not how to produce efficiently in order to accrue a surplus value, but rather how to spend or expend effectively in order to maintain the dynamic stability of a system. For Bataille, excess is the problem, not scarcity (and in this he gives us a position diametrically opposed to Malthus and classical political economy). Put another way, we might say that for capital the problem is the ongoing management of excess (pushes toward consumption of immaterial goods and services; conspicuous consumption; consumption of affects).

It is tempting to push this idea even further and to suggest that from Bataille's perspective of expenditure, *the problem for capital is not only managing excess, but taking over the production of excess in itself.* In the biotech industry, genomics and bioinformatics illustrate this tendency, for in them we see a twofold tendency. On the one hand, there is a push toward generating an incredible amount of information, a "tsunami of data" related to genomes, proteomes, SNPs, spliceosomes, transcriptomes, biopathways, and so forth. The first phase of the effort to map the human genome (by both Celera and the IHGSC) is the most well known of such productions of excess. On the other hand, the production of such excess has as its primary aim the subsequent development of mostly computational tools for analyzing and "making sense" of all this information. The proliferation of bioinformatics software, DBMSs, genome sequencing computers, and other technologies has largely followed upon "Big Science" efforts such as the HGP. We have a kind of push-pull relationship between these tendencies: the tendency toward an excess of data and a tendency toward the management of that excess (at which time it no longer becomes excess as such).

Using Marx's terms, we can say that in fields such as genomics and bioinformatics, information technology subsumes the excess information (which it has produced). There is no such thing as "too much information" for genomics, bioinformatics, or related fields such as pharmacogenomics. There is never too much data, only the production of an excess that serves to trigger further development of tools for the management of this excess. The productive capacity of this excess can be seen in recent patenting activity surrounding genetic material: up until 2001, patent applications for genetic material were being granted without a full knowledge of the function or specific application of the molecule in question. Genes (or rather gene derivatives) could theoretically be patented without an understanding of what the genes did. Only in the U.S. PTO's *2001 Federal Register Report* was it finally specified that patent applicants must demonstrate "both utility and application" in their files. Yet an ambiguity still remains concerning knowledge of the function of any patented compound: "The utility of a claimed DNA sequence does not necessarily depend on the function of the encoded gene product."

In the network between a laboratory BAC culture, a file in a genome database, and, perhaps, an application to the U.S. PTO, there is both an effort to regulate the relationship between labor (or biomaterial labor) and capital, as well as an effort to manage excess information. We can ask Bataille's question of the biotech industry: Is there unproductive expenditure in biotechnology? In one way, the answer is yes, for the production or even overproduction of information is increasingly being seen as one of the major obstacles for bioscience research. In a different light, the answer is no, for any excess in biotechnology is not only a production of excess (as in genomics), but an eventually useful or valuable excess. If this is the case, then the general economy of genomics is one in which excess biological "noise" is constantly managed and transformed into useful or valuable biological information.

The tension between labor and capital in the biotech industry is therefore a tension between the biomaterial labor of genes, enzymes, and cells, on the one hand, and the economic imperatives behind those instances of biotechnology research, on the other hand, where profit is the primary motivating force. Clearly, this is not the case with each and every instance of research in fields such as genomics, proteomics, or bioinformatics. And just as the mere existence of private-sector funding for research does not immediately imply a problematic relation between labor and capital, the mere existence of public or federal funding also does not immediately absolve those funding bodies of economic imperatives. The difficulty is initially in understanding how the tension between labor and capital (or between biomaterial labor and the biotech industry) takes formation in the so-called biotech century. From this understanding, the next step is to identify those elements that tip the delicate, fragile balance between medical and economic interests to one side or the other. These tensions are often played out at the technical level of BAC libraries, gene-finding algorithms, and XML standards. Sometimes the prioritization of economic interests create blockages in bioscience research, as in the case of proprietary data formats, databases, and software tools in bioinformatics. At other times, efforts toward the creation of flexible standards and open source environments can effectively aid in the research process.

We might wonder whether in this biomaterial labor that is transformed into capital there is still something that escapes, evades, or bypasses altogether the political economy of genomics. If fields such as genomics and bioinformatics are involved in the production of an excess, that excess is in turn transformed into useful information, and that information is also valuable information that can be exchanged. Use and exchange values are coordinated at this level, but use values also diverge into the potential medical benefits of drugs or therapies. Yet even drugs and therapies require further information and the development of technologies for diagnostics. As discussed in chapter 2, this vision of a future, genomics-based medicine means that the body of the patient is touched only minimally, configured in a way commensurate with the informatic systems that sample, scan, and profile.

To be fair, however, this overview is far from being the whole of genomics and bioinformatics research. A number of so-called alternative approaches have been pursuing the distributed, networked properties of gene expression, protein-protein interactions, cellular signaling, and metabolic pathways. This "systems biology" also shows some overlap in spirit with research in biocomplexity at the Santa Fe Institute, for instance.[65] Although the technologies and methodologies are new, the basic idea is not: as Evelyn Fox Keller reminds us, the history of molecular biology and biochemistry already pointed to the inherently "complex" and networked property of "life itself" at the molecular and genetic levels.[66] Alternative approaches merely remind us of what is implicit in the history of biological thought itself: relationality, flexibility, robustness, adaptation. Even the "official" HGP cannot help but to acknowledge certain excessive elements in the genome: a relatively modest number of genes (some 3 percent of the genome), an extremely small difference between the genomes of one individual and another (approximately 0.1 percent), an incredibly large number of "short tandem repeat" sequences, and entire "continents" that seem to have no known function at all (more than 50 percent of the total genome). In a sense, despite the biotech industry's efforts to manage the genome, the genome itself (like "life itself") appears to be nothing but *excess information*, with all the contradictions that the phrase implies. Bataille proposes that an organism always produces more than it needs and that, from a systemswide ecological perspective, abundance, not scarcity, is the rule of life. With regard to fields such as genomics, proteomics, and bioinformatics, it is not difficult to begin to wonder whether the human genome is a noneconomic relation between labor and capital, between expenditure and accumulation.

II

Recoding / Distribution

In all these cases, and they are all historical, it seems that distribution is not structured and determined by production, but rather the opposite, production by distribution.

—KARL MARX, *GRUNDRISSE*

4

Biocolonialism, Genomics, and the Databasing of the Population

The Map, the Territory, and the Population

When in the early 1990s the U.S. government–funded Human Genome Diversity Project (HGDP) drafted plans for a large genetic database of distinct ethnic populations, it was met with a great deal of controversy and criticism.[1] Led by scientists such as Luca Cavalli-Sforza and Allan Wilson—both pioneers in the field of population genetics—the HGDP's original aim was to collect DNA from individuals of genetically isolated populations around the world in order to reconstruct the evolutionary path of humankind.[2] But the initial planning sessions, which were funded by the U.S. National Science Foundation, the National Human Genome Research Center, the National Institute for General Medical Sciences, and the DoE, quickly became mired in disagreements over which populations to choose for sampling and the best way to define what *population* means in this context, a context that is as much cultural and anthropological as it is scientific. In addition, groups such as the Rural Advancement Foundation International (RAFI) and Third World Network voiced their concern over the possible implications of such research for those largely indigenous communities who would have their genetic material housed in cell lines in U.S. institutes. Although HGDP committee members denied that there is any economic motive behind their proposal, critics expressed concern over the way in which indigenous communities were effectively cut out of the planning process.

The controversy was further fueled in 1993 by related but independent incidents. U.S. scientists, with NIH funding, had applied for patents on the genetic material of a Guyami woman from Panama, only with the woman's "verbal consent." A similar case was found with a cell line derived from a Hagahai man from New Guinea. Such instances seemed to RAFI and other groups as an explicit example of colonialism, but now practiced on human genetic material instead of on territories or on natural resources. Native American groups in the United States accused the HGDP of practicing *bio-colonialism*, a term that has been used frequently in describing the practice of sampling genetic material for potential genes of value (be it medical or economic value).[3] The governments of Panama and the Solomon Islands separately protested to the international community over the patents (one of which was subsequently dropped), and indigenous communities, nongovernmental organizations, and South Pacific governments lobbied for a "lifeforms patent protection treaty."[4]

At the same time, in the midst of mounting tensions over the HGDP, the World Council of Indigenous Peoples publicly criticized the HGDP, dubbing it the "Vampire Project." Such concern and dissent forced the HGDP committee to draft up a set of "model ethical protocols" (MEPs), and in 1993 the HGDP formed the North American Regional Committee, whose aim was to come up with acceptable criteria for the collection of genetic material from human populations.[5] Written in consultation with leaders from indigenous communities in the United States (because Native American tribes were one major target of the HGDP), the MEPs stressed "informed consent" and the importance of cultural as well as scientific understanding. Since that time, however, the HGDP has been conspicuously silent, and despite the flurry of news items and press releases relating to the various genome mapping endeavors around the world—both government and corporate sponsored—there has been relatively no news or updates on the HGDP's progress.[6]

Much of the HGDP's curious disappearing act has to do, certainly, with the bioethical ties in which the HGDP has been involved and with the combination of criticism being voiced by groups such as RAFI and the HGDP's having been "marked" by the media.[7] However, although the HGDP as an organization may have slipped from science headlines, the issues and problems associated with it have not. Parallel developments within biotech and genetics have emerged, more or less taking up the "diversity problem" that the HGDP dealt with in the 1990s: bioinformatics and genomics. As we have seen, bioinformatics involves the use of computer and networking technologies in the organization of updated, networked, and interactive genomic databases being used by research institutions, the biotech industry, medical genetics, and the pharmaceutical industry.[8] Bioinformatics politically signals an important development in the increasing computerization of "wet" biotech research, creating an abstract level where new information technologies can establish the foundation for a new research paradigm of "in silico biology."

One of this chapter's arguments is that these two trends—the decrease in the HGDP's visibility, and the rise of bioinformatics—are inextricably connected. In fact, this twofold trend is striking: on the one hand, the relative disappearance of the HGDP and, on the other, a string of "population genomics" projects, the most visible being the IHSD managed by deCODE Genetics, and the government-sponsored U.K. BioBank. Such projects draw our attention to the ways in which race, genomes, and economics are mediated in complex ways by information technologies. Genomics—the technologically assisted study of the total DNA, or genome, of organisms—currently commands a significant part of the biotech industry's attention. In economic as well as scientific terms, genomics has for some years promised to become the foundation on which a future medical genetics and pharmacogenomics will be based.[9] This chapter attempts to draw out some of the linkages between the biotech industry and the emphasis within genomics programs on diversification and the collection of particular populations' genomes into computer databases. Such research programs, which highlight "genetic difference," demonstrate the extent to which culture and biology are often at odds with each other and the extent to which both ethnicity and race are compelled to accommodate the structures of informatics and information.

Blood, Sex, Data

With the sheer quantity of material being generated by such globalized science endeavors as the human genome projects, the need for a biotech-infotech science has become the more prominent, if only for organizational and managerial reasons. Biotech corporations such as Affymetrix, Incyte, and Perkin-Elmer not only specialize in research and development, but also emphasize product development.[10] Bioinformatics promises to deliver the tools that will make genomic science an information science and propel the human genome projects into the next phase of "post-genomic science."[11] With the aid of bioinformatics technologies, the "public" genome project, originally cast by the NIH as a 15-year endeavor, was shortened to a 3-year effort, with a "working draft" presented—not without a great deal of fanfare—during the summer of 2000.[12] Biotech corporations such as Celera and Incyte have initiated their own corporate-framed and privately run human genome mapping projects, claiming the ability to outdo the international consortium (IHGSC) in providing a more "medical" approach (i.e., mergers providing direct biotech-to-pharmaceutical access of information). The link between biotech and infotech is no longer a supplementary one; it is, for contemporary genomics, a current necessity.

The recent rise in genomics projects, especially those geared toward unique gene pools and genetic difference, has implied the hybridization of a new practice of statistics and medicine, combining studies in population genetics and new techniques in genomic mapping. This application of the genetic study of the genomes of specific populations—what has been called *population genomics*—brings together a lengthy tradition in the study of populations, hereditary patterns, and inherited characteristics with the contemporary development of large-scale genomic sequencing and analysis for clinical medicine.[13] Such endeavors stand in contrast to the more universal "human" genome projects, which attempt to construct "the" human genome, a universal signifier that will contain the average of all specificity. For instance, Celera's human genome database contains sequences from DNA sampled from approximately seven individuals, all anonymous, but taken from a range of genders, racial backgrounds, and ages. Where differences exist between individuals, the most common sequence or base pair, statistically speaking, is taken, and the alternatives become part of a different type of dataset comprising minute genetic differences. The field of population genomics thus contrasts itself to the universal human genome projects. Its focus is an entity whose definition is still in some dispute—the "population." Its emphasis is on those genetic elements that make a human population distinct from "the human" itself: genetic markers, STSs, haplotypes (HAPs), and SNPs (figure 4.1).[14]

Biotech companies such as deCODE, Myriad Genetics, and Gemini Genomics are focusing on the genomes of populations with histories of a low degree of migration and a low frequency of hybridity (Icelandic, Mormon, and Newfoundland communities, respectively).[15] Other companies, such as DNA Sciences Inc., are focusing on building a volunteer-based genetic health database to aid in the fight against disease. DNA Science's Gene Trust databank

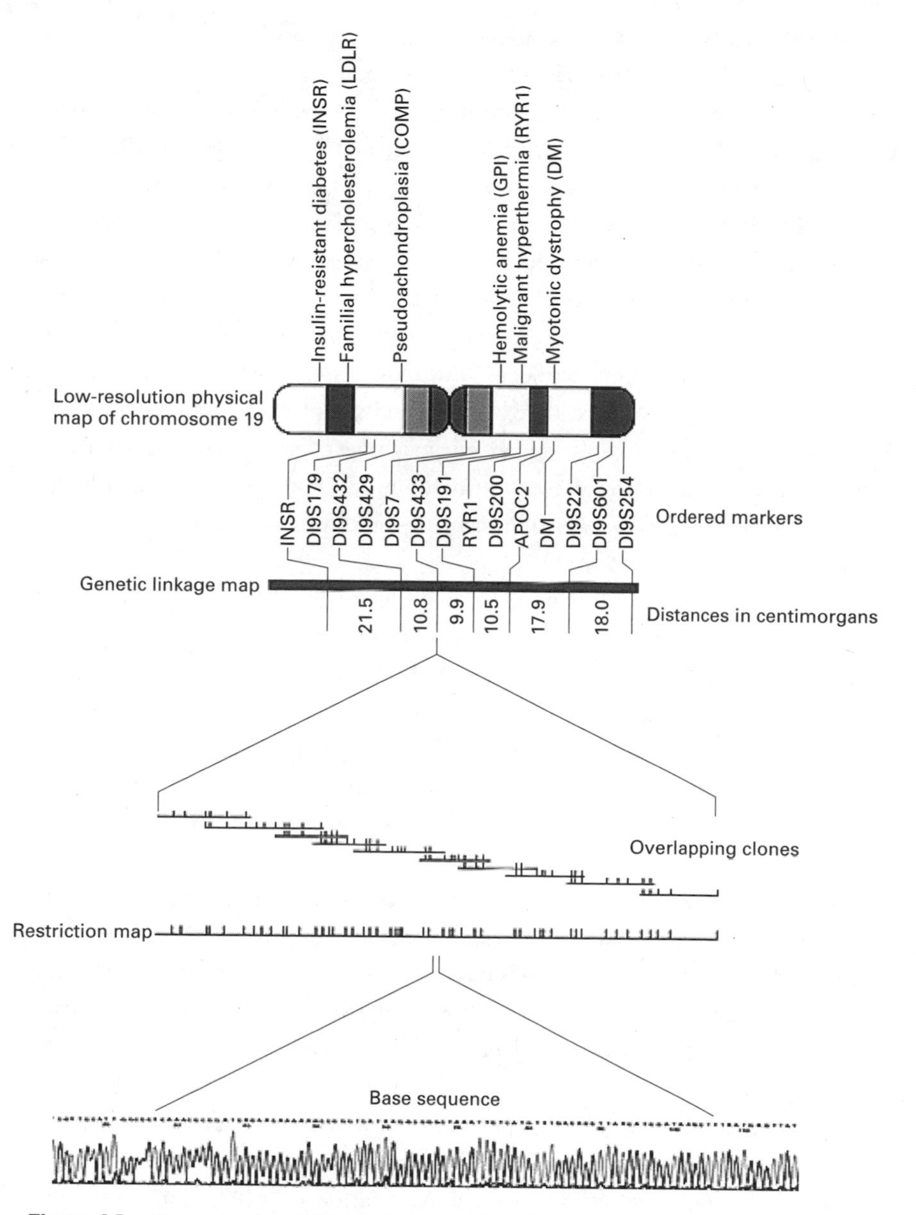

Figure 4.1 Map or territory? "Genomic geography" of chromosome 19. Image courtesy of the U.S. DoE Genome Program, http://www.ornl.gov/hgmis.

uses the GenBank model to archive medical, genetic, and health-related data (GenBank holds the IHGSC's human genome data).[16] A number of government-sponsored projects are also under way, the most noteworthy being the U.K. BioBank, which aims to collect DNA samples, medical histories, and medically relevant "lifestyle information" from more than half a million individuals. Still other companies and research labs are focusing on the minute genetic sequence differences between individuals—polymorphisms, SNPs, and HAPs—which may be the keys to individual genetic susceptibility to disease and, by extension, to pharmacogenomic approaches to drug development.[17] In most cases, such projects include genetic sampling (DNA extracted from blood samples), analysis (sequencing using plasmid libraries and sequencing machines), and the construction and maintenance of computer databases (including data analysis and search tools, as well as utilities for networking databases with other types of data such as medical records).

These and other population genomics projects (see table 4.1) have met with a great deal of controversy, most notably from the populations that are the object of study and data agglomeration. For instance, in Iceland a group of scientists, physicians, and citizens called Mannvernd continues to contest the ethical implications behind the IHSD.[18] Contentious ethical issues have also arisen in relation to the Tonganese genome (where Autogen had aimed to begin a genomics project), the government-sponsored Estonian genome, and efforts to establish a Korean genome (headed by Macrogen).[19] Critics of population genomics projects often raise concerns over issues of consent, information privacy, and the ethics of transforming a nation's health into a commodity for sale. Such concerns have prompted many population genomics projects to take on responsibility for ethical issues. For instance, the U.K. BioBank has established an independent ethics and governance council to oversee ethical, policy, and legal issues that will inevitably arise with such broad genetic sampling endeavors.[20] In addition, the issue of population genome databases has attracted the attention of the international health community. In 2002, the World Medical Association (WMA), along with the WHO, announced the "Declaration on Ethical Considerations Regarding Health Databases," which would, among other things, give priority to the individual patient's right in such databasing ventures.[21]

This emphasis on genomic differences is also, then, an emphasis on biotechnologies' ability to drive the diversification of databases, drug-development

Table 4.1 Population Genome Projects

Institute	Subjects
Estonian National Gene Bank Project (Estonian Genome Foundation, Estonian government)	Estonian volunteers
The Gene Trust (DNA Sciences, Inc.)	More than 100,000 U.S. volunteers via the Internet
Genomic Research in African-American Pedigrees (G-RAP; Howard University)	African American volunteers
Icelandic Health Sector Database (IHSD, deCODE Genomics)	More than 280,000 Icelanders
Korean Gene Bank (Macrogen)	Korean volunteers
Mormon Gene Bank (Myriad Genetics)	Mormon communities (United States)
Newfoundland Gene Bank (Newfound Genomics, a spin-off of Gemini Genomics)	Half a million Newfoundlander volunteers
P3G (Public Population Program in Genomics)	Includes U.K. BioBank, CARTaGENE (Quebec), Estonian National GeneBank Project, and GenomEUtwin Project (Finland)
Swedish Gene Bank (UmanGenomics)	More than 250,000 Swedes from Väterbotten
Tonga Gene Bank (Autogen)	More than 180,000 Tonganese in Australia
U.K. BioBank (Wellcome Trust, Medical Research Council, and the U.K. Department of Health)	Half a million U.K. volunteers

programs, and, it is hoped, the biotech industry itself. But we should also be careful to note distinctions between population genomics projects. For instance, some projects are funded with government money and for this reason are often represented in terms of national heritage. Other projects, by contrast, result from government partnerships with the private sector or have simply been undertaken with a community's consent. But perhaps the most noteworthy distinction is not between population genomics projects, but between recent projects that emphasize finding specific-population genomes and the earlier "biocolonialist" projects such as the HGDP. Whereas the HGDP proposal was an example of Western scientists gathering or appropriating genetic material from mostly indigenous populations around the world, the newer population genomics projects are by and large in-house operations. Population genomics projects such as the IHSD or the U.K. BioBank are examples of "national" genomics programs exercised reflexively: Icelandic scientists sampling Icelandic citizens to gain knowledge about the Icelandic genome and, perhaps, to aid in the health of common diseases affecting the Icelandic body politic (or so the story goes). If this is biocolonialism, it is arguably a very different sort of exploitation: government sanctioned, driven by national industry, and completely voluntary. As Michael Fortun notes, the study of such projects must therefore "trace how these rhetorics of exoticism and national difference are deployed not only by 'foreign' commentators and media outlets, but also have more 'domestic' origins."[22] What is produced in such projects is not only a database, but, in a sense, a new concept of what *population* may come to mean in the context of a genetics-based medicine and health-care paradigm. Fortun adds that such population databases "are like value-added export products designed to circulate in a global rhetorical economy."[23]

Indeed, in spite of the differences between specific population genome projects, one common thread runs through all of them: the central role of information technologies and the artifact of the computer database. In one sense, population genomics is a novel redefinition of *population* through the lens of the computer database. Yet, in another sense, this process is itself very old and can be traced back to the early-twentieth-century debates between Mendelian and population genetics, or even to the emergence of demographics and statistics in the eighteenth and nineteenth centuries. In this sense, population genomics is part of a much broader process constituting what Michel Foucault calls "bio-history," or "the entry of life into history." For Foucault, new scientific techniques, along with specific governmental modes of intervention and regulation, culminate in the inclusion of "phenomena peculiar to the life of the human species into the order of knowledge and power, into the sphere of political techniques."[24]

How is the bio-history of population genomics constituted? Foucault, as is well known, argues that the nineteenth century saw a passage from a bio-history centered around blood (kinship) to one centrally concerned with sex and sexuality. As Foucault notes, the blood relation was closely tied to "a society in which the systems of alliance, the political form of the sovereign, the differentiation into orders and castes, and the value of descent lines were predominant." Although many of these elements did not disappear completely, the nineteenth century marks, for Foucault, a shift in emphasis: "Through the themes of health, progeny, race, the future of the species, the vitality of the social body, power spoke of sexuality and to sexuality." The emergence of new scientific disciplines (psychopathology, germ theory, evolutionary biology) as well as a set of new social concerns (deviancy, hygiene, poverty, homosexuality) culminated in this shift "from a symbolics of blood to an analytics of sexuality."[25]

In the case of population genomics, it is clear that the concerns of "health, progeny, race" within a national context are still relevant. The particular challenge to population genomics endeavors—a challenge that marks them out as being unique—is that they must redefine the genomic "population" in ways that are congruent with the technical paradigm of the computer database. This redefinition requires, among other things, a prior understanding of a genomic population in informatic, or at least statistical, terms. And, indeed, the population genetics work by Cavalli-Sforza and others can be seen as an attempt to do exactly this: to utilize mathematical and statistical methods to study the genetic and evolutionary effects on particular human populations. Does population genomics, in its novel integration of biology and informatics, put forth another transition, from Foucault's "analytics of sexuality" to what we might call a "biopolitics of information"? If this is the case, then the governmental and medical concern over the health of the population will shift emphasis from sex and sexuality to a notion of population defined in genetic and informatic terms: *blood*, *sex*, *data*.

In order to investigate these claims, it will be helpful to consider how *population* is defined within population genetics, paying particular attention to the biohistorical and biopolitical aspects of which Foucault speaks.

The Genome Race

Many of the discomforts with population genomics and DNA sampling become clearer when we look at how *race* and *population* are articulated according to modern population genetics.[26]

In order to begin to isolate an object of study, a set of assumptions are needed to define what counts as a population for genetic study. In population genetics, these assumptions revolve around a predicable pattern of distribution within the population at a given time; in other words, the genetic stability of a group is a key defining factor of a population. Within this genetic stability, a profile of the population's gene pool can be assessed. The gene pool contains the totality of that population's genetic parameters and polymorphisms; it thus delineates the differences within a collectivity.

Luca Cavalli-Sforza, who originally headed the HGDP and is a leading researcher in population genetics, provides us with the following statement concerning race: "A race is a group of individuals that we can recognize as biologically different from others. To be scientifically 'recognized,' the differences between a population that we would like to call a race and neighboring populations must be statistically significant according to some defined criteria."[27]

Consider this definition in light of a similar one given by early population geneticists such as Theodosius Dobzhansky: "Races may be defined as Mendelian populations of a species which differ in the frequencies of one or more genetic variants, gene alleles, or chromosomal structures."[28] As historians of biology note, part of the debate in the early twentieth century between American Mendelians and British population geneticists lay in the disagreement over how to define a population.[29] Were observable patterns of phenotype in a given group the result of genetic variation, and, if so, did such patterns constitute or define a group as a particular "race"? These questions are still being debated today in the domain of genomics, but several aspects of these definitions highlight the way in which race is defined for modern genetics. When race can be approached from the perspective of genetics, difference becomes bifurcated along genotypic (genetic code), phenotypic (visible characteristics), and informatic (statistical) lines.

The first thing to note is that in this formulation race is, implicitly, biologically determined: as Cavalli-Sforza indicates, a race is a human collectivity defined by biological properties (be those phenotypic or genotypic). Race, for population geneticists, is biological, the substrate from which social customs, ethics, and culture emerges (and, indeed, Cavalli-Sforza speculates about applying the techniques used in population genetics to linguistics). It would seem that for anthropology to begin, population genetics or archaeology has to take the first analytical steps.

On the phenotypic, or visible, level, the concept of race has long been tied to a biological determinism that initially grew out of early colonial travel narratives to Africa and the Americas. The development of modern biology and natural history classificatory systems further contributed to the need to discuss ethnicity and cultural difference under the rubric of biological race. In some instances, the biological study of races and populations only bolstered racial stereotypes, providing legitimized, authoritative research to back up social and cultural presuppositions. Nineteenth-century sciences such as phrenology, Darwinism, hereditary studies, and eugenics contributed to the social marginalization of certain human types, from the criminal to the pervert to the primitive.

In many ways, it is difficult to discuss the concept of race without some reference to the role that modern science has played in the legitimizing articulation of racial and ethnic difference. Many early modern travel narratives depicted bizarre natives and races from Africa or the Americas and often relied on the discourse of the fantastic or the monstrous, describing others who were both debased and repulsive.[30] Darwinism and natural history during the nineteenth century helped to give such fantastic accounts a certain basis in scientific fact.[31] They also helped to explain, through evolutionism and classificatory biology, the basis and reasons for such differences. This basis in visible characteristics is such a core element of our everyday notions of race that it is difficult to think otherwise. Through the sciences, elements such as skin color became the set of characterizing morphologies that defined a given race and that explained that race's evolutionary development according to such factors as natural habitat and climate.[32]

The biological basis for defining race has several consequences. It makes race a biological substrate upon which all other formations depend, including the social and cultural dimensions that define ethnicity. This explanatory power not only links race to the natural environment, but also serves as a delegitimizing tool between colonizing and colonized collectives. Especially in First World contexts of colonization, it serves ironically to remove any accountability for difference from the colonized race because difference is biological and not socially contextualized or culturally chosen. A biological definition of race makes use of an abstracting biological science to explain away the social and cultural dimensions of race.

This also means that race must be scientifically defined, which is my second point. That race is something recognized is in itself significant; that race is scientifically defined, or recognized through science, means that the possibilities of other forms of communal recognition are excluded. Cavalli-Sforza's statement about race not only proposes that race is biological, but adds to this that race is not something self-evident to everyone; though we may see plenty of differences among and between human collectivities, race can be defined, in the final instance, only through scientific methodology.

On the genotypic level, or the contemporary level of genetic codes and sequences, race has been further articulated along its biological determinist lines. This articulation has not only added further scientific "truth" to the difference of races, but also brought to light differences that are exclusively genetic (that is, which are not seen but exist only as differences in genetic code). Early systems for blood-type detection are an example of this, and more current methods focus on population genetics and SNPs.

In one sense, the genetic basis for racial identification further extends the explanatory power of Darwinism and modern biology, but it also initiates a new type of biological determinism with regard to race. If, in projects such as population genome databases, race can be identified genetically—that is, according to sequences of genetic code—then the urgency of clear phenotypic markers between races is simply displaced by an emphasis on genetic markers. These genetic markers of difference can be visible and nonvisible (expressed and unexpressed), but with this unique mode of racial identification race becomes not only biologically determined but informatically determined.

Without a doubt, this approach is still biological determinism; the emphasis on abstract genetic code in no way liquidates the role that race and racism play in modern science's study and classification of populations. If anything, a genetic-based approach to race provides a mathematical proof (an informatic proof, as in DNA fingerprinting) of racial identification. It also provides a new means of quantifying racial hybridity, according to varying gene frequencies (a method used in population genetics but also applicable to individuals). A genetic-based approach to race thus informs the biological definitions of race already established by modern biology, and it does so through a basis in genetics (a basis in differential units of DNA) and informatics (a basis in new methods of quantifying and analyzing DNA patterns). It is, in many ways, a biological determinist conception of race that can be computed, mapped, and even, in the right conditions, predicted.

This scientific approach implies a methodological approach, which is that race is articulated through numbers. That scientific recognition is a bioscience of mathematics. More traditional population genetics approaches focus on frequencies of traits, demographic distribution of traits, and probability of gene migrations or inheritance of traits. But, as Cavalli-Sforza shows, the rise of genomics, new mathematical methods, and computer technology means that populations and races become informatic objects of analysis: "Within genetics itself, we want to collect as much information about as many genes as possible, which would allow us to use the 'law of large numbers' in the calculation of probabilities."[33] Calculation of gene frequencies, correlation of large datasets, multiple gene-factor probability analysis, and other techniques allow for a more statistical, more informatic approach to the recognition of visible and invisible differences on the genetic level.

From the population genetics perspective, particular populations—be they objects of archaeological study or of the study of Third World collectivities—are tied to nationality not only through social and political linkages, but also through biological ones. According to the biological determinism of early race theories and the discourses of monstrosity, if one was born into a certain nation, one was also born into a certain race and even into a certain culture extending from that race.[34] This connection between biological race and the geopolitics of nationality can become the opportunity for a range of situations, from colonial expansion to First World aid to economic globalization.

From Population Genetics to Population Genomics

A number of constraints are thus placed on a defined population because most collectivities we can think of accommodate a range of circumstances that mitigate against absolute purity, absolute isolation. Population geneticists traditionally look to a theory called the Hardy-Weinberg equilibrium to assess the degree to which a population can be defined as such, and the degree to which knowledge about that population will be minimally affected by the noise of other influences. The Hardy-Weinberg theory, first developed in 1908, makes the broad claim that although genes may shuffle and recombine within a population, the total genetic makeup of the population remains the same.[35] It also goes on to posit five basic assumptions:

- The population should be infinite and evenly distributed, which creates the conditions for a large enough group to be studied for patterned characteristics, as well as for an even distribution of possible mating matches.
- There should be little or no mutation within the population—that is, no error mutations that would cause epidemic or a radical decline in the birth rate.

- There should be little or no genetic migration in or out of the population; that is, the gene pool may display internal dynamics, but there should be no new input of genetic material.
- All mating and reproduction within the population should be random; that is, reproduction should not be constrained by genotype.
- The totality of genotypes should be equal; despite genetic reshuffling, the genetic frequencies within the population remain stable.

In short, the Hardy-Weinberg theory isolates all those elements that deter a population from maintaining a homeostasis with its overall gene pool or genetic frequency. A population thus defined is not only entirely removed from environment, external influence, or context, but such a definition also recuperates biological dynamics within an overarching static framework. This emphasis on information sciences such as statistics gives rise to specific techniques, a few of which can be mentioned here.

Population geneticists use the concept of "genetic distance" to study human collectivities and their evolutionary histories.[36] For modern evolutionary biology, genetic alterations happen for one of four reasons: mutation, or a random alteration in the genetic sequence, which is passed on to subsequent generations (and which may or may not affect the organism); natural selection, or changes in the organism (genetic and otherwise), usually in response to environmental conditions; genetic drift, or the complex fluctuations of particular genetic traits or genes within a stable population; and genetic migration, or the geographical movement of peoples out of one collectivity and into another.

Because evolutionary biologists and population geneticists take populations and races as biologically determined, they consider changes in the organism to be always either responses to the natural environment (in which there is strong teleological factor, form fits function) or the result of randomness (in which there is a strong probability factor). Again, this assumption implies that societies, cultures, and ethnicities form because, first and foremost, populations and races form. Although this emphasis on the contingency of the biological body can operate as a critical tool, it can also "explain away" other cultural formations equally connected to the body, but not necessarily within the paradigm of evolutionary biology. The lack of this distinction has led scientists such as Cavalli-Sforza to assume that geographical distance is equal to the degree of genetic change; genetic distance would thus be the statistical analysis of these two correlative changes, and wherever one sees a rift in the gene pool, one can assume a significant moment of genetic and geographical diversification.

Within the concept of genetic distance is a strong linkage between territory, population, and biology. Scientists such as Cavalli-Sforza explain race in terms of biological determinism and biological necessity. The reasons Cavalli-Sforza gives for migration—overpopulation, food shortages, low reproductive rates—are based on a biological pragmatism. But can this teleological, determinist framework adequately explain genetic variation, especially when at the genetic level many variations appear to have no connection to biological survival?

With this in mind, we can see that the field of population genetics has left several areas unconsidered, areas that at first glance are nonbiological, but that on closer inspection are intimately related to the concerns of population genetics. For instance, if a great deal of genetic change occurs via unexplained or random processes, do not these warrant some radically new approaches to the study of population, instead of accounting for random variation simply by mathematical analysis? If a great deal of our polymorphisms in SNPs reside in "junk DNA," does this not this suggest a more complex network of genetic differences, working to produce a wide range of qualitatively different polymorphisms?

Furthermore, if for a moment we accept Cavalli-Sforza's biological determinism, changes within populations would be the result of a response to environmental changes (climate changes for instance). This conclusion suggests that the human organism has a great deal of adaptability at the genetic level, and, indeed, much of evolutionary biology supports the idea (though through a long-term, gradualist approach). But what about horizontal biological adaptability, not through generations, but between individuals and collectivities? To some extent, we adapt all the time in new contexts, new environments. Are these adaptations simply elements of synchronic genetic adaptability?

In this perspective, genetic variation, biodiversity, and gene recombinations may occur; indeed, they even define the internal character of a population. But they are constrained by the fact that they should maintain the population's overall genomic makeup. From a biosocial perspective, a pure population, or a healthy population, is one that maintains a regularity in its genetic makeup (and assumedly in its phenotypic makeup), allows for conservative internal recombination, and can be studied using statistical and probabilistic means—isolated, static, and ahistorical.

But the important thing to note about the Hardy-Weinberg theory and population genetics is that they are part of a statistical science—or, in more contemporary terms, an information science. The question that traditional population geneticists and contemporary population genomics ask is how a heterogeneous, dynamic, adaptive collective of people can be measured through information science. This use of information science is not in itself necessary; early studies on hereditary patterns, evolution, and species dynamics did not rely extensively on statistical methods. But we also need to remember that population genetics in its modern form has developed out of modern molecular genetics, in which the central importance is the genetic code, the kinds of sequences that make up genes, their gene-protein relationships, and their different types of polymorphisms.

Each and Every

Thus far, we have seen how the controversies surrounding biocolonialism stem in part from the way in which genetic science inscribes *race* biologically and from the way in which population genetics and genomics reinscribes *population* mathematically. If these fields view their object of study—the "population"—through genetic and informatic lenses, then what social and political effects might such approaches have?

One result is that population genomics, in the way it stitches together genomes, governments, and corporations, forms a novel type of biopolitical power, one in which the biological—and social—population is reconstituted through the high technologies of genomics and bioinformatics. This biopolitical dimension to population genomics is not simply the exercise of sovereign power, as if genome projects are an attempt to instill top-down apparatuses of power. Rather, what is at stake is "the biological existence of a population," the point at which the health of a population becomes isomorphic with the health of a nation and the wealth of the nation.[37]

For Foucault, *biopolitics* was this technoscientific incorporation of the "life" of a population into a set of political and economic concerns. In its most assertive forms, biopolitics was not a negative, repressive power, but a power that "exerts a positive influence on life, that endeavors to administer, optimize, and multiply it, subjecting it to precise controls and comprehensive regulations." At the very least, biopolitics was a political relation in which state and economic forces took up the "task of administering life."[38] In his analyses of the politics of medicine and health care during the eighteenth and nineteenth centuries, Foucault cites a number of key instances of biopolitical power: the "medical police" plans active in eighteenth-century Prussia, birth-control programs in England and France, urban hygiene efforts in Paris, hospital reforms in London and Paris, the English Poor Laws, inoculation and vaccination efforts in Europe, and the emergence of a science of statistics, demographics, and "political economy."

Biopolitics, then, implies some political and economic incorporation of scientific and technological notions of the "life of the population." Foucault distinguishes biopolitics from "anatomo-politics," the latter that type of power relation that dealt primarily with individuated bodies of subjects, which were rendered docile within a range of social institutions. Whereas anatomo-politics worked through institutional disciplinarization (in the prison, the military barracks, the school, the hospital), biopolitics "focused on the species body, the body imbued with the mechanics of life and serving as the basis of the biological processes: propagation, births and mortality, the level of health, life expectancy and longevity, with all the conditions that can cause these to vary."[39] Whereas the disciplinary mechanisms of anatomopolitics were addressed to the individual, anatomical body, the regulatory mechanisms of biopolitics were addressed to the species body, or to the population. "Their supervision was effected through an entire series of interventions and regulatory controls: *a biopolitics of the population*."[40] The difference between these two types of power relations—anatomo-politics and biopolitics—is thus also a difference in two approaches to the "power over life." For Foucault, they are historically overlapping, rather than opposed to each other: "After the anatomo-politics of the human body established in the course of the eighteenth century, we have, at the end of that century, the emergence of something that is no longer an anatomo-politics of the human body, but what I would call a 'biopolitics' of the human race."[41]

This shift in power relations had a number of consequences. One is that a biological and medical notion of "population" became the primary concern for state and economic interests.[42] Though the "life" of a body politic is an implicit concern in all political thinking, the emergence of biopolitics, for Foucault, is a specific concern with the population as a biological entity, such that "biological existence was reflected in political existence."[43] The notion of a biologically defined population gained much currency in the eighteenth century from political economists such as Thomas Malthus and, later, Adam Smith, David Ricardo, and John Stuart Mill. Malthus's *Essay on the Principle of Population* (first published in 1798, then modified in 1803) put forth the famous and largely discounted argument that a population grows at a geometrical rate (1, 2, 4, 8), but the resources to sustain that population grow only at an arithmetical rate (1, 2, 3, 4). If this argument is accepted, then the population, if unchecked, will outgrow the resources that can sustain it. For Malthus, the concern was that this discrepancy would lead not only to overpopulation, but potentially to an increase in poverty and subsequently moral degradation.[44] For this reason, Malthus acknowledged the negative controls of nature (e.g., famine), while also promoting the positive controls of "moral restraint" (e.g., birth control). In combining mathematical analyses of population growth with a social and moral concern for the health of a population, he formulated a biopolitical concern of political economy. The population is not only biological at its basis, but, as a social entity, it is constantly threatened by "misery and vice," or by overpopulation, unchecked reproduction, and promiscuity.[45] Furthermore, the population is not just a biological or social entity, but, more important, a political and economic one. In his *Principles of Political Economy*, Malthus acknowledged a relation between population and wealth, but that "population alone cannot create an effective demand for wealth."[46] For Malthus, production is not always the same as reproduction; the practical application of political economy in this period was thus to regulate, modulate, and control the population, rather than simply to promote its growth as a source of production itself. "The question really is, whether encouragements to population, or even the natural tendency of population to increase beyond the funds destined for its maintenance, will, or will not, alone furnish an adequate stimulus to the increase of wealth."[47]

In addition to this biological—but also political—definition of *population*, a second consequence follows from biopolitics: the emphasis on mathematical and informatic-based approaches to studying the population. In his discussion of biopolitics, Foucault makes frequent reference to the fields of statistics, demographics, and other mathematical methods used in political economy to account for birth and death rates, the spread of epidemics, and the monitoring of citizens' well-being by health officers. Ian Hacking, in his history of statistics during the nineteenth century, suggests that the rise of such mathematical fields occurred in conjunction with a medical view of human norms as regulated by "laws of dispersion" rather than with a deterministic view of "human nature." The development of statistics and probabilistic methods to measure risks, dangers, and tendencies was part of the long process whereby "society became statistical."[48] The statistical assessment of the health of a population was among the first areas to be affected by these methods. Hacking argues that statistical and probabilistic methods played a central role in the expansion of information regarding populations. Whereas such information for much of Great Britain prior to 1815 had been limited largely to births, deaths, and marriages, by the mid–nineteenth century a number of professional committees and societies dedicated themselves to articulating the "statistics of sickness": that biological processes such as health or illness could be accounted for through mathematical techniques. "Statistical law was on the march, conquering new territory," such as the role of occupation, poverty, age, and frequency of illness.[49] As Hacking notes, "a new type of law came into being, analogous to the laws of nature, but pertaining to people."[50] In this process, the indeterminacy of natural selection, combined with strict laws of biological selection, would produce a tension-filled zone in which statistical and informatics-based methods reconfigured the population as a nondeterministic yet regulatory entity. Hacking's historical context is nineteenth-century industrialism, but his comments resonate with contemporary population genomics: "The avalanche of numbers, the erosion of determinism, and the invention of normalcy are embedded in the grander topics of the Industrial Revolution. The acquisition of numbers by the populace, and the professional lust for precision in measurement, were driven by familiar themes of manufacture, mining, trade, health, railways, war, empire."[51]

Finally, a third consequence of biopolitics is that this combination of a biological definition of a "population" and a set of statistical and informatic approaches for studying (indeed, producing) the "population" culminates in a flexible, variable, differentiated body politic. On the one hand, Foucault's description of biopolitics places emphasis on the mass quality of population, the result of "collective phenomena which have their economic and political effects, and that they become pertinent only at the mass level."[52] Yet—and this is equally important for Foucault—this massification of the population also implies a set of techniques for differentiating within the population. This "biologization of the state" involves approaching the population on the biological level as a particular kind of species with defined characteristics, for Foucault a significant move away from earlier notions of the state grounded in territory.[53] As a defined unit, the population-species can not only be studied and analyzed (for health/medical reasons), but also be extrapolated, its characteristic behaviors projected into plausible futures (birth/death rates, etc.). The proto–information sciences of demographics and statistics provided a technical ground for a more refined, mathematically based regulation and monitoring of the population (and thus of the state's prime resources). The sciences of statistics and demographics are tools "to intervene at the level at which these general phenomena are determined," to inscribe a specificity within a generality.[54] More important, this internal differentiation is the point at which biological notions of race become relevant: by folding biological notions of race (difference) onto biological notions of population (sameness), biopolitics treats "the population as a mixture of races, or to be more accurate, to treat the species, to subdivide the species it controls, into the subspecies known, precisely, as races." In its most extreme cases, this biopolitics of race results in forms of "race war." Thus, "racism makes it possible to establish a relationship between my life and the death of the other that is not a military or warlike relationship of confrontation, but a biological relationship."[55]

Foucault is clearly thinking of the use of medicine in the service of racial purity and of ethnic-cleansing programs, and, as such, his statements may seem extreme in the case of population genomics projects. But there is also a more "liberal" (or, perhaps, neoliberal) side to biopolitics in this regard: by creating internal differentiations within the population, biopolitics opens up new ways of monitoring and regulating the political and economic health of the population. David Arnold notes this tendency in the nexus between colonialism and biomedicine when he states that "colonial rule built up an enormous battery of texts and discursive practices that concerned themselves with the physical being of the colonized," attempting to use this medicalized body "as a site for the construction of its own authority, legitimacy, and control."[56] This hegemonic position is also, importantly, in attune with an all-encompassing, even diversifying approach to the body, for this modern form of governmentality, far from a reductionist homogenizing strategy, is predicated on a dual approach, which both universalizes and individualizes the population. In this gradated approach, populations can exist in a variety of contexts (defined by territory, but also by economic or class groupings, ethnic groupings, or social institutions)—all within a biomedical framework that analyzes the fluxes of biological activity characteristic of the population. Such groupings do not make sense, however, without a means of defining the unit of grouping—not just the individual subject (and in Foucault's terminologies *subject* is also a verb), but a subject that can be defined in a variety of ways, marking out the definitional boundaries of each grouping.

This dual character of governmentality directly applies to current approaches to genetic difference in the biotech industry, especially in fields such as genomics, bioinformatics, and population genomics. Although the various efforts to map the human genome have been concerned with constructing a universal, representative genome (in which specifics as to race are left out of its description), genomics has also become a thriving business in terms of genetic differences, population genomes, or polymorphisms. According to some reports, pharmaceutical companies are realizing that the real money to be made and the most powerful discoveries to be made are not within the universal, single human genome, but in the minute markers that distinguish different human genomes from each other. This pharmacogenomics, or "personalized medicine," promises a tailor-fit drug program in place of the "one size fits all" methods traditionally used.

Thus, to summarize, Foucault's concept of biopolitics, although historically situated in the eighteenth and early nineteenth centuries, can reveal a great deal about the larger political questions that population genomics raises. As a novel form of biopolitics, population genomics redefines (and arguably produces) a notion of "population" that is genetic in its basis. It does this through a set of informatic techniques and technologies, including genome sequencing and bioinformatics analysis. Finally, this combination of genetics and informatics makes possible a simultaneous massification and internal differentiation of the population, which is, it may be argued, a redefinition of biological "race."

Decoding the Population

If population genetics is an extension of the informatic and biological aspects of Foucauldian biopolitics, then how might this connection change the definition of *population* in genomics projects such as those in the United Kingdom or Iceland? In most cases, what population genome databases promise to provide is an extensive, computer-driven, data-mining analysis of the genetic basis of disease as well as the development of treatments, cures, and preventive practices. A database such as deCODE's IHSD has become a paradigmatic example from a scientific point of view in that it brings together three types of data: phenotypic data and health-care information, genotypic data (genomic sequence information), and genealogical and hereditary data (gene pool and statistical information).[57] In 1998, the Icelandic parliament approved deCODE Genomics's proposal for the development of the IHSD, an agreement that would give deCODE exclusive rights to control the access privileges to the IHSD for up to twelve years. Under this agreement, even Icelandic physicians would be forced to pay a fee to access health data that was previously freely available. At the same time, deCODE also established a multimillion-dollar alliance with pharmaceutical giant Hoffman-LaRoche for carrying out basic drug R&D on the IHSD.

The IHSD provides us with a situation in which a widespread sampling-, sequencing-, analysis-, and research-based product and services development takes place. The IHSD is both highly specified in its object (a genetically isolated population) and widespread in its coverage (containing national health records, genealogical records, and genetic information). In addition, the IHSD, as part of deCODE, is a business endeavor as well as a health-care one, and deCODE uses this product—here the product is information or database access—to forge productive relationships with governments (Icelandic national genealogical records) and other business (database services, drug partnerships).

However, it is precisely this combination of medicine, privacy issues, and economics that has many scientists, activists, and citizens worried about the implications of such population genome projects. At the same time that deCODE's proposal was approved, a collection of concerned scientists, activists, and citizens gathered together to voice dissent over many of the ethical questions raised by the IHSD. This group, known as Mannvernd, and the Icelandic Medical Association continue to emphasize the importance of ongoing, truly democratic debate, not just on the IHSD, but on the trend of genomics databases generally.[58]

The main criticisms of the IHSD are fourfold. First, there is the issue of the commodification of health-care data. That deCODE has exclusive rights to control database access means that individual patients' rights might be pushed to the background in the interests of economics. Second, there are also concerns over information privacy and the confidentiality of medical data; although deCODE controls database access, it does not claim any responsibility for how patient information would be used by third parties. Confidentiality—the cornerstone of trust in the medical encounter—would be compromised by the unregulated distribution of genetic sequence, medical history, and lifestyle information. Third, there is the more scientific controversy over the efficacy of such population genome projects: many scientists claim that they are of benefit only regarding direct disease-causing genes (e.g., in cystic fibrosis or Huntington's disease), where single-point mutations or variations in a population can be targeted for the development of drugs. Other so-called complex diseases—polygenic diseases, with each gene expressing itself in variable degrees, along with environmental factors—would not be so easily detectable in a population genome (or indeed in "the" human genome). Fourth, a central area of dispute is deCODE's notion of "presumed consent": that, by default, Icelanders will be included in the database unless they voluntarily opted out. The assumption is that a visit to the doctor is as good as signing a consent form, an assumption that groups such as Mannvernd strongly contravene. Population genome projects thus offer four central challenges: commodification, privacy, consent, and surveillance. We can examine each briefly.

In any science-based project where human biological materials are being taken by a corporation, the issue of commodification is foremost in the social debate and ethical considerations. Although it is true that projects such as the IHSD are ostensibly aimed at providing positive models for the future of health care, the organizations that run them, such as deCODE, are also, when it comes down to it, businesses. This connection thus involves a consideration of what a particular biotech company's product is. In genomics, the product is most often, not surprisingly, information. But in genomics and in biotech generally, the rise of informatics has shown that many types of valuable information are available on the biotech market—universal genome data, gene pool data, disease profiling, SNP databases, and so on. A company's success depends on "having" not only the most valuable data, but also the most accurate (this is Celera's claim over the public consortium), the most technically sophisticated (thus the importance of tech companies), and the most articulate, or data that are shown to have a direct, immediate impact on health care and the fight against disease (the search for pragmatic genetic data). The challenge put forth by those critical of corporations and businesses handling health-related issues is that the research conducted and the projects undertaken be in the best interests of public health and not be determined by commercial concerns.

However, even when public projects attempt to assemble biological databases, there is still discomfort over the very process of sampling, extraction, and utilization of one's own body for medical research. In the case of genomics, this is a very abstract process, but also a very simple one, moving from a blood sample to a DNA archive in a computer. Contemporary issues of privacy have such a resonance in biotech, however, because of the central importance of new computing technologies in the biotech and biomedical lab. There have been disputes concerning the ownership of one's own DNA, in which, for instance, a company such as deCODE develops novel patents based on research done on individual human DNA samples. For this reason, many companies require complex disclaimers, and they also make an important further distinction between a person's own lived body and health-related data generated from a person's body. This reduction of the debate to a distinction between blood and data may solve the patenting and ownership issue, but it still does not address that fundamental gap: the ontological difference between one's own, proper body and the genetic data extracted from one's body. Many of the debates concern the handling of genetic data (for some projects, such as Celera's, individual donors are anonymous, and for others, such as deCODE's, individual identification is required to correlate genealogical with medical data).

Because these two issues of privacy and commodification are present in virtually every endeavor to create a genetic database, it is common for differing levels of informed consent to be an imperative for the research carried out. Informed consent mandates became especially important in so-called biopiracy cases, either involving the appropriation of Third World natural resources or the sampling of genetic material from indigenous populations around the world by First World science organizations. Similar issues apply to genetic sampling within a nation or community. Although there are different levels of consent, ranging from individual informed consent to the broadly applicable "presumed consent," they all contain clauses that permit certain types of activities and enable others. The IHSD, for instance, requires three levels of consent because there are three basic levels of appropriation of genetic material: consent for the blood/DNA sample extraction, consent for the genotyping, and consent for the analysis and diagnosis of the sequence and of its potential range of uses in research and clinical practice. As may be guessed, the ambiguities in different consent clauses have led to disputes over issues such as patentability, the buying and selling of genetic data, and so on.

As a unique technoscientific approach to race and ethnicity, population genomics also undertakes a unique type of biological or genetic surveillance. First World big science projects (such as the HGP) that harvest biological materials from Third World communities are an explicit form of *biocolonialism* in the more common usage of the term. The HGDP's controversy surrounding the extraction of cell lines from a New Guinea community is an example. Even with the now-mandatory clause of informed consent (again, there are many, often ambiguous levels of informed consent, including "presumed consent"), the appropriation of biological materials by First World science more often than not feeds into First World health-care economies (principally drugs). This situation, it should be said, becomes more complex when we consider the level of involvement or compromise of those who are the object of biocolonialism. Whereas some instances do involve the veritable plunder of biological materials, with little or highly falsified information given, other instances, in which the colonized community demands some economic reimbursement, are more complicated because such negotiations play into the commodification of the body.

But biocolonialism is also a phenomenon within First World countries, where the pharmaceutical industry stands to gain the most returns. This "endocolonization"—not only of the body but of medical practice itself—focuses on the ways in which the biological body can be turned into a value generator, either in drug development or through novel medical techniques such as gene therapy.[59] With a very straightforward pipeline of capital investment, R&D, products and services, and finally medical application, the biotech industry has transformed the First World into a giant molecular laboratory. The human genome database is only the most explicit manifestation of this literal data mining of precious, valuable genes. The patenting of biological materials and new technologies feeds back into an economy that operates largely at the level of finance capital.

What these four issues illustrate is that the question of genomics has much to do with institutional control and the level of biopolitical practices within societies, be those practices handled by governments or corporations. Some projects, such as DNA Science's Gene Trust, operate on a volunteer-only basis, using the rhetoric of altruistic humanism as the primary motivation (although, incidentally, DNA Sciences accrues immediate monetary gains). Other projects, such as Celera's human genome database, are exclusive, selling access only to major pharmaceutical companies for drug development. And still others, such as the IHGSC databases, are technically public-access databases, but are used almost exclusively by research institutions and universities.

The scope and ambition of projects such as Celera's or deCODE's are, of course, made possible by advances in computing technologies, most of which are exorbitantly expensive and inaccessible to nonspecialists, and which have a high learning curve. These restrictions, combined with the illegibility of genetic sequence data to nonspecialists, makes the potentials of genomics extensively out of the reach of the general health-care public and makes any informed critique or public debate challenging as well. Reasons for this difficulty includes the facts that current molecular biotech research builds on previous and equally opaque research and that the revolution in computer technologies challenges traditional modes of research and scientific practice.

For instance, in articles written by Kari Stefansson, CEO of deCODE, the potentials of informatics technologies transform science research from a linear, hypothesis-driven approach to a semiautomated data mining agent that completes computations far beyond what was possible prior to the use of parallel processing computers and data mining algorithms.[60] Companies such as deCODE and Human Genome Sciences have stated that they are in the business of information and discovery. The intersection of business approaches to genomics and the use of informatics-based tools means that science research is based on combinatorial techniques; the best pattern-recognition combinations will be of the highest value, the greatest assets.

In considering this complex of life science business models, new computer tools, and a genomics-based approach to populations and disease, we can actually differentiate several strategies within contemporary genomics. One strategy has to do with the utilization of a universal reference in genomics. Generalized human genome projects, such as that undertaken and first completed by Celera, emphasize their universality as models for the study of disease, for treatment, and for a greater understanding of life at the molecular level. They also highlight the backdrop against which all genetic difference and deviations from a norm will be assessed. Indeed, part of the reason why genomic projects by Celera, Incyte, and the public consortium have received so much attention is that they are in the process of establishing the very norms of genetic medicine. Their practices and techniques themselves are the processes of establishing a genomic norm, what will or will not exist within the domain of consideration for genetic medicine, and what will or will not be identified as anomalous, excessive, or central to genetic knowledge and pharmacogenomic drug design. A company such as Celera, though it assembles its sequences from a number of anonymous individuals, constructs one single, universal human genome database. That database becomes the model for all sorts of research emerging from genomics—proteomics, genetic drug design, functional genomics, population genetics, gene therapy, and genetic profiling.

The utilization of a universal reference also implies a high degree of individualization within targeted population groups. At the opposite pole of Celera's universal model of the human genome is a field of research that deals with the minute, highly specific base-pair changes that differ from one individual to another.[61] However, many SNPs are phenotypically nonexpressive; that is, they are base-pair changes that do not effect the organism in any way—they are simply differences in code sequence. Many large-scale genome projects, such as Celera's, also contain information on polymorphisms, and many bioinformatics applications are designed to provide analysis of polymorphisms. But specific projects, such as Genaissance Pharmaceutical's HAP series of database tools and the Whitehead Institute's SNP database, focus exclusively on these minute base-pair changes.[62] For Genaissance in particular, the study of single-case base-pair changes in themselves are less productive than a total chromosomal perspective of SNP positioning. Genaissance uses its proprietary bioinformatics tools to assemble the given SNPs within a given chromosome or gene, thus offering a more distributed map sample of the interrelationships of SNPs. These databases of individual point changes form linkages between variations within a gene pool and a flexible drug-development industry that operates at the genetic, molecular level.

From the combination of a universal reference and minute differences within the universal, population genomics, specific modes of integrating biology and race are woven. Whereas Celera's human genome database attempts to establish itself as a general, universal model for genetic research, other genomics projects are focusing on collectivities within a universal gene pool. The projects from deCODE, Genaissance, Myriad, and others focusing on genetically isolated populations move somewhere between the *universality* of Celera and the *individuality* of the SNP databases. Often combining the usual genotypic data with data from demographics, statistics, genealogy, and health care, these projects are both new forms of health-care management and paths toward understanding the effects of disease within genetically homogenous groupings. This is perhaps the main discomfort with projects such as deCODE's IHSD; they promise the ability to perform large-scale computational analysis on entire genomes, but in this they also threaten to abstract genetic data from real, physical communities. They take a genetics-based, or genotypic, view of race and make connections to the functioning of norms within medicine and health care. In doing this, they establish an intermediary space of negotiation between the universality of a human genome project (which claims a uniformity under the umbrella of a distinct species) and the high-specificity of SNP or HAP databases (which claim difference within a general category). This intermediary space is precisely the space of racial (mis)identification and ethnic boundary marking; it is the space where collectivities form, composed of individuals united under a common species categorization. These population genomics projects, with their primary aim as medical (specifically in drug development and genetic diagnosis), are involved in the rearticulation of race and ethnicity through the lens of bioscience research and corporate biotechnology.

In this schematic of different types of genomics projects, we can see an approach toward biological information that is far from a simplistic monocultural model in which difference is marginalized, silenced, or pushed out of the domain of serious consideration. From a scientific perspective, of course, difference—or diversity—is the cornerstone of traditional evolutionary theory, be it random or directed environmental influence. But in this scientific narrative, these mutations or evolutionary changes always have some end and are directed toward the establishment of a new and more adaptable phenotype, maintaining the same hierarchical dynamics. Likewise, from a business perspective, diversity not only is the key to creating more custom-tailored products (as in genetic drug design), but also enables a more thorough production of knowledge about the population. This is a kind of niche biomarketing, in which information is extracted from a heterogeneous population, then selectively organized and rerouted into research and product development.

Projects such as deCODE's IHSD are prototypes for scientific-business practices that are not separate from a consideration of race and ethnicity, which are themselves considered from a molecular and informatic perspective. They do two things: they work toward establishing new, more flexible sets of norms, both within biomedicine and in business strategies, and in doing so they form new methods of population management and regulation.

"On the Surface of Life, to Be All Open Wounds"

In the specific context of molecular biotechnology and biomedicine,[63] biocolonialism brings together the discourses and practices of contemporary bioscience with a colonialist approach to specific bodies and populations.[64] This approach involves the application of the ideologies of expansion, technological development, and appropriation of natural resources to modern medicine. It also brings in the critical perspectives of postcolonial theory, especially as it pertains to the complexities of the relationships between colonizer and colonized, of hybridity, and of the relationship between conceptions of race and the body.[65]

However, a stance critical of biocolonialism is not, it should be said, simply an antiscience perspective. Biotechnology is too diverse and complex a field to be simply denounced wholesale, and it promises to affect materially too many bodies to be dismissed as ineffectual. We should recognize several things about biotech in relation to biocolonialism. To begin with, biotech is founded on Western bioscience research and Western medical practice. As a technical practice, it emerges from a discontinuous tradition of Western science. This also means that it has a more or less defined position on what counts as legitimate scientific research (that is, which kinds of research will form linkages to medical practice).

But as is well known, biotech is as much a business as it is science research. Along with its growth in the finance capital sector and its use of government funding, biotech is partaking in the broad processes of economic and technological globalism. The result is that the gap between the knowledge of the patient and the black-boxing of biomedicine becomes greater and greater. This global perspective has meant a series of problematic scientific and political relationships between First World and Third World countries. The expansion of "Big Pharma" to other countries, the sampling and patenting of biological materials from Third World populations, and the influence of Western medical practices are all instances of an asymmetrical conflict of interests.

This appropriationist approach to Western medicine is applied equally within technologically advanced sectors of the First World, especially in the United States. The globalization process of biotech means, first and foremost, an expansion and reconfiguration of the (mainly U.S. and British) medical and health-care landscape. The agglomeration of genetic databases, high-tech drug development, and high-pressure clinical trials combine to transform the United States into a biotech-infotech laboratory.

If, generally speaking, *colonialism* refers to the process whereby one collectivity (usually geographically and nationally defined) forcibly or coercively appropriates the land and resources of another collectivity, then some important issues can be brought up concerning the current influence of global biotechnology. However it is not exactly accurate to speak of biotech as a colonial instance, despite explicit examples of biological sampling and patenting. Rather, we should ask how biotech, as a technology-driven practice networked into a globalizing economy, approaches, transforms, and encodes different kinds of culturally specific bodies. In particular, we need to consider the relationships between the science of biotechnology and the concept of race as manifested within the biosciences.

Bearing in mind Foucault's emphasis on the population as a biologically defined entity and on the proto-informatic forms of regulation of the population, we can discern three primary issues with respect to biocolonialism and population genomics: informatics, biodiversification, and the notion of "genethnicities."

First, informatics is a key factor in considering contemporary power relationships in biotech because it works as a medium for transforming bodies and biologies into data. But those data are understood in many different ways, not simply as the liquidation or the dematerialization of the body. The case of population genomics suggests that databases are anything but "just data." Networked into drug development, medical records, and health-care services (including health insurance), genome databases are literally information that "matters." Genomic, population, and SNP databases are just some examples of this variability of biological data that are both digital and material. As a technical mode of analysis, knowledge production, and application in medicine, informatics at its roots brings up philosophical questions (What is the body if it is essentially information?), but preceding these questions at every point are political questions.

Second, and in a similar vein, *biodiversity* is a term most often reserved for debates concerning the preservation, conservation, or sustainability of natural resources, which depends a great deal of natural diversity, as opposed to the advantage of natural diversity that transnational corporations take to produce monocultures as product.[66] In the context of genomics, biodiversity becomes a signifier for genetic difference and for the ways in which genetic difference gets translated into cultural difference. Biopiracy (the attack on biodiversity) is not simply about the destruction of natural resources; it is about a complex reframing of "nature" and the use of diversity toward commercial ends. As Vandana Shiva argues, the discourse of biodiversity is actually less about sustainability than it is about the conservation of biodiversity as a "raw material" for the production of monocultures.[67] The same can be said of biotechnology, especially in the case of genome projects, genome databases, cell and tissue banks, DNA sample collections, and other instances of accumulating biological information.

A third point of controversy regarding many such projects is the issue of genetic discrimination, which is, in the case of bioinformatics and genomics, a kind of data-based discrimination. With molecular genetics, a unique type of identification and differentiation has come about in which individuals and populations can be uniquely analyzed and regulated. Discussing recent genetic screening of African Americans in the United States for sickle-cell anemia, Troy Duster observes how the argument for genetic screening changes when dealing with specific racial or ethnic groups. Whereas general medical tests can be proposed as being "transparently in the general public interest," this "ceases to be the case when screening is aimed at a specific population with its 'own' health problem, and when the cost effectiveness of such screening is assessed by those not in the screened population."[68] This twofold process of molecular genetics (genetic essentialism) and informatics (population databases) paves the way for a new type of data filtering, or, in some cases, discrimination. Such discrimination is based not on race, gender, or sexuality, but rather on information (genetic information). Bioinformatics—an apparently neutral technical tool—thus becomes manifestly political, negotiating how race and ethnicity will be configured through the filter of biotech, constituting a unique type of *genethnicity*.

These biopolitical features of population genomics—informatics, diversification, genethnicities—point to the ways in which the object of such practices is the molecular body itself. If colonialism historically involved the forced or coerced appropriation of people, land, and economy, *bio*colonialism presents us with a situation in which the bodies of the colonized *are* the land and economy. In this case, "population" morphs into territory and resource, and in some of the more economically motivated ventures, these things in turn translate into biological *value*. The appropriation of land meant the acquisition of territory for expansion, and the acquisition of this territory intersected with a like expansion of the colonial economy. So far the biotech industry has displayed little interest in the kinds of colonial expansionism historically demonstrated by European nation-states. Despite the millions of dollars flowing into corporate biotech, the majority of companies, excepting the large pharmaceutical corporations, are smaller upstarts in the process of making large promises and taking significant financial risks in R&D. What then justifies the "colonialism" of biocolonialism?

Biocolonialism takes the molecular body and biological processes as the territory or the property to be acquired, *but insofar as this body can be modulated at the level of informatics*. When governmental regulations so stipulate (as in the case of the U.K. BioBank), this acquisition is made through some type of informed consent, so that the individuals and community whose biological materials are being acquired are informed about the reasons and future uses of their bodies. At other times, this acquisition is handled without such formalities, resulting in either biopiracy (simple appropriation without any consent from or reimbursement to the community) or patenting (based on cell lines or genes that are minimally modified). In order to conduct successful research and possibly to turn over a profit, the first thing needed is a large resource pool, which in this context means a biological sample suited for the particular type of research being conducted. If we are to take seriously the concept of biocoloniaism, then we will have to understand the ways in which the territory appropriated by the population genomics project is the biomolecular body of a "population."

Postgenomic Collectivities

Given the controversy that the HGDP caused in the early 1990s, the work of population geneticists is also, Cavalli-Sforza states, a statement against racism, through the disciplines of population studies and evolutionary biology. Although this defense of Western science is an interesting political move, the logic behind it is deeply problematic.

Cavalli-Sforza takes race as a biologically determined entity, but he discusses racism as a cultural manifestation with biological roots. He traces this connection to the inclusivity of an exoticized, prehistoric tribalism threatened by a rampant xenophobia. For Cavalli-Sforza, racism uses the biological differences that science studies, but for antagonistic reasons. In other words, he acquits science of any involvement in racism—the problem is in the use of science's results, not in science itself.

But many of these issues dealing with racism, culture, and biology are left fundamentally truncated in Cavalli-Sforza's textbooks as he explains the science of population genetics. It is not clear whether racism can be approached on a cultural level or if the way in which "race" itself is defined by Western science is problematic. Although he does point to the complexities in visible and nonvisible genetic variations, he does not connect these complexities to his initial comments against racism.

The justifications Cavalli-Sforza gives for population genetics and projects such as the HGDP follow a common universalist logic: "we" all are descended from a common race; thus, this science and this knowledge are for everyone's benefit in understanding the lineage of *Homo sapiens*; and in the more near term, such knowledge promises to contribute to the improvement of Western medicine, especially in fields such as pharmacogenomics. It goes without saying that Cavalli-Sforza does not take up the issues of patentability, profit making, and the rigidity of Western medicine, and the same can be said of related science articles and textbooks.

The most important issue brought up by population genetics and related fields is the future of genetic difference. Is this a new form of racism for the "biotech century"? With the widespread use in science of informatics and computers, will biotechnical forms of racism also become informatic forms of discrimination?

As a way of addressing this question, we can point to three possible directions that a biocolonial critique might initiate. A starting point would be a consideration of the population within a temporal—and not just spatial or statistical—framework. A population should be taken as *dynamic*. Although traditional population genetics aims to study an "ideal" population—static, closed, predictable—human collectivities of all sorts are embedded in heterogeneous, fluctuating, and dynamic environments. As such, they cannot help but also be dynamic and, above all, adaptive. Bioscience has yet to consider seriously this notion of an adaptive, flexible population. Doing so would mean integrating, in a transdisciplinary manner, other perspectives—distributed pattern-recognition systems, autopoietic theories, systems biology, nonlinear dynamics, and molecular epidemiology.

A second guideline would take into account the way in which populations are dynamic; that is, a population should be taken as a *hybrid*. Again, the ideal population for study is coherent; it forms a closed system in a vacuum, with no new genetic material coming in and the overall gene frequency remaining stable, though internal fluctuations may occur. But—as this guideline points more directly to the notion of race—populations not only are dynamic, but can be incredibly diverse, even to the point of internal fragmentation. This hybridity is not a deviation from a biologically pure ideal; it is the very substance that composes the population. A mode of articulating and understanding difference at the biological and genetic level is needed, one that does not define difference as deviation, but, in a more constitutive manner, as something without which a population cannot form.

Finally, although there are a number of problematic points concerning the conception of race and ethnicity in population genomics, a critique of such points should not be taken as antiscience. The biological should be integrated into the cultural, the social, and the political. Population genetics—and biotech generally—cordons off the biological determinism of bioscience from cultural processes, but we might question, from the point of view of the biosciences, whether the biological is all that containable. Our definitions of the biological domain also need to be more flexible and more articulate. The biological should not simply determine or explain race; it should be a core part of the cultural experience of race. This quasi-dialectical move would also require that cultural issues be a central part of science projects such as the HGDP. It is through this perspective that non-Western viewpoints on race and population can be seriously considered.

Given these starting points, we can reconsider now the apparent contradiction with which we began. Recall that I noted the simultaneous disappearance of the issues associated with the HGDP and the rapid rise of bioinformatics and the use of computer technologies in genomics generally. So, then, what has become of the original issue put forth by the critics of the HGDP? Part of the problem is that the issues dealt with in this criticism have been handled in the same way that criticism of genomic mapping and human embryonic cloning have been handled: they have been filed under the worrisome category of "bioethics" (which commands a mere 3 percent of the HGP's budget). As postcolonial critiques have pointed out, the HGDP came to a relative standstill because it could not reconcile Western scientific assumptions and intentions with non-Western perspectives toward agriculture, population, medicine, culture, and so forth. The sheer gap between the HGDP's bioinformatic colonialism and those predominantly non-Western cultures who were to be the source of biomaterial for the HGDP database illustrates the degree to which *global* once again means "Western" (and, increasingly, "economic"). *One of the meanings of the decrease in the presence of the HGDP and the rise in bioinformatics developments and applications is that the issues of ethnicity and cultural heterogeneity have been sublimated into a paradigm in which they simply do not appear as issues.* That paradigm is, of course, one based on the predominance of information in understanding the genetic makeup of an individual, population, or disease. When, as geneticists repeatedly state, genetic information is taken as one of the keys to a greater understanding of the individual and the species (along with protein structure and biochemical pathways), the issue is not ethnicity but rather how to translate ethnicity into information. In such propositions, ethnicity becomes split between its direct translation into genetic information (a specific gene linked to a specific disease) and its marginalization into the category of "environmental influence" (updated modifications of the sociobiological imperative).

The biopolitics of genomics is that of an informatics of the population in which cultural issues (ethnicity, cultural diversity) are translated into informational issues (via either a universal, generalized map of the human genome or individualized maps of genetic diversity). As with numerous other science-technology disciplines, the apparent neutrality of abstract systems and purified information needs to be questioned. As Evelyn Fox Keller, Donna Haraway, and others have pointed out, information is not an innocent concept with regards to issues of gender and race.[69] The question that needs to be asked of bioinformatics, online genomic databases, and genome mapping projects, is not just "Where is culture?" but rather "How, by what tactics, and by what techniques is bioinformatics reinterpreting and incorporating cultural difference?"

Coda: Fanon's Database

As a closing note, I should point out that despite the increasing visibility of the more "neutral" technical fields of genomics, bioinformatics, and population genomics, the HGDP's broad initiative has not disappeared. In fact, it can be observed alongside the high-tech fields of genomics in contemporary media culture. For instance, consider the 2003 *Nova* documentary *The Journey of Man*. Narrated by population geneticist Spencer Wells, *The Journey of Man* attempts to show how all races—no matter how different culturally, economically, or politically—are united by a common genetic heritage. This heritage is, of course, "the" human genome, which genetic archaeologists claim can be traced back to the earliest human beings on the African continent. Using both high-tech genomics and "old-fashioned" ethnography, *The Journey of Man* shows us cultural difference at the same time that it posits genomic unification. *The Journey of Man*, in its ethnographic representation of Wells visiting a range of cultures, even purports that this common genetic heritage can be the basis for a greater cultural understanding. The question is, of course, Understanding for whom and dictated by what criteria? *The Journey of Man* makes a number of alarming claims, such as Wells's closing comments that "we are all literally 'African' under the skin." And the documentary is made more volatile by the fact that Wells is a former student of Cavalli-Sforza.

Frantz Fanon's essay "Medicine and Colonialism" offers a counterpoint to what may be the new face of population genomics. Fanon's text situates the tensions between colonizer and colonized within the framework of medicine. Although more recent developments in postcolonial studies have complicated and tempered Fanon's position—one thinks of Homi Bhabha's work on the colonial encounter, or Gayatri Spivak's work on the subaltern—Fanon's essay still has great import in thinking about the problem of biocolonialism. Writing in the midst of the Algerian revolution, Fanon's essay is instructive for colonialisms of all kinds, for it attempts to do two things: remain decisive in a critique of colonialism and, at the same time, remain open to the transformative and empowering aspects of a science and technology that serve the people or "the population." Fanon is adamant about the impermeable barrier between a colonized society and the colonizing one, "the impossibility of finding a meeting ground in any colonial situation." In addition, the biopolitical concerns of a population's health make the colonial imperative immune to critique: "When the discipline considered concerns man's health, when its very principle is to ease pain, it is clear that no negative reaction can be justified."[70] For Fanon, the introduction of European medicine into the colonies is part and parcel of the colonial program. Not only are local knowledges and practices delegitimized, but, in the case of colonial medicine, the gift of health care always leads to an indebtedness.[71] The figure of the colonial doctor is, for Fanon, a figure that represents the most insidious form of colonialism because it is precisely "life itself" and bodily health that cannot be questioned. At the end of the day, Fanon sees political and economic interests behind this imperative of health. As he notes, "in the colonies, the doctor is an integral part of colonization, of domination, of exploitation." Furthermore, "in the colonial situation, going to see the doctor, the administrator, the constable or the mayor are identical moves."[72] Thus, "this good faith is immediately taken advantage of by the occupier and transformed into a justification of the occupation."[73]

Despite his militant position, Fanon also notes a number of complexities that arise in the colonial encounter as a *medical* encounter between doctor and patient. For one, he notes a fissure within Algerian culture in the figure of the native Algerian doctor. On the one hand, the native doctor understands the people and culture much better than does the colonial French doctor; on the other hand, the native doctor has been trained and schooled in Western medicine—the colonizer's medicine—and is therefore a subject of mistrust. Not only this, but the native doctor has denounced all other modes of medical treatment as superstition, as primitive, or as irrelevant to the domain of rational scientific inquiry. The native doctor is thus put into a difficult position, at once taking on the burden of cultural understanding and yet facing resistance and mistrust from patients precisely because of this burden.

If in colonial medicine the native doctor is effectively alienated, this situation is reversed by the condition of revolutionary conflict. Fanon notes how in the context of Algeria's war of liberation the native doctor, nurse, and technician went from being an ostracized third party to becoming a constitutive part of the anticolonization movement. Certain key events served as the impetus for this transition, such as the French government's decision, throughout 1954–55, to place an embargo on all medicines entering Algeria for the Algerian people. Although the people were barred not only from seeing the European doctors, but also from receiving vaccines and medicines, the native medical professionals, occupying an intermediary position, were able to create an infrastructure for the flow of medicines and medical knowledge into Algeria. Fanon describes the Algerian National Liberation Front's controversial medical program:

These medications which were taken for granted before the struggle for liberation, were transformed into weapons. And the urban revolutionary cells having the responsibility for supplying medications were as important as those assigned to obtain information as to the plans and movements of the adversary. . . . It [the National Liberation Front] found itself faced with the necessity of setting up a system of public health capable of replacing the periodic visit of the colonial doctor. This is how the local cell member responsible for health became an important member of the revolutionary apparatus.[74]

The preventive practices, hygiene plans, and routine examinations that were shunned by the Algerian people were suddenly now put into place with great fluency. Native medical professionals, previously alienated as constituting part of the problem rather than a solution, were now brought into the fray of colonial struggle. "The war of liberation introduced medical technique and the native technician into the life of the innumerable regions of Africa."[75] There are many examples like these, but what Fanon shows us in the case of Algeria is the degree to which politics is biopolitics and how colonialism is never far from biocolonialism.

Of course, Fanon's largely oppositional account sidesteps a number of frustrating compromises. For instance, the relationship between medicine and politics is complicated in the colonial situation, to say the least. Medicine may ideologically be seen to constitute a part of the colonial apparatus, but the experience on the ground may be quite different, where a mere guilt by association may in fact prevent much needed medical treatment to colonized peoples. In addition, although Fanon psychologizes large social groups—the colonizing nation, the colonized people—he does not consider the ways in which resistance emerges from specific individuals, such as colonial French doctors sympathetic to anticolonial sentiment. Finally, Fanon's injunction of medicine as part of colonialism really points to a larger issue, which he leaves unmentioned: the relationships between medicine and poverty in the colonial context. A wide range of factors—from diet to urban planning to labor conditions—affect the "health of the population" of which Foucault speaks. These and other issues further complicate Fanon's oppositional stance regarding colonial medicine.

Thus, although Fanon's generalizations may be questioned on particular historical grounds, the general lesson he points to is worth thinking about in the context of biopiracy and biocolonialism. Fanon is militant about the irresolvable conflict between colonizer and colonized, but he adopts a more complex approach on the issue of biopolitics—the "health of the population," of the body politic. The example of the native Algerian doctor, nurse, and medical technician is an example of reappropriating the benefits of medical practice and knowledge, but without the political and economic imperatives set on them by colonial interests. Political, economic, and cultural issues will certainly come up in such situations, and, indeed, we are seeing such issues play themselves out in the ongoing efforts to offer American governmental and corporate financial aid to combating AIDS and tuberculosis in Africa (most notably, from the Gates Foundation). Writing in the midst of Algeria's colonial struggle, Fanon offers the possibility of a deeply committed critique of colonialisms of all kinds, along with an equal commitment to the social and political empowerment of medical knowledge and practice. However, Fanon's overall point still retains a degree of ambiguity, despite his vehement voice against colonialism. As he notes, "science depoliticized, science in the service of man, is often non-existent in the colonies."[76] The trick, or course, is knowing how to separate the benefits of medical knowledge and practice from its political and economic motivations—or, indeed, if such a separation is possible at all. It may be that the immediate and future contestations in developing nations over agriculture, health care, and genetics will decide on this point.

In the case of contemporary biocolonialism and biopiracy, we see a situation significantly different from the examples of Algeria, India, and other sites of colonial struggle. As I have noted, it is not the territory that is at stake, but rather the biological population—the population is the territory in biocolonialism. Moreover, this population is configured as a biological and a genetic entity, and in this sense the body of the population is separated from the body of a culture. Blood, protein, and DNA samples do not contest their status as items of property, especially when they are extracted and abstracted from the particular bodies of individuals. This biological-genetic definition of the population is coupled with an informatic approach to the population, as we see in the discipline of population genetics and in the techniques of population genomics. Both approaches—the genetic and the informatic—are approaches that Foucault implicitly points to in his discussions on biopolitics.[77] Finally, biocolonialism is also unique in that, by shifting the emphasis from territory to population, it makes possible layered forms of "endocolonization": population genomics and prospecting within a given population or nation (for example the U.K. BioBank, or the Gene Trust in the United States). Thus, biocolonialism raises not only cultural issues (as in the case of the HGDP), but also, in its endocolonial form, issues pertaining to medical surveillance, the privacy of health-related information, and economically driven health-care practices in technologically advanced countries.

Although the potential medical benefits from fields such as population genomics are still being debated, the economic and civil liberties issues have prompted many individuals and groups (such as RAFI) to speak out against biocolonial practices. Is there a space, within the *biocolonial* encounter, for negotiation? As Fanon notes,

> Specialists in basic health education should give careful thought to the new situations that develop in the course of a struggle for national liberation on the part of an underdeveloped people. Once the body of the nation begins to live again in a coherent and dynamic way, everything becomes possible. The notions about "native psychology" or of the "basic personality" are shown to be in vain. The people who take their destiny into their own hands assimilate the most modern forms of technology at an extraordinary rate.[78]

For Fanon, the political and military program of occupation is also a medical program of occupation. But his twofold position—a critique of colonial imperatives and an openness to tactical innovation—can be seen as one example of how to avoid simply demonizing science and medicine en masse. This model—difficult as it may be to carry out—can be of great benefit for confronting the issues raised by, for example, population genomics. In fact, such biobanking efforts demand at the minimum a sort of "critical genomic consciousness," or, as Fortun notes, a "genomic solidarity": "experimenting with the idea of genomic solidarity and working to imagine and invent its social practice seems increasingly necessary to me. One way or another, we are going to find ourselves grouped into some kind of population that has some kind of value, commercial or otherwise. . . . So why should any of us become scientifically and commercially consolidated, but remain socially and politically isolated?"[79] In the colonial context of Algeria, Fanon notes that "the Revolution and medicine manifested their presence simultaneously."[80] Perhaps, in the genome era, we can complicate Fanon's terms without losing his critical acuity: revolution and biotechnology manifest their presence simultaneously, as that which is always about to arrive.

5

The Incorporate Bodies of Recombinant Capital

Cut-and-Paste Bodies

In 1973, biochemist Herbert Boyer and molecular biologist Stanley Cohen published a paper in the *Proceedings of the National Academy of Sciences*, outlining their procedures for producing "recombinant DNA."[1] Working at the University of California, San Francisco, Boyer had developed a method for using "molecular scissors"—an enzyme known as a restriction endonuclease—to snip bacterial DNA at precise locations. Cohen, working separately at Stanford University, was simultaneously experimenting on the use of bacterial plasmids—short loops of DNA in bacteria such as *E. coli*—as a kind of delivery system for novel segments of DNA. Together, Boyer and Cohen, along with other researchers, were central to the development of what became known as genetic engineering. Put simply, genetic engineering is based on the premise that the DNA of an organism can be reshuffled and engineered; the tools of genetic engineering are the organism's own biology, and its techniques are based on the logic of cutting (restriction enzymes), pasting (DNA ligase enzymes), and replicating (in the cellular divisions of the bacterial plasmids). Boyer and Cohen had demonstrated not only that DNA could be artificially "collaged" into new hybrids, but also that this cut-and-paste logic could operate across species. They had provided a technical logic for engineering at the molecular and genetic levels, and they had provided a set of powerful tools—many of which are still used—for carrying out genetic engineering research.

This research had a number of consequences. Several years after Boyer and Cohen published their recombinant DNA findings, both researchers patented their techniques for producing recombinant DNA, and they put together a business plan for a company to market these techniques. That company, called Genentech, garnered some $35 million from its IPO in 1980 and is often cited as the first true biotech company. Genentech—its name an abbreviation for "genetic engineering technology"—would focus on the production of unique drugs and therapeutics through recombinant DNA techniques. It went on to produce a number of highly valuable products, including human insulin and human growth hormone, and many industry analysts point to the emergence of Genentech as the beginning of the first wave of the biotech industry.[2]

But beneath this apparent success story are some important intersections between bodies, technologies, and systems of value. Boyer and Cohen's recombinant DNA experiments signified an important scientific shift within molecular biology research. Whereas postwar developments in molecular biology were geared primarily toward producing knowledge about the organism on the genetic level, the Boyer-Cohen experiments were focused on the development of techniques for working on and intervening in the organism. In a sense, such genetic engineering experiments were the first explicit examples of a "bio-technology": the technical design, engineering, and application of biological life. Although there is a long history of selective animal breeding, food preparation (e.g., fermentation), and agricultural techniques (e.g., ecological optimization of crops), the Boyer-Cohen experiments were the first such instances effectively to combine modern life science research with an engineering perspective, a paradigmatic look toward the technological control over the body as nature.

However, the steps in this narrative of genetic engineering also illustrate the degree to which "pure" research is never pure, but always networked through economic, institutional, and disciplinary zones: research and discovery, publishing and patenting, business start-up and IPO, product development and medical application. These intersections are what Bruno Latour has described as "translation," or the production of cultural, scientific, social, and technological mixtures and hybrids.[3] The networks produced from such translations are often condensed into hybrid objects such as, in our case, recombinant DNA or transgenic organisms. Such hybrid articulations between organisms and technologies come to reflect and eventually to express the adaptability of the biotech economy, in turn affecting technology development and science research itself. In other words, if we consider the scientific implications of recombinant DNA research—that the body can be "reprogrammed" on the molecular level—we must also consider the ways in which that capability to reprogram is spliced with a business plan and an emerging industry. The dual events of recombinant DNA techniques and the founding of Genentech are indicative of the ways in which the *biotech industry* has been keenly aware of the ability to engineer biological matter so that it can generate both medical and economic value. In such an instance, a "cut-and-paste" body is less a cause for bioethical alarmism and more an example of the ways in which bodies and economic value systems can accommodate each other.[4] The cut-and-paste body of recombinant DNA technologies is a mode of flexible accumulation, the transformation of certain biomolecules into "wet" factories for the generation of a range of custom-tailored proteins. Recombinant DNA can not only surpass nature by providing the molecules nature does not, but also forms a novel technological infrastructure in which the body's "natural" processes (for instance, the production of insulin) can be recontextualized in a kind of upgraded, biomolecular black box.

These connections between bodies, technologies, and economies are the driving force of the current "biotech century." The example of Genentech not only points to the complex interrelationships between science, industry, and society, but also serves as a moment that illustrates the "encoding" of an economic logic directly into the biotechnological organism. As the new millennium gets underway, the biotech industry maintains many of the elements it did during the 1970s and 1980s, but new factors have also transformed it into something quite different. This chapter aims to explore these factors. In particular, of interest is the possible transformations of the notion of "labor" as an expressly *biological* notion in the biotech industry. Affecting not only life science research, but drug development, medical diagnostics, and the health-care sector, the redefinition of "life itself" at the core of the biotech industry is key for understanding how the industry balances the imperatives of medical and economic value. The relation between life, labor, and capital is thus key for the success of the biotech industry in medicine; or, put another way, the biotech industry aims to insert "life" into the relation between labor and capital in a way that purports to provide gains in both medicine and economics.

Industrial, Postindustrial, Biotech Industry

Since the 1970s and the founding of Genentech, the term *biotech industry* has taken hold as a descriptor of a large and diverse set of practices and subindustries. The qualifier *industry* seems to imply that biotechnology is related to industrialism in some way. In the same way that industrialism utilized advances in science and technology to transform the production process (in the familiar examples of textile factories or automation), the biotech industry utilizes advances in molecular biology and genetics to transform the way that drugs, therapies, and diagnostics are produced. Although this comparison is not untrue, it is also important to note the differences that mark biotechnology apart from industrial or, indeed, postindustrial modes of production.

Industrialism, for sociologists such as Daniel Bell, is marked by the production of material goods, most often through developments in technologies in which "energy has replaced raw muscle and provides the power that is the basis of productivity—and art of making more with less—and is responsible for the mass output of goods."[5] Industrial societies are centered around the way that "energy and machines transform the nature of work," necessitating new modes of hierarchical and bureaucratic organization. The challenge for industrial production is, "[H]ow does one extract the greatest amount of energy from a given unit of embedded nature (coal, oil, gas, water power) with the best machine at what comparative price?"[6] From the steam engine to automated "large-scale industry," from the nineteenth-century textile factory to the twentieth-century automobile plant, industrialism implies a sort of physics, an economics of force, motion, power, and energy.

By contrast, what Bell famously termed "postindustrial society" was formed, in part, by global political changes and by developments in new technologies, such as information and communications media. "If an industrial society is defined by the quantity of goods as marking a standard of living, the postindustrial society is defined by the quality of life as measured by the services and amenities."[7] What Bell designates as "services" include a range of activities that, strictly speaking, produce no goods: transportation, tourism, entertainment and media, insurance, education, real estate, and health care. Bell—along with David Harvey, Scott Lash, and others—has traced these shifts into a postindustrial or postmodernized social structure in the West during the 1970s. The shift from products to services, from class to status, from practical to theoretical knowledge, from planning to development, and so forth can be seen to revolve around the emergence of new modes of cultural, social, political valuation based on information, knowledge, and signs. As Bell notes, "what counts is not raw muscle, power, or energy, but information." Furthermore, "information becomes a central resource, and within organizations a source of power."[8] Information is, therefore, not only a new kind of production, but also what is produced. Manuel Castells concurs when he notes that the "emergence of a new technological paradigm organized around new, more powerful, and more flexible information technologies makes it possible for information itself to become the product of the production process."[9] Michael Hardt and Antonio Negri more recently have described this sort of work as "immaterial labor," a concept to which I return later.

Where does biotechnology fit in such scenarios? On the one hand, biotech companies and pharmaceutical corporations still employ many aspects that would characterize them as industrial (e.g., assembly-line plants for bottling drugs). On the other hand, the biotech industry—at all levels—is becoming increasingly integrated with new information technologies (e.g., online genome databases, global operations, sophisticated public-relations campaigns). In addition, most accounts of the so-called postindustrial society rarely mention biotechnology, an industry that has, as we have seen, been in existence since the 1970s.[10] All technology is arguably a form of biotechnology, at least in the sense that each technological system demonstrates some type of interrelation between human beings and their environment. Discussing the particular qualities of industrial technologies such as the spinning machine, Marx notes how Darwin's theory of evolution had already laid the groundwork for understanding biology as a technology:

> A critical history of technology would show how little any of the inventions of the eighteenth century are the work of a single individual. As yet such a book does not exist. Darwin had directed attention to the history of natural technology, i.e. the formation of the organs of plants and animals, which serve as the instruments of production for sustaining their life. Does not the history of the productive organs of man in society, of organs that are the material basis of every particular organization of society, deserve equal attention?[11]

This comment comes in chapter 15 of the first volume of *Capital*, in which Marx discusses the emergence of industrial technologies and the differentiations between tools, machines, and "large-scale industry." Many things can be deduced from this brief observation, especially in light of the then-current fields in natural history, germ theory, and evolution. Marx's observation suggests, among other things, that the newly emerging biological sciences can be considered from the perspective of political economy. That is, biology can be understood not only as the adaptation of organisms to their environment, but also as the complex means through which organisms use their bodies and their productive capacity continually to ensure their survival and their existence. In this sense, the organism and its organs continually work on the environment and other organisms; organisms continually produce themselves and their modes of existence, and their bodies become indistinguishable from technologies that ensure the continuation of their biological being. When this happens, the sciences of life or of the human body also become engineering sciences, or technologies of production, the "productive organs of society."

Marx was not the first to wonder whether the sciences of biology could also be understood as a technology, but his later writings on labor commonly make use of biological metaphors, from the description of the labor process as a "metabolism" between the human being and nature to his description of the raw material of production as "one of the organs of his activity, which he [the laborer] annexes to his own bodily organs," to his description of automated large-scale industry as an "objective organism."[12] In addition, in the often-referenced "fragment on machines" in the *Grundrisse*, Marx configures industrial technology in organic terms, speaking of "mechanical and intellectual organs," "living (active) machinery," and the general "social metabolism" of industrial capitalism.[13] This intersection between nature and artifice, biology and technology, takes on a particular flavor in Marx's analysis of industrial modes of production. However, despite the fact that Marx was looking specifically at the shift from manufacture to large-scale industry in the nineteenth-century, his points also have much to say about the kind of labor specific to biotechnology. Thus, it may be helpful to consider briefly Marx's analysis on industrial technologies before going on to consider the equally unique aspects of the biotech industry.

In *Capital*, Marx describes a machine as having three parts: a "motor mechanism," or that part that provides the driving energy or motive force; a "transmitting mechanism," or that part that performs the harnessing of energy or processing of force; and a "working machine or tool," or that part that actually performs modifications on an object of labor.[14] Marx notes that it is in this third part, the working machine or tool, that human labor power is transformed by the introduction of technology into the production process.

Marx begins by addressing the issue of what happens when machines enter the production process in industrial capitalism. At first, the labor power of the human body may be augmented by tools ("simple machines"), in which case the tool serves to facilitate the labor power inherent in the human body of the worker. Marx mentions spades, hammers, chisels, but also the early weaving machines as examples of tools of this type. According to Marx, a key transformation occurs when we pass from tools ("simple machines") to machines ("complex tools"). The weaving machines and machines driven by steam or by electromagnetic or hydroelectric force are all examples of a new level of production, a shift from mere "manufacture" to the development of organized factories and "large-scale industry." Machines now derive their motive force not from the human body's labor power, but from some force of nature that has been artificially harnessed (e.g., water, air, electricity).[15]

The key element in this shift is that the labor power of the human body goes from being a primary motive force to being a supplementary and managerial force. As Marx notes, "as soon as man, instead of working on the object of labor with a tool, becomes merely the motive power of a machine, it is purely accidental that the motive power happens to be clothed in the form of human muscles; wind, water or steam could just as well take man's place."[16] The result is that the role of human labor power becomes inverted. Instead of technologies extending the labor of the body, the body now acts in a more regulatory, managerial mode: setting the machines on and off, maintaining and repairing, and so forth. When this happens, "labor no longer appears so much to be included in the production process; rather, the human being comes to relate more as watchman and regulator to the production process itself."[17] A new mode of cooperation and further division of labor results in human bodies surrounding the labor power of machines in the factory. Instead of the earlier manufacture model, in which the labor power of the human body appropriates natural resources and tools, now, with industrial production, "the worker has been appropriated by the process; but the process had previously to be adapted to the worker."[18]

This appropriation results in the familiar image of the oppressive, dystopian industrial factory, which haunts much nineteenth-century British fiction, but also a great deal of twentieth-century science fiction.[19] The threat becomes not only the disenfranchisement of human labor power, but, more specifically, a technologically driven mode of production that exhibits a strange kind of vitalism: "But, once adopted into the production process of capital, the means of labor passes through different metamorphoses, whose culmination is the machine, or rather, an automatic system of machinery . . . set in motion by an automaton, a moving power that moves itself; this automaton consisting of numerous mechanical and intellectual organs, so that the workers themselves are cast merely as its conscious linkages."[20] Despite such dramatic passages, Marx did not quite announce the end of human labor power, nor did he ever forecast the coming end of human labor in the face of technological advance. The condition that results from this shift is a complex relation between technological and scientific advance, on the one hand, and the varying roles that human labor power plays in the production process, on the other. Thus, instead of simply reading Marx's analysis in antitechnological terms, we would be more accurate to say that, in his analyses of new technologies of production, he notes how the very notion of labor undergoes a qualitative change. In the case of increasingly automated industrial modes of production, this change is, for Marx, a shift from human labor power based on direct physical work to one based on the monitoring, regulating, and controling of sophisticated machinery. As he notes, one of the results of this process is a move from the collective living labor of workers to a new integration of living labor into technology, a "living (active) machinery" or "organism."[21] The machine becomes alive or vital, a kind of living labor of technology.

Thus, in discussing industrial production, Marx points to several ways that transformations in technology are connected to transformations in human labor power. First, human labor power is configured as both biological and economical. Marx begins with a consideration of labor as a productive mode of existence, in which labor produces direct use values. Labor power in this guise is described as "the aggregate of those mental and physical capabilities existing in the physical form, the living personality, of a human being, capabilities which he sets in motion whenever he produces a use-value of any kind." Labor power in this sense is corporeal, physiological, biological—a biological capacity for performing certain activities over a certain span of time. "Labor power exists only as a capacity of the living individual."[22] Second, the biologically inflected labor power is not an essence of labor, but is always bound up with modes of technical production. Labor power is as much technological as it is biological. It is a medium. In this process, it is also a means, a tool that can be applied to a range of tasks in return for a wage. In order for this to happen, the human subject must fulfill two conditions: "He must constantly treat his labor power as his own property, his own commodity"; and "the possessor of labor power, instead of being able to sell commodities in which his labor has been objectified, must rather be compelled to offer for sale as a commodity that very labor power which exists only in his living body."[23] The result is an ambivalent intersection of labor as simultaneously biological, technological, and, of course, economic: "In machinery, objectified labor confronts living labor within the labor process itself as the power which rules it; a power which, as the appropriation of living labor, is the form of capital. The transformation of the means of labor into machinery, and of living labor into a mere living accessory of this machinery, as the means of its action, also posits the absorption of the labor process in its material character as a mere moment of the realization process of capital."[24]

In this and other passages, Marx makes a distinction between "living labor" and "dead labor," the former being akin to labor as the "living, form-giving fire" of human activity and the latter best represented by the fixed capital of technology. Labor power is described in *Capital* as the "capacity of the living individual."[25] But Marx also notes that this definition does not necessarily presuppose a notion of labor power outside of capital—a presupposition that is, at best, conjecture. Thus, living labor comes to mean a number of things in Marx's writing, and therefore the term harbors an ambiguity to it. Sometimes living labor is indeed a notion of labor outside of capital, which is then appropriated by the labor-capital relationship.[26] In this guise, living labor is always a potential for labor, labor that is living because it is the constant potential of the worker's living body. Wage labor is thus a demand placed on living labor to become labor power, or a potential for labor that can be rendered quantifiable in money terms and that can be sold and purchased as a commodity itself. At other times, living labor is that which capital seeks to extract through the use of technology, or fixed capital.[27] If living labor is circulating capital, it is vulnerable to any number of changes (changes in the worker's living conditions, changes in wages, changes in worker demands, etc.). With the introduction of new technologies, living labor can be redefined and stabilized, while the bulk of labor can be performed by the fixed capital of machinery. In this guise, living labor stands directly opposite dead labor.

Conventional readings of Marx's distinction between living labor and dead labor have understood the former as the human labor of workers and the latter as the technological labor of machines. Tensions arise when capitalism attempts to replace the former with the latter (hence the neo-Luddite interpretations of the information society). But in doing so, capitalism not only indirectly loses its capacity to sell (because workers are unemployed), but it also loses the ability to extract surplus value from wage labor (because labor is performed by machines). But, as we have seen, Marx moves beyond this simple dichotomy in the organicist, vitalist metaphors he deploys in describing the way that labor is redefined in the context of technology: the human-machine interface forms a complex "social metabolism" for capital, as "living (active) machinery."

This is where Marx's more ambiguous comments on living labor come into play. At points, Marx seems to refer to living labor as an essence of human subjectivity, as something that exists outside of and prior to capital, which is then appropriated and turned into labor power.[28] But I would take a more nuanced view and suggest that *living labor is the fissure that exists between labor and capital*. In a sense, living labor is the surplus of the biology of labor power. As Antonio Negri and others in the autonomist tradition note, labor is that which capital is forever striving to subsume totally, a process from "formal to the real subsumption—this passage entails the effective, functional and organic subjugation of all the social conditions of production and, concomitantly, of labor as an associated force."[29] But, by definition, capital cannot totally achieve this subsumption, for if it could, then there would be no labor power, no replenishment of labor in the relationship between capital and labor. Living labor is akin to what Negri calls an "antagonistic tendency," that which is defined by its resistance to capital—at the same time that it is incorporated into capital.

In this sense, *living labor is not an essence of labor, but rather an immanence to labor power*. Living labor—"the form-giving fire"—does not disappear in labor power, nor does it disappear in the "dead labor" of machinery or fixed capital. Living labor is that which is immanent to every instance of labor, dead or alive (or both, as Marx's hybrid organicist and mechanist language suggests). Keeping in mind our own context—that of the biotech industry—I suggest that Marx's distinction between living labor and dead labor be taken quite literally. Living labor in the biotech industry is, quite simply, "life itself." But this "life itself" is not life in general, but a biological notion of "life itself" that is supported by developments in fields such as molecular genetics and biochemistry, genomics, bioinformatics, drug development, diagnostics, and other fields. This does not, of course, say that genetics and its related fields are capitalist endeavors, but it does note the key role that science plays in the development of modes of production in the biotech industry.

Biology: The New Dotcom

In Italian autonomist thought and practice, the shift from an industrial to a postindustrial society is concurrent with transformations not only in capital, but also in the composition of labor and the working class.[30] For every transformation in the mode of production (from products to services, from goods to information), a set of corresponding changes occur in the way human labor defines itself and is defined by capital. Capital struggles to encompass the totality of labor (both in its actuality and in its potential as future labor), whereas labor continually struggles to define its own needs and to survive vis-à-vis capital. As Marx points out in the *Grundrisse*, "capital stands on one side and labor on the other."[31] Autonomist writers often highlight the dynamic relationship between labor and capital, as noted by Nick Dyer-Witheford:

> Far from being a passive object of capitalist designs, the worker is in fact the active subject of production, the wellspring of the skills, innovation, and cooperation on which capital depends. Capital attempts to incorporate labor as an object, a component in its cycle of value extraction, so much labor power. But this inclusion is always partial, never fully achieved. . . . Labor is for capital always a problematic "other" that must constantly be controlled and subdued.[32]

In fact, as Antonio Negri suggests, labor is always antagonistic with respect to capital. "Labor can therefore be transformed into capital only if it assumes the form of exchange, the form of money. But that means that the relation is one of antagonism, that labor and capital are present only at the moment of exchange which constitutes their productive synthesis, as autonomous, independent entities."[33] Far from presupposing an ideal state of labor outside of capital, Negri argues that any potential politics of living labor must come only from the labor-capital relation itself. Living labor is at once that which is always escaping from capital and that which is always being subsumed under capital. He posits antagonism, but also autonomy. The autonomy of living labor—of "life" as living labor—stems from the fact that it is bound up in a complex with capital. At stake is the very subjectivity of individuals in their laboring capacity, inasmuch as "the opposition determines subjectivity." This subjectivity bears the marks of the exploitative difference between labor and capital, it is also the potential for a kind of "general intellect" (to use Marx's phrase). "Here use value is nothing other than the radicality of the labor opposition . . . the source of all human possibility."[34]

Science and technology play a significant role in this ongoing, tension-filled relation. I should be careful, however, to say that despite this role, science and technology do not necessarily determine the labor-capital relationship. Raniero Panzieri notes in his critique of "objectivist" accounts of new technologies that "technological development presents itself as a development of capitalism: as capital."[35] As he observes, this presentation does not mean that the development of science and technology are themselves completely determined by purely economic forces; new patterns of consumption, the redefinition of the work-leisure boundary, and new modes of cultural production all play a part in complicating the labor-capital relation. Panzieri echoes Marx, however, in noting how forms of "life" or living labor are often positioned between labor and capital, a "more and more complex and sophisticated attempt to adapt the planning of living labor to the stages [of capitalist development] progressively attained."[36] In the relation between labor and capital, there is, then, also a tension between living and dead labor. Or, put another way, *it is "life" that is situated between labor and capital*—in particular, the "life" of living labor.

But this relation is not, at I have noted, a static one. One of the significant changes brought on by so-called postindustrial societies has been a shift from labor-producing material goods to labor-producing services, communication, and information. Maurizio Lazzarato refers to this postindustrial, postmodernized form of labor as *immaterial labor*. Lazzarato defines immaterial labor as "the labor that produces the informational and cultural content of the commodity."[37] Immaterial labor is a qualitative change, both in the nature of what is produced (services, communication, information) and in the nature of the labor process itself (cultural-, entertainment-, or communications-based labor). Examples include "audiovisual production, advertising, fashion, software," and so on, as well as labor linked to networks, the "interactive labor" of problem solving and analysis, and the cultural labor of "the production and manipulation of affects."[38] For Lazzarato, "immaterial labor forces us to question the classical definitions of work and workforce, because it results from a synthesis of different types of know-how: intellectual skills, manual skills, and entrepreneurial skills."[39]

Furthermore, immaterial labor is two sided in its effects. On the one hand, it serves as an ongoing expansion of capital into media, culture, entertainment, and the "production of affects." In this guise, immaterial labor is the urgency for communication ("always connected, always on") in the service of more complex forms of commodification. Lazzarato is direct is his critique: "The communicational relationship . . . is thus completely predetermined in both form and content; it is subordinated to the 'circulation of information.' "[40] However, this "opening" of immaterial labor also demands a new level of cooperation and coordination between multiple points and many subjects. In a sense, immaterial labor already bears within itself the seeds of its own form of "self-valorization" in the political uses of technologies. Hardt and Negri highlight this aspect of immaterial labor: "The novelty of the new information infrastructure is the fact that it is embedded within and completely immanent to the new production processes."[41] Thus, immaterial labor, as Lazzarato defines it, "constitutes itself in immediately collective forms that exist as networks and flows," which create the conditions for productive "antagonisms and contradictions."[42]

Although autonomist Marxism has not dealt explicitly with the biotech industry, the terms it uses—immaterial labor, living labor, and the tension between labor and capital—are helpful for beginning to understand the biotech industry and all its complications. I can start by providing a general overview of the biotech industry, paying particular attention to those sectors related to medicine and health care (genomics, proteomics, drug development, bioinformatics). As I do this, it is worth keeping in mind a number of points put forth by autonomist thinkers. First, inasmuch as the labor-capital relation changes historically, it follows that biotechnology, with its emphasis on "life itself," will constitute another qualitative change to the notion of living labor as "life itself." Second, as Negri notes, new relations between capital and labor, especially those driven by new technoscientific advances, force us toward an "ontological broadening" of the very notion of labor itself.[43] Given that the biotech industry employs both new informational technologies and engineered *life forms* (from microbes to transgenic mammals), it may be that what we are witnessing in biotechnology is, in a sense, the production of a new type of labor.

Thus, when we speak about the biotech "industry," this description is, in some ways, misleading. Though biotech in many ways certainly is an industry, it has rarely been about industrial modes of production; even the early developments in chemical engineering and food biotechnology have been built on modifications of preexisting forms of life, not on products (that is, in contrast to standard production models, biotech foods are "genetically modified").[44]

Research in molecular biotechnology is more explicit on this point. Although on the research end there is no shortage of ideas, experiments, and novel techniques, on the production end "Big Pharma" has yet to deliver on this potential of biotechnology to spark a genetic medicine "revolution." Obstacles such as the excessive amount of clinical data that must be analyzed and correlated before products can be developed are in part responsible for this lag between research and product development. Research labs and biological supply companies are able to output large quantities of needed cells, genes, DNA segments, sequence clones, and other tools of research, but large bottlenecks have occurred at the product development and consumer ends. But if biotech is not an "industry" proper, if it is not based in the development of products and services utilized in medical practice, if it is not about the development of novel pharmaceuticals—if biotech does not overtly "produce" anything for medical practice—what then is it? I can begin an answer to this question by outlining three main kinds of companies within the biotech industry.

First, there are the "pick-and-shovel" companies, mostly in the technology and laboratory supply sector, which provide the tools for research. Like the original pick-and-shovel business during the California gold rush, such companies implicitly believe that although the actual genome may not yield any profits, the need for research technologies on it will. These companies are generally the lowest risk takers, though radical new tools such as microarrays can be risky, unless they quickly take off (as has happened with microarrays). An example of such a company is Affymetrix, which is one of the leading suppliers of microarrays, or DNA chips for large-scale, efficient sequencing.

Second, there are the software and service companies, which operate mostly on the level of computer technology, software, and network applications. These companies often provide a counterpart to the pick-and-shovel companies by providing the software tools necessary to complete the work done by pick-and-shovel hardware. Such companies can offer software packages (such as Incyte's "LifeSeq" sequencing and analysis software package), data analysis services (mostly computer and network business), or access to a database on a subscription-only basis (as the private genome companies such as Celera are doing).

Third, there are what we might call the product makers, those companies, usually large pharmaceutical corporations, that take the information generated by the second group (say, the information generated by Celera on the human genome) and transform it into an array of products, services, and practical techniques. The most prevalent among these are Big Pharma and the emphasis in such companies on drug development and gene therapy–based drug treatments (or "pharmacogenomics").

A consideration of these types of companies not only illustrates the degree to which biotech has become infotech, but also suggests that the future success of the biotech industry is dependent on the ability to generate value out of the data collected from biological material. All of this is predicated on the assumption that biological bodies—tissues, cells, molecules, chromosomes, genes—can be unproblematically translated into data. Such a move indicates the degree to which biotech relies on the notion of a stable "content" in the genome, irrespective of its material instantiation (be it in cells or in computer databases).[45] Recombinant capital demands that everything within biotech has an informatic equivalent; it does not, like biotech research, demand that everything be translated into information, but it does demand a direct link between genetic bodies and relevant data.

Recombinant capital touches the population, not through direct genomic database management, but more indirectly through the commodification of such databases. As biotech becomes increasingly privatized, the database corporations such as Celera or Incyte will become the main biocommerce brokers. At issue is not the buying or selling of databases, but the generative potential of genetic data; in such a case, ethnic population genome databases, individualized SNP or genetic screening databases, and various animal genome databases important for human medicine will become sources of a biopolitical management.

"Of Living, Breathing, Yea Ev'rie Move"

One of the major dynamics that characterizes the biotech industry presently is the privatization of molecular biology research, encapsulated in the biotech start-up company.[46] This so-called corporatization of biotech is illustrated by looking at the HGP. Originally a 15-year endeavor, the HGP was initiated in

1988 and is currently headed by various research institutions in the United States and abroad.[47] It was and still is a "public-domain" research project. It is supported by two large governmental bodies: the National Human Genome Research Institute (NHGRI, part of the NIH) and the DoE, in partnership with the Wellcome Trust in the United Kingdom.

This publicly funded genome project was joined in 1998 by a small group of privately funded genome projects, which positioned themselves in direct contrast to the publicly funded HGP. Celera Genomics (which at the time became involved as the Institute for Genomic Research [TIGR]) burst onto the biotech scene by claiming that it was initiating its own human genome project and that it would complete the map of the human genome for less money and in less time than the HGP. Celera was followed by Incyte Genomics, which also claimed to be initiating its own human genome project. In constructing their own genomes, both companies are adopting a subscription-only access model to their data, most of which is accessed by pharmaceutical companies for drug development. In addition, Human Genome Sciences, although not taking on the sequencing of the entire human genome, has instigated a series of patent claims for techniques and products based on genes sequenced within the human genome. Each corporation and its strategy—Celera's "whole genome shotgun sequencing," Incyte's use of in-house microarray technologies, and Human Genome Sciences' short-cut strategy of gene patenting—offers a different approach to the molecular biology research that we have seen with the HGP; research does not happen without a business infrastructure.[48]

We might differentiate the biotech industry further by pointing to several trends. As is evident by looking at corporations such as Perkin-Elmer, Pfizer, Genzyme, or Millennium Pharmaceuticals, both biotech start-ups and Big Pharma are participating in the development of complex corporate bodies within the economic framework of global capitalism. A driving economic force is finance capital, bolstered from within by a wide range of "future promises" from biotech research (gene therapies, genetic drugs, and so on). What we are witnessing now in "digital capitalism," to use Dan Schiller's term, is an intersection of economic systems with information technology.[49] As Michael Dawson and John Bellamy Foster show, this trend leads to an emphasis on a "total marketing strategy" that is highly diversified: consumer profiling, individualized marketing, "narrowcasting," "push-media," and so on.[50] Such trends are transforming medical research as well. More often than not, a research field within biotech can flourish or perish in the future depending on the tides of stock values. In turn, those stock values are directly tied to the proclaimed successes or failures of clinical trials or research results. Most of the stock value of the biotech industry is an example of what Catherine Waldby calls "biovalue": with companies either being able to produce valuable research results that can be transformed into products (such as genetic-based drugs or therapies) or being able to take research and mobilize it within a product-development pipeline (mostly within the domain of the pharmaceutical industry).[51]

However, the most significant factor for our concerns here is that the biotech industry, along with biotech research itself, is becoming an information science, an informatics. As Ben Rosen, chairman of Compaq has stated, "Biology is becoming an information science . . . and it will take increasingly powerful computers and software to gather, store, analyze, model and distribute that information."[52] The advent of advanced computer-based tools, the Web, and, soon after, a proliferation of e-commerce models has meant that computer technology offers the exemplary site for the continued extension of biotechnology as a model for future research and as a model for business.

Within biotech research such as the HGP, the generation of endless amounts of genetic sequence data and of genes to be analyzed has necessitated the incorporation of computer science into scientific analysis in the field of bioinformatics. As we have seen, the most prevalent example of bioinformatics has been databasing—where, for instance, up-to-the-minute information on the HGP can be accessed by researchers worldwide via a Web site and a central database. Bioinformatics also includes complex analysis systems, molecular and protein modeling applications, "gene discovery systems," data mining tools, and the development of simulation environments in order to study further the function of genes and pathways.[53]

In biotechnology, the bioinformatics of database access is inextricably connected, to software and subscription models for research, which is where bioinformatics intersects with biocapitalism, or genetic bodies are integrated into an advanced capitalist framework. Discussing the free-floating dynamics of late capital, Fredric Jameson notes that the self-referential feedback loops of finance capital propel it into a zone of "autonomization," a viruslike epidemic that forms a speculation on speculations.[54] For Jameson, this autonomization has resulted in "the cybernetic 'revolution,' the intensification of communications technology to the point at which capital transfers today abolishes [*sic*] space and time and can be virtually instantaneously effectuated from one national zone to another." It is this instantaneousness and total connectivity that has driven many labs to incorporate advanced computing and networking technologies (such as Celera), and it is this integration of biotech with infotech that has brought companies such as IBM, Compaq, and Sun Microsystems into the life sciences. If, in the biotech industry, finance capital and laboratory research are interconnected, how does this connection transform the "wet," biological materials in the lab, the molecular bodies of life science research?

One response is to suggest that *it is in the unique, hybrid objects of the biotech industry—genomic databases, DNA chips, automated protein analysis computers—that genetic bodies and an immaterial labor intersect.* In other words, the correlations between bodies and capital, which enable a biotech industry to exist at all, are currently mediated by computer and information technologies. The use of such technologies is predicated on the assumption that a range of equivalencies can be established between, say, a patented genetic sequence and the marketing of that sequence through genetic-based drugs.

The opening example of Genentech illustrates this point in the correlation between the molecular "tools" of the cell (the enzymes for cutting and pasting genes) and the novel molecules produced through those tools (Genentech's human insulin product). The example of Genentech and recombinant DNA technology is also helpful for another reason. The science behind Genentech—recombinant DNA technologies—is based on the ability to insert, reinsert, and recombine genetic sequences strategically to attain a desired end (for instance, the production of a novel protein). Genentech is an example of a constant modulation of the generative potentials of both science research and capital accumulation; it is hypermarketing and product management infused with laboratory technique. Boyer and Cohen's challenge in forming Genentech was not about making "health" or health care marketable; it was about creating a context in which both science research and capital accumulation would be reciprocally sensitive to each other, with a common goal of maximizing profits (medical and economic). Current biotech companies have realized this connection, which is one explanation why the biotech industry is focusing so heavily on the production of "silver bullet" drugs and is sidestepping more complex, "messy" approaches such as systems biology.

Because this logic of correlating bodies and value inserts itself at multiple sites within the biotech industry (just as a recombinant gene can insert itself in multiple ways within a strand of DNA), we might do better to refer to this fusion of corporate biotechnology and information technology as *recombinant capital*. Recombinant capital is not just the prevalence of online stock fluctuations, speculations, and losses; it is a diverse business model that deals with the informatics of biotechnology. At their roots, both recombinant capital and bioinformatics begin with the worldview that everything is information—or, to be more specific, that the world can be reduced to informational pattern, a pattern that identifies the "essence" of, in this case, the human body. Such a process involves a level of technical sophistication (for example, in converting DNA samples into computer databases), but, more than that, it requires a constant process of correlation between medical value and economic value. In an example such as the genome databases of Celera or Incyte, the process of converting DNA samples into databases is also a process of creating a relationship between medical value (a possible gene target for a cancer drug) and economic value (the profit potential of such a gene when it is patented). Whether involving a sequence of DNA in the lab or a computer-generated stock profile of a gene, medical value or stock value, information and information technologies enable genetic sequences to be decoded and enable online interactions within the market to fluctuate and interact. Thus, we might say that the balancing act in the biotech industry between medical and economic value is mediated by information technologies, which can establish modes of correlating different value systems by converting everything into data.

This mode of informatic abstraction means that, technically speaking, both recombinant capital and bioinformatics operate according to an encoding-processing-decoding protocol. For example, a bioinformatics approach to database management for one of the research centers in the HGP proceeds by developing a way to translate genetic information into computer information. Once that genetic body is encoded as computer information, it can then be reconfigured and organized into a database structure.

Likewise, recombinant capital also encodes the genetic body, but in a much more diluted manner. Many online hubs for the biotech industry list side by side a daily update of research results or discoveries and stock quotes.[55] As may be expected, glowing reports on Phase III clinical trials for a gene-based drug will show significant rises in stocks, whereas a failed merger between two companies might show a decline. For example, the announcement in March 2000 by President Clinton and Prime Minister Blair that genomic sequence data should be "freely available to all" triggered a panicky downward slide in biotech stocks, especially in those companies such as Celera and Incyte, which have built their entire stock portfolio on a genomic database subscription model.[56] The Clinton-Blair announcement is important not for what it said (it said nothing new, of course), but for what it demonstrated politically: a certain point of crisis in the discourse of biotech in the correlations between medical and economic value systems.

This is one level of recombinant capital—that of online stock fluctuations, which is often in direct correlation with either laboratory research results or offline scientific discourse. But stock value must first be developed, or cultured, so that an influx of stock investment can occur in the first place, and this is the level where laboratory research becomes more fully integrated with economic interest. For a biotech start-up, there must of course be something to "sell," something to garner potential investors' attention and competitor companies' envy. This something can be a product—"owning" the genome will certainly be an asset—but it can also be a service or technique—such as genetic analysis or genetic profiling. It can also be related to technology development, such as silicon-based microarrays or DNA chips. Bioinformatics carries this next step, as the traditional biological "wet" labs become more computerized, with the advent of "labs on a chip." That is, once genetics is successfully translated into a problem of informatics, programming, and digital encoding, working with biological materials in the lab becomes abstracted into working with code on a monitor.

As with other restricted-access e-commerce models, the most in-demand data will be the most valuable and even the most costly to access and download. This will create a real-time value-assessment model in which each click by a researcher will correspond in some way to the fluctuations of value in the data accessed and of the overall value of the company hosting the database. In this situation, the database-subscription business model is simultaneously research and real-time stock and pricing adjustment. The end result is that the most expensive and most costly research will be dictated first by economic negotiations, second by data-access privileges, and last by a consideration of the disease or gene being studied. The trend clearly foreseeable here is that research done on certain genes or gene-based drugs will prosper, only because they yield from the beginning a low-risk, cost-effective guarantee of success in clinical trials, U.S. FDA approval, and market utility. Research of more complex diseases such as AIDS or the study of more complex biochemical pathways (in which genes are *not* the central component) will be assessed as high-risk endeavors, highly unpredictable, and unlikely to turn any returns in the near future. A shift in research will then be encouraged, from knowledge-based medical models of healing to the full-fledged investment in temporary, genetically specific drugs that have their own built-in obsolescence. Does the biotech industry really have anything to sell to begin with?

A Biovalorization of All Values

As a way of looking at this network between bodies, data, and value, consider the recent strategy employed by some biotech companies to develop participatory models for DNA sampling. During the summer of 2000, DNA Solutions, Inc., announced the creation of the Gene Trust, the first large-scale, organized DNA sample bank.[57] It is based on a volunteer model, and interested individuals can register and send in blood samples for DNA analysis and archiving. As an information business, the Gene Trust will gather genetic samples from a range of individuals with a range of medical conditions. Such information will help pharmaceutical companies and research laboratories to speed up the process of therapeutics, drug development, and clinical trials.

Although on the research and development side the Gene Trust is exemplary of many biotech endeavors, on the consumer-patient side it is at the forefront of producing a new type of relationship between medical practice and the biomedical patient. In short, the patient or consumer—who supposedly takes part in and ultimately benefits from the Gene Trust—enters into a unique relationship with his or her own information. The medical patient becomes a *data patient*.

The data patient is characterized by several constraints, all of which relate to the relationship between patient body and the medical use of computer technology. First, the physical, biological body of the medical patient must be translatable into computer-based information. Anything that cannot be encoded will be either designated as supplemental or fully excised from the data patient's profile. All medical diagnosis, treatment, and communication in the telemedical context will then take place via information and information technology. This means that even personal consultation may involve remote physicians or technicians. One of the top priorities of medical treatment will therefore be the dataset relating to the patient. In addition, the data patient will constantly be mirrored by an electronic double, his or her digital profile. In a sense, the real patient will have to negotiate constantly with this data double, so that they correspond to each other. This means changing the data in computers, especially when those data are used to make important health-related decisions.

The cycle starts when the medical patient's genetic body is sampled and encoded. From there, a loop is initiated in which the patient's biological body must always first answer to his or her telemedical profile. The types of drugs needed or the types of therapy best suited to this patient will be decided at the level of informatics first and foremost (this is the basic logic of pharmacogenomics). The patient's biological body will have been diagnosed and treated with no mention of the phenomenal condition of embodiment, the blatant gap that currently exists between "wet" genetic molecules and digital codes, or the dynamic and flexible quality of the body.

The point here is not simply to go back to those precomputer days of Galenic bedside medicine, endlessly listening to the patient's testimonies. But we should also be very astute in assessing the ways in which this complex of biological science and information technology structures a certain narrative of progress. The data patient in genetic therapeutics and telemedicine presents us with a classical case of Baudrillardian simulation, where the simulacrum not only verifies the real (what is real comes to be what can be perfectly simulated), but also threatens to blur the real/simulated distinction.[58] Because patient data cycle back on and ultimately affect the physical, real patient, there is a confusion of where to locate the object of medical attention; in a sense, the real patient is preceded by the data patient.

As its name indicates, the Gene Trust represents itself as a form of voluntary, altruistic investment. As a long-standing legal concept, a "trust" involves a relationship between a property owner and a property manager, where the latter is entrusted with the property of the former, with benefits going primarily to the property owner (in the same way that one entrusts one's finances or one's business to someone else).

An investment trust (also called a closed-end trust) involves a financial organization that gathers the funds of its shareholders and then invests them in a diversified securities portfolio. Simply put, securities (such as stocks) confer ownership of something that is not in the owner's immediate possession; it is a means of rendering property and materiality virtual. The owner of a stock can, theoretically, demand the right to receive the property designated in the stock, or, of course, the stock can be traded. As such, stocks can take on a free-floating fluidity, as informational value units, which significantly, if only momentarily, detaches them from any material substantiation. Most investment trusts hand over the majority of the management and control of the funds to the financial organization. With a fixed number of shares, the value of the portfolio will depend on the status of the supply and demand for certain types of shares in the market.

We are already witnessing a massive virtualization of finance capital (the actual technological implementation in e-trading is perhaps only its latest phase). What happens when we consider an investment trust based on the body? As patenting debates in the past have illustrated, one of the first transformations is that the body becomes, more specifically, biological property. Bodies as objects certainly have their own convoluted history (in labor, sexuality, cultural stereotyping, etc.), but in the case of the Gene Trust the body is simultaneously genetic and informatic. It is genetic because what the Gene Trust is interested in is not personal testimony, patient-described symptoms, or even past medical records. The Gene Trust is centrally interested in a molecular pattern that is commonly "read" as a sequence of letters. In contemporary biotech research, this reading of molecules also means that DNA is translated into information, for computer-based analysis and diagnostics. This translation is a kind of body, framed by informatic and economic concerns, which we might call a "bioinformatic body." As the basic logic of the legal trust indicates, the ownership of this biological property is defined by its absence (the stock holder does not actually have the property designated in the stock).

The Gene Trust's central asset will be its databases and the diagnostic and medical information it can tease out of that genetic data. What the Gene Trust volunteer donates is less his or her body and more bioinformatic value. Under a broad category of medical altruism, the Gene Trust promises a future of humanistic returns (the DNA sample you supply now could save a life in the future). In the meanwhile, the more short-term circulation of biodata enables the Gene Trust, as a privately funded venture, to gain immediate economic returns.

The Gene Trust, though it may communicate the best of intentions, actually has a twofold agenda, in that it uses *investment* in two different ways. First and foremost, the Gene Trust presents itself through a rhetoric of an altruistic, biomedical humanism. The investment alluded to here is made possible by the overarching category of collective human well-being. At its basis is a biological essentialism, which implies that because we all are the same genetically, volunteering for the Gene Trust is an investment in the future of human health. In fact, the sound-bite advertising phrases on the Web site push forth micronarratives (the young woman whose aunt has Alzheimer's and who is thus gaining a sense of "participating in history" by helping researchers combat disease). All of us have had some experience with the often traumatic effects of disease and the death of those close to us. Despite this, however, neither the human genome projects nor their spin-offs such as the Gene Trust have mentioned the fact that scientific research still does not know for sure whether genes "cause" devastating diseases such as cancer. And researchers have yet to consider both the complex effects of biological networks and environmental influences in a serious manner.

What are the dynamics of change when a novel technology is introduced, via recombinant capital, into biotech research? The introduction of biologic-informatic hybrids as part and parcel of biotech research has meant something different; it has meant that the data produced through biotech research always have a direct, linear, and causal relationship to the development of systems for generating biovalue (mostly with pharmaceutical corporations and genetic drug development). We can see this not only in the primary interests of DNA chips (efficiency, cost-effectiveness, instantaneous results), but in the kinds of data they produce, suited for their adaptive utilization in genetic drugs and preventive medicine.[59] In this sense, the data produced by DNA chip and related technologies resemble demographics or statistics: a selected range of analytical categories (gene sequence, gene frequency, iterative patterns) composes a genetic profile of a DNA sample. Once a greater number of genes are "discovered" to have direct and indirect relationships to a wide range of "disorders," it is only a step toward seeing something such as DNA chips as sociobiological surveillance systems—biopolitical modes of regulating and accounting for individuals and populations.

Thus, one of the primary issues in projects such as the Gene Trust, is the ways in which recombinant capital can adaptively insert itself into a range of contexts, from hard lab research to the highest ethers of finance capital. In doing so, it scans, cuts, and recombines the resources inherent in fields such as genomics, articulating them as information-based solutions to an ideology whereby disease is fully accounted for through genes. In recombinant capital, sociobiological perspectives allow for the quick development of commercial health-care models based on a simple model of mechanical addition or blockage through genetic drugs.

Biomaterial Labor, or Living Dead Labor

It will be helpful to summarize some of my main points thus far. In the biotech industry, we see—in instances such as genetic engineering (e.g., Genentech), patenting, genetic testing, and gene banking (e.g., the Gene Trust)—a continual negotiation between economic and medical value. How is "life itself" positioned in this tension-filled zone? In the case of banking endeavors such as the Gene Trust or Celera's high-tech genome project, we see biological "life itself" inserted at multiple points: in biological samples from patient volunteers, in extracted DNA in test tubes or plasmid libraries, in digital DNA in computer databases, in bioinformatics software analyzing DNA, in molecular modeling applications, in genetic screens utilizing DNA chips, and so forth. It seems, then, that "life itself" is constantly positioned between medical and economic value: life situated between labor and capital. Again, I want to note the twofold tendency of biotechnology in this regard: on the one hand a thoroughly informatic view of biology and on the other hand an informatic view of biology that is nevertheless *not* immaterial. If we take all this into consideration, what is the role of "life itself" in the nexus of life, labor, and capital?

We can now begin to address this question by considering Marx's notion of living labor elaborated earlier and the notion of immaterial labor in the Italian autonomist tradition in relation to contemporary biotechnology. As seen previously, Marx makes a number of important contributions to our general understanding of the role of technological development in the relation between labor and capital. In particular, he sets up two relationships that take on new meaning in the context of biotechnology. The first relationship is that between living labor and dead labor, with the latter represented by machinery. Although this relationship certainly can take the more familiar Luddite form of a replacement of humans by machines, Marx also notes how it qualitatively changes human labor, in effect opening onto activities defined by control, regulation, and intellectual labor.

However, Marx also generates some ambiguity in his distinctions between living labor and labor power as well. On the one hand, living labor is always conceived of by capital as potential labor power, labor that is a commodity, labor that is made amenable to a quantitative exchange in wages. Yet, on the other hand, capital never totally subsumes labor, if only because labor is defined temporally and is always potential. This tension, highlighted by

Italian autonomist Marxism, is a tension between labor and capital. In the context of the biotech industry, we can see how living labor as "life itself" is inserted in between labor and capital.

Marx's formulation has taken on new dimensions in the wake of discussions over postindustrialism, globalization, and empire. We have seen how Maurizio Lazzarato has used the term *immaterial labor* to describe these qualitative changes in labor focused on the communications, computational, and service-based industries. Immaterial labor involves not just the transformation of corporeal, living labor into labor power, but the transformation of living, intellectual labor into terms set by capital. It is for this reason that Hardt and Negri describe immaterial labor as biopolitical.

However, the biotech industry shows us yet a further transformation in the nature of labor (living labor, labor power, and dead labor). Consider a range of technologies in the biotech industry, all of which make use of some biological process as the core of their mode of production:

- PCR (polymerase chain reaction), a technology developed in the early 1980s at the Cetus Corporation for efficiently and rapidly copying any desired DNA sequence (see figure 5.1). PCR makes use of simple DNA base-pair complementarity to replicate thousands of copies of any desired segment of DNA for laboratory research. PCR is common in any molecular biology lab today (and requires a license, depending on the type of cycler purchased).
- Bioreactors, laboratory devices that re-create the internal conditions of cells (temperature, moisture, growth medium, etc.). By simulating the biological conditions of a cell, bioreactors enable various molecular interactions to take place in the lab that would "naturally" take place in the body.
- Cell lines, or the use of molecular biology techniques to create a viable cell culture from a single cell sample. The use of cell lines is common in the construction of biological banks or libraries of cells (e.g., the American Type Culture Collection), and is a common resource for laboratory research. Cell lines that are able to replicate indefinitely through the modification of chromosomal telomeres are often called "immortal" cell lines.
- The burgeoning field of regenerative medicine, which has attracted attention for the most part from research into stem cells, or those cells that exist in an undifferentiated state and that can, with the right prompting, be controlled in a way that directs their differentiation process so that they become bone, cartilage, or muscle cells.

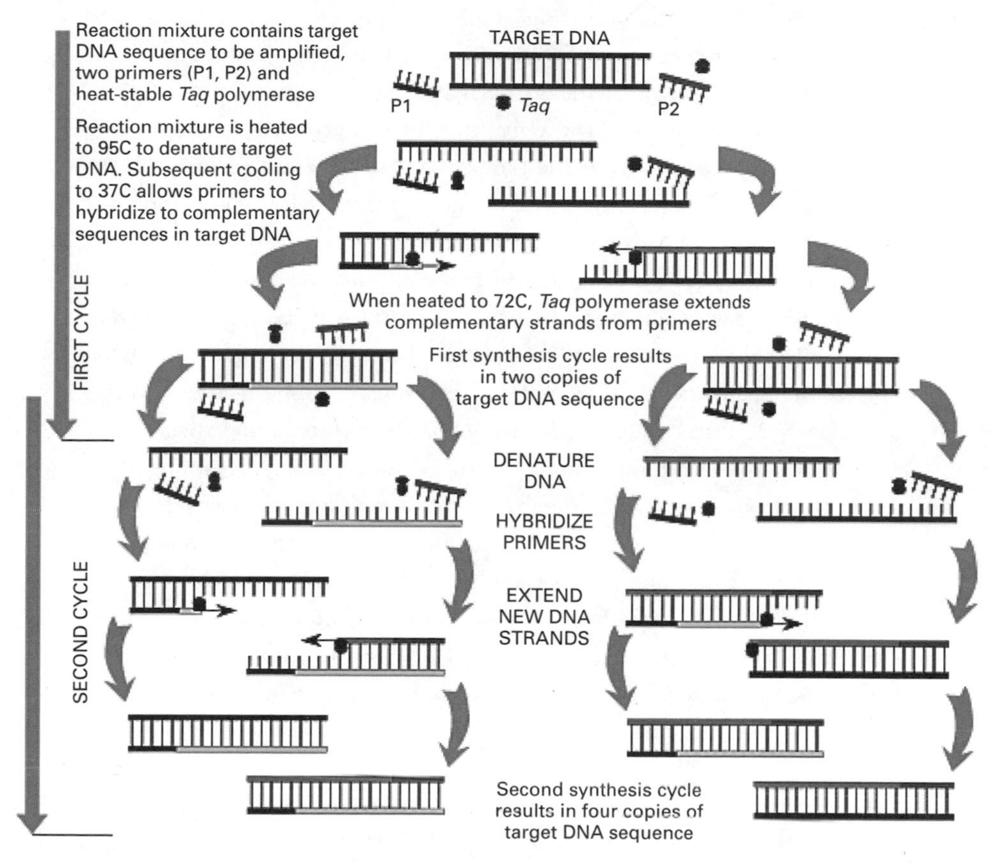

Figure 5.1 The "labor" of DNA in PCR. Image courtesy of the U.S. DOE Genome Program, http://www.ornl.gov/hgmis.

In each of these cases, we see plenty of technology at work—thermal cyclers, incubators, computers, and so forth. But more important, we also see a range of technologies that have some biological process at their center—cellular replication, cellular differentiation, DNA annealing and ligating. In other words, without these core, biological processes—processes that occur routinely in the living body—there would be no biotechnologies, no regenerative medicine without cellular differentiation, no PCR without DNA base pair bonding, no bioreactors without cellular metabolism. In the short history of the biotech industry, the well-known examples of the production of human insulin, HGH, and other genetic engineering feats are a demonstration of the ways in which "naturally occurring" biological processes are harnessed and made to work in the context of a biotechnology industry. Transgenics, GM foods, and regenerative medicine are currently following suit.

But this is certainly not the only kind of activity that takes place in the biotechnology lab. Within the biotech industry, we can identify several different kinds of labor. There is, first, modern factory labor involving the production of either goods such as prescription drugs or tools such as gene sequencing computers (intersecting with the information technology industry). The worker who performs this labor might formally correspond to what Negri calls the "mass worker."[60] Second, there is the immaterial labor surrounding the marketing, promotion, design, and public-relations factors of the pharmaceutical industry, focusing mostly on drugs and diagnostics. However, the role of information technologies is increasingly becoming a key part of this labor as well (e.g., databases, software design, data services, programming), as the fields of genomics, proteomics, and pharmacogenomics are attempting to show. Along with these two types of labor, there is also the more highly skilled labor that takes place in labs, be they at universities, research institutes, or biotech companies. These jobs include principal investigators, administrators, professors, lab technicians, student assistants, work studies, and so forth, formally akin to Negri's "professional worker." Both in the lab (research) and outside of the lab (grant writing, affiliations, patent applications), these jobs require a combination of education and skilled experience. Finally, as a special category, we might also mention those workers in the medicine and health-care sectors who are often the endpoint of biotech research (e.g., prescribing new drugs, requiring tests, analyzing biological samples, nursing, and coordinating health insurance). Although these jobs are also skilled jobs, they exist apart from the basic lab research we see in genomics, proteomics, or bioinformatics. However, in the case of the medical application of biotechnology, they are connected because they serve as the point of contact between a biotechnology and an individual patient.

Clearly, there are many types of labor involved in the biotech industry. However, in some sense, none of this is new, for every industry arguably has this sort of breakdown—a combination of conventional and contemporary forms of labor. What is it, then, that makes the biotech industry unique? One possible answer is that all these types of labor in the biotech industry revolve around a more fundamental type of labor: the specific type of labor performed routinely by cells, proteins, and DNA. In other words, biotechnology is unique in that it bypasses altogether the division between humans and machines, or between human living labor and the dead labor of machinery of which Marx spoke. Certainly, in any molecular biology or genetics lab, one will find an array of technologies, including familiar tools such as pipettes, the centrifuge, or cell cultures, as well as more modern tools such as thermal cyclers, oligonucleotide synthesizers, and entire labs full of networked personal computers. Therefore, to state simply that the biotech industry ignores technology is inaccurate. What is important to note is that each of these technologies is predicated on a deceptively simple principle: that the "naturally occurring" processes of biology at the molecular and genetic levels can be recontextualized in such a way so as to transform them into an instrument, a technology.

If we accept Marx's points concerning the labor-capital relationship and the more contemporary formulation of "immaterial labor," then we might ask how the situation has changed in the current biotech industry, an industry that produces novel artifacts—biotechnologies—that seem to fulfill Marx's description of a mechanical "organism" and yet at the same time *frustrate any clear division between living and dead labor*. Right away we can note a number of differences between Marx's characterization of industrialism and the biotech industry. In the biotech industry,

- Biology is the motive force, the method, the medium. Biology is what drives production. Biology is the source material.
- Biology is the process of production. Biology is not replaced by machinery, but it replaces machines. Biology is the technology.
- Biology is the product, the endpoint, and the aim. Biology does not aim to produce a material good or a service, but, above all, more biology.

The result is a vision of biotechnology in which biology is the technology. What characterizes the current biotech industry is that the technology that biology is, is seen to be informatic. In the biotech industry, biology is the technology, and that technology is informatic, without ever becoming purely immaterial. Instead of an immaterial labor that is concentrated in the computer and information industries, we see a *biomaterial labor* that brings together the emphasis on informatization with the novel ways in which biological "life itself" is radically recontextualized in genetic engineering, genomics, and bioinformatics.

Within the biotech industry, the existence of this uncanny, "living" labor of cells, enzymes, and genes is an indicator that the relationship between life, labor, and capital is undergoing some significant changes. In biotechnology, labor power is produced through cells lines, transgenics, recombinant DNA, PCR, and genome databases. The results appear to be products and services like any other: prescription drugs, genetic diagnostics, cell- or gene-based therapies, and novel in vitro and in vivo means of producing antibodies and other compounds for medical use. The notion of biomaterial labor is meant to indicate these significant shifts in the labor-capital relation within the biotech industry (which does not exclude government-funded or university-based research).

What does this literal biotechnology produce? On one level, it produces information—test results, experiment results, assay results—which is then used to perform more experiments, leading (it is hoped) to the development of an application such as a biomarker for drug targeting. On another level, however, this biotechnology simply produces biology—it produces itself, or, rather, it generates the conditions for the subsequent instrumentalization of biology. The interest in biotechnology from a scientific view is that it is not simply "used up" in one experiment, but forms a whole panoply of techniques and experimental systems for research. The development of cell cultures is one example, in which specific foreign DNA can be maintained in culture indefinitely.

The biotech industry produces not just products or services (though it does produce both, certainly), but above all biology: drugs, therapies, cells, tissues, organs, and information. The core of the production process in the biotech industry is that it is fully constituted by biology: *biology is what produces, it is how production occurs, and it is what is produced.* Biology is the technology in biotech, but that technology is, at its core, biological.

The core of the biotech industry is biological "life itself," or, put another way, the economic uses of "life itself." As an industry, biotechnology is markedly different from nineteenth-century industrialism or early-twentieth-century automation, but it is nevertheless an industry, one built around the ability to harness biology (see table 5.1). The key to understanding the biotech industry as an economic endeavor is this basic approach of providing the conditions in which "life itself" can become productive, living labor. In the case

Table 5.1 Biomaterial Labor

Period	Industrialism	Postindustrialism	Biotech industry
Type	Material labor	Immaterial labor	Biomaterial labor
Modality	Dead labor	Vital labor	Living dead labor
Products	Material goods	Information; services	"life itself"
Resources	Raw materials	Information	Bioinformation
Science	Physics; thermodynamics	Computer science; informatics	Biology; genetics
Affect	Alienation	Signification; communication	Life as technology; life as instrumental
Aim	Surplus value	Informational valorization; circulation of information	Biological valorization
Level	Economic base	Culture industry	Medicalization
Examples	Weaving and textiles; automobile factory	Internet Service Providers; computer and software industry; management; design	Drug R&D; database services; patents
Form	Physical, corporeal, manual labor	Intellectual, communicational, affective labor	Biological, cellular, enzymatic, genetic labor
Extreme cases (science fiction)	Robot workforce (*R.U.R.*; *Metropolis*)	Intelligent computers (*Colossus*; *Tron*)	Clone army (*Brave New World*; *Blade Runner*)

of biotechnology, living labor actually means the work performed by biology: mammalian bioreactors, transgenic lab animals, immortalized cell lines, lab-grown tissues and organs, and the bioinformatic labor of cells, enzymes, and DNA. But if this is the case, then what is the difference between contemporary genomics and ancient practices of agriculture and fermentation? The history of molecular genetics provides us with one important difference: the role played by information in relation to "life itself." If one of the primary innovations of molecular biology was the emphasis on sequences, codes, and messages in DNA, then it follows that this informatic understanding of biology can be further expanded in plasmid libraries, genome databases, and genetically engineered animals. What differentiates the living labor of the biotech industry from earlier forms of classical biotechnology is the emphasis on an informatic—*but not immaterial*—understanding of biological "life itself." This emphasis is historically accompanied by the way in which the biotech industry combines the "immaterial labor" of the computer and information technology industries with the newly emerging biomaterial labor of cells, enzymes, and DNA.

At the same time, biomaterial labor raises a set of complicated questions. For instance, in biomaterial labor, biology performs work, and yet there is no worker, no subject; biomaterial labor is strangely nonhuman. Who sells the labor power of biomaterial labor? We have a situation in which there is living labor without subjectivity. If biomaterial labor is a labor power in the biotech industry, and if there is no human subject or wage, is there still exploitation and alienation? Is biology always already biomaterial labor? In biomaterial labor, has this twofold change—the nonhuman scale of genes and the role of informatics—altered the quality of human subjects' labor in the biotech industry?

Another question: Does biomaterial labor configure biological "life itself" as more natural or more artificial? Consider the arguments for and against patents in the biotech industry. On the one hand, patents for, say, transgenic plants, animals, and microbes are seen to be safe and useful, in part because they are fully biological and thus bear some close relation to "nature." Yet, in order to fulfill the criteria for patentability, scientists must show that their inventions are "new, useful, and nonobvious"—in short, that they are artificial. Biomaterial labor—and by extension the biotech industry—presents us with the paradoxical notion of a "life itself" that is both transparent and yet totally mediated.

Finally, at the level of ontology, biomaterial labor points to a qualitatively different type of estrangement, a biological estrangement (indeed, biology generally can be understood as the estrangement of "life itself"). The very fact that your "own" body parts (cells, DNA) can be patented without your knowledge implies a unique type of separation between biology and embodiment. Biomaterial labor gives us a paradoxical life without agency, a "life itself" that is always already objectified in the lab, even as your cells outside you are transplanted back into your body (as in tissue engineering).

Coda: Nonorganic Life and Its Discontents

The motto of biotechnology, as an industry, is "nature does it better." What this means is that biological knowledge plays a key role in articulating what exactly nature (or biology) does, or how, exactly, biology can be understood as a productive and value-generating process.

The rhetoric of "DNA as information" is now a given in biotech research, as it is in textbooks and even popular culture. Despite many researchers' concerted efforts to state that the individual is not, in fact, determined by his or her DNA, a powerful cultural mytheme has developed around the centrality of DNA in forming the individual and collective (species) subject. Many critics have brought up the problems inherent in this information-based view of genetic life, and, as Evelyn Fox Keller and Lily Kay have shown, the notion of a genetic "code" emerged out of a complex set of historical, institutional, and disciplinary contexts during the postwar era.[61] One of the central tensions in this consideration of a genetic code or a genetic program is the dichotomy it sets up between organic and nonorganic matter. Science fiction has given us many intelligent machines, but there is still a basic division between carbon-based and silicon-based systems, the former assumedly existing "in nature" (that is, supposedly preexisting technology) and the latter an invention by human beings.

In an essay on "nonorganic life" (a term borrowed from Gilles Deleuze and Félix Guattari), Manuel De Landa suggests that recent research into chaos and self-organizing systems (from cloud patterns to wave solitons to "chemical clocks") forms a significant paradigm shift in the traditional notions of what is meant by living and nonliving.[62] De Landa shows that such systems, which display complex patterning and dynamic interactions with their environment, are more than automated reactions. They show how dynamic processes such as bifurcations, attractors, and so on can create the condition for self-organization and emergence; in effect, these "nonorganic" systems display evolutionary patterning. As De Landa states, "what is important for our purposes, though, is that both periodic and non-periodic oscillations and coherent structures are among the unexpected possible behaviors of 'inert' matter, behaviors of which we have only become recently aware. Matter, as it turns out, can 'express' itself in complex and creative ways . . ."[63]

When this research is combined with research into cognitive science and artificial intelligence, the boundary between the living and nonliving becomes even more blurred. The direction De Landa takes, however, does not simply culminate in the anthropomorphic, humanist subject. Instead, systems such as geological formations, crowds, insects, highways, sand dunes, computer networks, cell metabolism, and the economy display unique sets of dynamic structuration that, from one perspective, might contentiously be called "life."

I might then suggest that what I have been calling biomaterial labor—"informed" by recombinant capital—is a symptom of a more fundamental crisis concerning the boundary management between the living and nonliving. The biotech industry needs effectively to separate the human and technological, the living and nonliving, in order that biotechnologies can be clearly seen as simple, nonthreatening tools or medicines in the service of the betterment of human health. Although a noble cause in which motive is not the issue, this kind of separation is simply not possible if we begin to consider the actual techniques, technologies, and discourse of informatics that pervade current research.

The confusion between medical value and economic value can thus be seen as a confusion between living and nonliving systems. The current integration of information technology and biotechnology has brought us the challenge of rethinking the still-active vitalist ideology in the life sciences. The question is not whether or not vitalist-based approaches to life are in themselves good or bad; the question is whether the ideological division between living and nonliving that vitalist thought highlights is the best approach to dealing with contemporary biotech. This is especially so in biotechnology research, where blood and tissue samples, DNA sequencing computers, "immortalized" stem cells, online genomic databases, and an array of biological-technological hybrids are in the process of redefining the conventional distinctions between humans and machine, nature and technics.

Biomaterial labor asks us to consider Marx's concepts of "living labor" in a technoscientific way. Whereas Marx's early writings experimented with underappreciated concepts such as "species being," and the "inorganic body" of human labor, the later Marx employed the frequently used metaphor of "social metabolism" in *Capital*. Perhaps this is the moment in which to revise Marx's biologically inspired concepts with an eye toward the biotech industry. Biomaterial labor asks us to consider the labor-capital relationship not just at the level of factory work or the service sector, but at the core of biotechnology's production process.

In light of Italian autonomism's emphasis on the resistant, creative capacities of labor, biomaterial labor also encourages us to ask how the nonhuman domain of cells, enzymes, and genes displays a different sort of "resistance." Can the "complex" properties of biological "life itself"—metabolic networks, biopathways, single-point mutations, immunoknowledge, protein folding—offer a resistance to the genecentric and reductionist approaches taken by the biotech and pharmaceutical industries?

Finally, biomaterial labor asks us to rethink the relation between the human and the nonhuman in the cycles of life, labor, and capital. It asks us to innovate new ways of thinking about the connections between a sequence of DNA, a genome database, a gene patent, a genetic test, an individual patient, an illness, and the economics of health care.

6

Bioinfowar: Biologically Enhancing National Security

Honorific Body, Horrific Body

The Crazies (1973, retitled *Code Name: Trixie*), an underrated film by the cult director George Romero, portrays the events that ensue when a top-secret military bioweapon is accidentally released in a midwestern American town. Once released, the bioweapon turns the townspeople into raving mad zombies who seem to devour everything in their path, including human flesh. As in many films in the "zombie" genre, *The Crazies* contains a number of familiar motifs: an incompetent military unable to control the outbreak, a small group of civilian "survivors," a doctor struggling against state bureaucracy to find a cure, and of course the "silent majority" of the zombies—this time in the guise of the sick and the ill, rather than the living dead.[1]

What makes *The Crazies* noteworthy is the way in which it portrays the network properties of what is now known as the biotech industry. An engineered bioweapon spreads among the townspeople via biological networks of infection. Once the military arrives, it establishes communications networks with government agencies to facilitate the disaster management; within the area of the town, the military and local police set up perimeters and quarantine (see figure 6.1), as well as chaotic emergency medical stations and holding pens (a high school biology lab serves as the site for the development of a vaccine). All of this is, however, to no avail, for, as in many of Romero's films, the zombies slowly, steadily triumph through a fleshy network of infection, cannibalism, and slow-motion pursuit.

Figure 6.1 The convolution of military and medicine in George Romero's 1973 film *The Crazies.*

Romero's film was made one year after the 1972 Biological and Toxin Weapons Convention (or Biological Weapons Convention, BWC), which prohibited nations from stockpiling, developing, or producing biological weapons, except for "prophylactic, protective or other peaceful purposes."[2] The film was also made at the same time that the first recombinant DNA experiments were being published, which would raise a host of ethical, social, and political questions in the larger public sphere surrounding biotechnology.[3] *The Crazies* picks up on the outbreak or "medical thriller" genre, but it also combines this genre with Romero's rich use of the zombie metaphor. We as viewers are never really told what the experimental bioweapon is, and the communications by the military to the government are as mysterious as the accidentally released bioweapon. In effect, both the nature of the bioweapon/epidemic and the nature of military communications are as alienating as the zombies themselves.

Compare this context to that of the 2002 release *28 Days Later*, another zombie-epidemic film, directed by Danny Boyle. Set between London and Manchester, the film also features a mysterious, experimental virus—the "rage virus"—that seems to be as much induced by violent media images as it is triggered by a biological agent. The film combines several subgenres: the apocalyptic "last man" scenario, the outbreak or medical thriller film, and of course the zombie film.[4] Like the zombies in Romero's films, the "infected" in *28 Days Later* are turned into rabid, lifeless, flesh-eating creatures, except that in the age of digital media the infection happens within seconds, and the infected move significantly more quickly than Romero's zombies (figure 6.2). But, like Romero's *The Crazies* (as well as *Day of the Dead*), the military is featured in a highly problematic light, at once establishing sovereignty over the crisis and authorizing violence of all kinds.

Released first in Great Britain, then in the United States, *28 Days Later* appeared the same year that the U.S. government passed the 2002 Bioterrorism Act.[5] Among other things, the Bioterrorism Act has provided the conditions for an increase in research into biological warfare and potential bioterrorist agents, while also boosting the regulation and oversight of biological materials (agriculture, livestock, microbes) distributed in the United States. A number of U.S. government legislations—from Project BioShield to

Figure 6.2 War, the plague, and the abandoned city in Danny Boyle's 2002 film *28 Days Later.*

subdepartments within the Department of Homeland Security—have furthered the "war against terror" by making funds available for the construction of "next-generation medical countermeasures" (new vaccines and therapies), a streamlined FDA approval system for such drugs, and upgrading current "disease surveillance capabilities."[6]

So, then, we have, arguably, two eras, two different kinds of war, represented in popular culture by two different films. If the BWC is scientifically contextualized by genetic engineering, then the Bioterrorism Act is likewise contextualized by genomics, bioinformatics, and computer-driven drug development. If the BWC implicitly supported "defensive" bioweapons programs (of which the U.S. program is by far the largest and most exhaustive), then the Bioterrorist Act supports broad government and even civilian "countermeasures." If the BWC still harbored within it the language of Cold War politics, the Bioterrorist Act actively identifies "a new enemy" in the distributed network of international terrorism.

In both cases, however, there is an important similarity: the establishment of political sovereignty in the "state of emergency."[7] This state of emergency is defined by an explicitly biological threat to the body politic, and in this sense all war is biological war. Indeed, one of the arguments of this chapter is that a biologically derived notion of political conflict is what defines this type of sovereignty. Michel Foucault refers to this nexus of biology, politics, and war as a form of "biopolitics." It not only results in a permanent state of emergency, but also induces widespread societal fear, an imperative on health surveillance, and a culture of body anxiety. Although there is no question that terrorism generally and bioterrorism specifically represent a threat to peace efforts, it is also important to understand the forms of power engendered by these threats and by this *biological security*.

In a sense, the hegemonic role of genetic science represents two sides of this current biopolitical condition. On the one hand, the beginning of the twenty-first century saw the completion of the sequencing of the human genome, which many lauded as a landmark in scientific endeavor. This is what we might call the "honorific body," the body—or rather genome—built up from the accumulation of knowledge and aided by technological advance. On the other hand, a string of bioterrorist attacks worldwide—several by unidentified terrorists in the United States, by the Iraqi government against the Kurds, by a religious cult in Japan—points to a "horrific body," a body that is suddenly and unexpectedly rendered vulnerable by invisible bioweapons.

On the one hand, we see the honorific body celebrated in the boom of the biotech industry—new discoveries, new tools, new drugs, and so on. On the other hand, we see the horrific body manifest itself in popular culture (e.g., the television serial *24*), military wargames (e.g., Homeland Security's *Dark Winter*), and actual terrorist attacks.

This polarized view of the role of science vis-à-vis politics and war has led a number of government agencies, nonprofits, and professional associations to talk about a "new type of threat." For instance, the WHO released a series of agenda items and memos in 2002 relating to the ways that local and international governments can strengthen their defenses against bioterrorism.[8] The WHO pointed to two main strategies: increased or tighter regulations on agriculture, food, and livestock, and the development of information systems for tracking outbreaks, be they natural or artificial. For some years, the U.S. Centers for Disease Control (CDC) have been developing the latter, including the National Electronic Disease Surveillance System (NEDSS), which gathers medical data on patients nationwide as a means of profiling worst-case scenarios and potential bioterrorist targets.[9] This concern for deterring potential attacks has also been expressed by the British Medical Association (BMA). In one report, published prior to September 11, the BMA authors noted the possibility that information derived from the human genome could be used to target specific populations.[10]

In short, there is a growing tendency to acknowledge the specificity of the threat that biowar generally and bioterrorism specifically represent. But at the same time that governments install preparedness protocols, surveillance systems, preemptive vaccinations, and defensive research programs, it is also important to ask how the "new threat" represented by biowar is changing the relationship between politics, war, and biology. If Foucault broadly describes biopolitics as the moment in which "biological existence is reflected in political existence," then it is also important to ask how bioterrorism, genomics, data networks, and health organizations constitute a type of biopolitics in which the twofold character of biology and informatics is foregrounded. As a step toward exploring this question, let us first consider biowar in all its complexity.

Weaponizing the Body: Levels of Biological Warfare

A point of clarification: throughout this chapter, I use the term *biowar* to refer broadly to all forms of biological warfare as well as to bioterrorism. My intention is to identify a term that emphasizes the way in which some knowledge of biology or biological thinking pervades and conditions thinking about war, especially in an instrumental sense. At other times, I distinguish quite sharply between *biological warfare* and *bioterrorism*. Each arguably has a different conceptual core—the former defined by what Foucault calls a "race war," and the latter by the vulnerability of biology or "life itself." There are, certainly, many other differences between them, one of which is that biological warfare has been traditionally defined as a war between nations, whereas bioterrorism is not a nation-to-nation war, but a war waged against nations by non-nation-state actors.

What makes biowar unique is that it is not simply another way to wage war or to destroy large groups of soldiers, civilians, or regions; it is the direct application of biology to the practice of war. In one sense, this is already apparent: the very definition of *biotechnology*—from premodern techniques of breeding to modern gene therapy—is the use of natural, biological processes for human ends. Unlike chemical warfare, biowar uses living elements and life forms, most often bacteria and viruses, as the agents or weapons themselves. These biological agents are effective, however, only when they act upon other life forms, other biologies. This is the unique character of biowar: *biology is both the weapon and the target*, biological life acting upon biological life (or, rather, biological "life itself" acting upon biological "death itself"). This means that the extent of the damage inflicted and the range of effects that can be achieved are significantly expanded. An effective, infectious agent can cover much more ground than a single explosion, especially when we take into account transportation technologies such as air travel. The presence of some biological agents have a longer incubation period, which makes it more difficult to detect biological agents when they are released. An outbreak may already be well under way before it is identified as a biological attack.

This is one of the greatest points of anxiety within all discussions of biowar and biotechnology generally: the irresponsible or maliciously motivated implementation of novel biotechnologies and the range of unforeseen, possibly devastating results that may arise from integrating engineered organisms into the human body, into human communities, into livestock, into plants, or into the environment. There is, to be sure, good cause to fear the potential effects of the new bioweapons, and their descriptions are often accompanied with a great deal of dramatic flair. Richard Preston, a writer known for his fictional works, describes a near outbreak of Ebola in a Washington, D.C., suburb in his book *The Hot Zone*. Again, the descriptions of real events are rendered in Preston's familiar, fictional style. Here is a description of an individual named Charles Monet undergoing the terminal stages of an Ebola infection:

> He appears to be holding himself rigid, as if any movement would rupture something inside him. . . . The intestinal muscles are beginning to die, and the intestines are starting to go slack. He doesn't seem to be fully aware of pain any longer because the blood clots lodged in his brain are cutting off blood flow. His personality is being wiped away by brain damage. This is called depersonalization, in which the liveliness and details of character seem to vanish. He is becoming an automaton. The higher functions of consciousness are winking out first, leaving the deeper parts of the brain stem . . . still alive and functioning. It could be said that the *who* of Charles Monet has already died while the *what* of Charles Monet continues to live.[11]

Preston's Ebola description, a fictional description of a nonfictional event, feeds on the deep-seeded body anxiety that biowar elicits. Death is not instant, but protracted and disturbingly transformative. The virus's only aim is to transform the human body into a viral replicator—in effect, to transform the body into a virus itself. It is this "nonhuman" aspect of biological agents that, perhaps, instills the greatest fear. Once a bioweapon is set loose, the intention is in a sense irrelevant. A wide range of factors—how good the bioweapon is, how infectious it is, how quick people are to respond, how effective countermeasures are, even the weather conditions—affects the largely chance outcome of a bioagent.

If nuclear arms were feared for their massive scale and expediency (at the push of a button), biological weapons are both more extreme and more subtle. Because biological weapons operate through microorganisms, they not only destroy or harm, but literally take control of and occupy the body. This microbial "occupation" of the population offers something more valuable than total annihilation. It offers a means of precise targeting and control of a given population's level of health—that is, their veritable dependence on health standards set by the nation possessing biological weapons (even if this nation is one's own). Quarantine and confinement here become both political and medical, just as demography and statistics become military tactics.

In 1999, the BMA published a report that outlined three general stages in the history of biowar: a first generation, which includes all premodern examples of biowar (mostly dealing with sabotage); a second generation, which includes chemical warfare in World Wars I and II; and a current, third generation, which may include molecular genetics and genetic engineering.[12] This historiography has been similarly echoed by publications from a number of U.S. government offices, including the Department of Health and Human Services, the CDC, and the Department of Homeland Security.[13] Biowar is understood to proceed in distinct stages akin to the progress of technology, and "doomsday scenarios" begin to read much like the speculative fiction of the medical thriller genre. Although such historiography is for the purposes of education, it also serves to draw out a narrative of scientific progress in the practice of war. Current documentaries, such as the *Nova* special *Bioterror*, locate bioterrorism as the latest phase in this progression, complete with a plethora of scare tactics, ambitious journalists, and a replaying of the Cold War tensions between the United States and the Soviet Union.[14] However, such histories make little mention of the long and extensive history of offensive biological warfare programs in the United States that extend back to the American Revolution. In addition, each instance of biowar takes place within a social, political, and scientific milieu and is not exclusively a technical narrative.

This blindness to the interconnections between political power struggles and scientific-technological research enables many popular accounts of biowar to read as a very one-sided story, in which the threat or the attack always seems to happen for no apparent reason and from without. This historical conservatism takes different forms. For instance, the account given by the BMA report is dedicated almost exclusively to biological warfare between nations, and the *Nova* special portrays biowar as a struggle between First World democracies (meaning the United States) and shadowy terrorist networks in the Middle East. The differences in context and the diversity of the ways that biology, war, and politics intertwine are often dropped in favor of historical narratives with a single punch line: the impeding threat of a war through biology.

In addition, it is important to recognize that the rise of biowar does not mean that nuclear arms are now simply out of fashion, just as the demonstration of "infowar" during the Kosovo crisis did not mean that all war simply became "virtual." If anything, the narratives of scientific and technological progress told by the United States create a picture of a military-industrial (and military-medical) complex that multiplies its forces and proliferates its means of security. Nuclear arms races, biological warfare, chemical warfare, infowar, and good old-fashioned air, sea, and ground combat are all at the disposal of these military superpowers. Thus, in thinking about biowar generally, we might do better to think about concurrent but historically differentiated levels of conflict that proceed through the knowledge and know-how of biology. Thus, we can outline several "layers" or "levels" of biowar, all of which are present to varying degrees in any event or identified threat.

First Level: Biological Sabotage

Accounts of early examples of biological warfare in antiquity already outline three main components of biowar: the use of substances that make the body ill, the sabotage of food and water resources, and attempts to create "modern" biological weapons.[15]

Examples include forms of sabotage of food, water, or animals among the Greeks.[16] The use of poisons directly or indirectly ("weapons" composed of venomous snakes or scorpions) was not uncommon in Greek and Roman warfare. Examples of the second kind are found in Thucydides' account of possible pollution of wells during the Peloponnesian War.[17] In his account of the outbreak of plague in Attica following the invasion of the Peloponnesian army, Thucydides notes the patterns of infection and the disastrous political effect that the plague had in the battle: "Athens owed to the plague the beginnings of a state of unprecedented lawlessness."[18] Although Thucydides' account concerning pollution of food and water is conjecture, what is relevant is that he consciously juxtaposes war and epidemic, as if the two become naturally coexisting phenomena (in this case, the latter determining the former).

The development of perhaps the first "modern" biological weapons is found during the first outbreak of the Black Death during the Middle Ages.[19] The adjective *modern* is in quotes because, although the Black Death did not result in a formalized, scientific knowledge of infectious disease, it did demonstrate a moment in which war was consciously thought of in terms of biological death. As is known, the Black Death first spread throughout Europe between 1347 and 1351, by some estimates destroying nearly half of Europe's population. Trade routes, trading posts and towns, religious conflict, and the use of military organizations in facilitating trade are known to have had a significant effect in the transmission of the plague.

One event is of particular note, and it is thought to have occurred around the early part of 1346. Historical records are lacking for this often-mythologized event, except for one Italian chronicler, Gabriele de Mussis, a lawyer from Piacenza, whose *Historia de Morbo* remains one of the important accounts of the early stages of the Black Death.[20] According to de Mussis, in September 1345 the Black Death crossed into European territory. How did this happen? At an Italian trading settlement in Caffa, on the northern coast of the Black Sea, a skirmish broke out between the Italian Christian merchants and local Muslims. The skirmish escalated into a gang war, involving a small Tartar army and military exchanges from both sides. The Tartar army attempted to siege Caffa but was hit with the Black Death, which had then been spreading throughout the Mongol region. Before retreating, the Tartar commander ordered troops to take soldiers' diseased corpses and catapult them over the walls of Caffa, where the Christian armies were entrenched. Days later the Black Death was reported in Caffa, and by 1351 it had traveled through Asia Minor, Greece, Egypt, Libya, Syria, and southern Europe.[21]

Historians continue to debate the accuracy of the events at Caffa and the degree to which it may be exaggerated. Even if exaggerated, the case of the Black Death is interesting for several reasons. First, it very literally demonstrates the weaponizing of the body, in which biology becomes both weapon and target, a propagator of disease and death. But more than this, the siege at Caffa demonstrates something that is at the core of biowar: the application of knowledge in the service of war. The very idea that a diseased cadaver could have biological and strategic effects beyond its own lifelessness is itself a significant moment in biowar thinking. In fact, even in contemporary contexts, the concurrence of disease and war is striking (bioterrorist threats alongside new infectious diseases such as SARS), and the events at the siege of Caffa illustrate the basic strategy of biowar: that, metaphors aside, disease is war.

These early examples of biowar place an emphasis on the uses of disease or toxins to affect an enemy or target indirectly; they did not yet include direct militaristic methods of attack, and certainly did not yet have access to the new technologies of genetic engineering. They made a rudimentary and fairly uncontrolled use of disease and toxins, most often as a means of sabotage. In contrast, the controlled sabotage of food and water systems is a top concern for the U.S. FDA, whose responsibility within the 2002 Bioterrorist Act is to monitor and prepare for possible terrorist attacks in the food and water supply.[22]

Unlike direct combat, sabotage occurs invisibly and in secret; its effects are often not immediately felt or are noticed only after a delay. Biological sabotage operates in this indirect manner, even more indirectly than the dispersal of a biological agent. Infection happens not directly through the air or blood, but through the metabolic process of food and water—the very substances that maintain the body. In addition, in our contemporary context, the preparation, distribution, and processing of food constitutes a complex network of farms, slaughterhouses, train cargo, food handlers, and so on, which can make the backtracking of sabotage a difficult task. It is for these reasons that biological sabotage continues to be one of the primary concerns in terrorist preparedness programs in the United States.

Indeed, in 1984 an attack such as this was carried out on a small scale within the United States. Followers of the Bagwan Shree Rajneesh cult living in Oregon contaminated salad bars in several restaurants with salmonella.[23] In an effort to thwart a local election, cult leaders had intended this act as a precursor to a more extensive act of sabotage that would be carried out at a later date. More than 700 cases of food poisoning were reported, some of which required hospitalization. In addition, early-twenty-first-century scares over the nonterrorist outbreaks of mad cow disease, bird flu, and monkey pox have further heightened fears about the possibility of a terrorist attack through biological sabotage.[24]

Second Level: Biological Weapons

Biological sabotage was made "more scientific" through the application of microbiology and germ theory during World War I. The antiplant and antianimal campaigns carried out in the two world wars are an important aspect of biowar, for they not only demonstrate the systematic application of the life sciences to war, but also show an awareness of the network properties of infectious agents, be they in food, water, or distribution systems.[25]

This second level of biological weapons extends from the scientifically driven sabotages of World War I to the emergence of recombinant DNA, genetic engineering, and a biotech industry during the 1970s. Here, a scientific knowledge of disease and lethal biological agents is more closely fused with contemporary tactics and strategies of war (including the chemical bomb or nerve gas bomb). The most common approaches were mobilizing pathogenic agents toward targeted areas, biological resources, and both the military and civilian populations.[26] A greater effort is made on this level to control the biological weapon and its desired impact (its target area, carriers, lethal rate and dose, infected perimeter, modes of protecting soldiers).

During 1915 and 1916, the German army initiated a number of antiplant and antianimal biological warfare campaigns against Allied forces.[27] The primary agents developed were anthrax and glanders, and the primary targets were grain stocks and livestock such as horses and cows. Pathogens were cultured in the lab, then distributed by German operatives within the United States to various distribution points, in which horses and other livestock would be injected with infected needles. In addition, some evidence also exists that the French also had an antianimal biological warfare program during the war.[28]

Though by most estimates the effects of these attacks were minimal, the alarm they caused, along with the specter of chemical weapons, led to the 1925 Geneva Protocol, which was, in effect, a "no-first-use" agreement between the signatory nations.[29] However, although the Geneva Protocol prohibited the use of chemical and bacteriological weapons, it did not prevent the further research, development, and weaponizing of biological weapons. This major weakness in the agreement left the door open to a number of offensive biological warfare programs, including those in the United States, Japan, Germany, France, Great Britain, and the Soviet Union.

One of the most harrowing examples of offensive biological warfare programs involves the Japanese experiments on Chinese prisoners during World War II. Known by the name Unit 731, this top-secret program began in 1936 in occupied Manchuria, under the leadership of Ishii Shiro.[30] Over the next four years, the respected scientists and physicians of Unit 731 would intentionally infect Chinese prisoners with a range of diseases, including anthrax, cholera, and bubonic plague. Other experiments involved the use of biological sabotage, bacteriological bombs, and insect disease vectors on the unsuspecting civilians of local Chinese towns. Historians estimate that some 10,000 people were killed as a direct result of Unit 731's experiments. As the war came to an end, Unit 731 members came into U.S. hands. The U.S. government brokered a deal with the Unit 731 members, granting them immunity from war crimes prosecution in exchange for the knowledge they had gained from their experiments.[31]

Following World War II, the awareness of the extent of Unit 731's program led a number of leading nations, including the United States, the Soviet Union, and Great Britain, to more aggressive research into offensive biological warfare. Much of this research centered around field tests, either in populated, civilian areas with nonlethal forms of a biological agent or in unpopulated areas with lethal agents and animal subjects.[32] In 1942 and 1943, the British government tested an anthrax bomb (N-bomb) on Gruinard Island off the coast of Scotland.[33] The most extensive of these activities was that of the U.S. biological warfare program, initiated in 1942 by the War Research Service.[34] Between 1949 and 1969, field tests led by the Committee on Biological Warfare in the Defense Department were conducted in more than 200 populated areas within the United States, totally unknown to the civilians who lived in those areas. Examples of such field tests include a 1950 *Serratia marcescens* and *Bacillus globigii* test off the shore of San Francisco; a 1951 *Aspergillosis* test at a shipping center in Virginia; a 1955 test of *Hemophilus pertussis* in the Gulf Coast of Florida; as well as urban field tests in Minneapolis (1953), St. Louis (1953), and New York City (1966).[35]

In the examples of Unit 731 and the field tests conducted in the United States, we see a noticeable shift away from an ad hoc, tentative deployment of biological sabotage (in World War I) to the development of specifically funded, government-mandated research programs. In addition, in the case of the U.S. program and a bit later in the Soviet germ warfare program, we also see the use of the civilian population as a kind of testing ground for the theoretical effectiveness of bioweapons.

This level of biowar might be said to close with the BWC, which was signed by the United Kingdom, the Soviet Union, Japan, and many other countries in 1972 and was ratified by the United States in 1975. Numerous reviews, policy modifications, and suggestions have been made to the original BWC since its inception date, including more stringent methods of verification. To this day, an agreed upon, workable protocol for biological weapons monitoring and verification remains one of the central weak points of the BWC.[36]

Third Level: Genetic Warfare

Whereas the biowar programs of the previous level were dedicated primarily to the analysis and experimental use of already existing biological agents, another level—that of genetic warfare—takes a further step into the possibility of engineering and designing novel biological weapons. The controversy over the Soviet germ warfare program is but one example. A 1979 outbreak of anthrax in the city of Sverdlovsk resulted in the death of approximately 70 civilians and the illness of many more.[37] It was not until 1992 that inspectors were allowed to visit the city, but their visit was presaged by the defection of a number of Soviet scientists such as Ken Alibek, who publicly testified to his and other scientists' government research into a genetically altered "superplague."

Thus, this layer of genetic warfare is dominated by the recent advances in molecular genetics and biotechnology, in examples such as the HGP and the HGDP. This level involves the use of techniques in genetic engineering, gene therapy, medical genetics, and genomics to design, for the first time, biological weapons that may be able to target specific regions, ethnic groups, populations, or biological resources. One hypothetical example is the use of the information from human genome projects and the HGDP, to develop novel pathogens to target ethnic groups, which would use a gene therapy–based carrier.[38]

However, the concept of engineering biological weapons has to be understood also in light of the history of eugenics in the United States and Germany. Modern eugenics follows upon the work of Francis Galton, who in the 1880s coined the term and had proposed applying Darwinian principles of artificial selection to human beings. Galton's eugenics took hold in a United States grappling with mass immigration, population growth, rising urban poverty, and a looming economic depression. The idea that science could be used to prevent social degeneration was formalized in a number of institutions, primary among them the Eugenics Record Office, founded and run by Charles Davenport, a respected biology professor at the University of Chicago.[39] The Eugenics Record Office generated an immense amount of survey data, including studies of "feeblemindedness." Such studies feed into the perceived social need to exercise a "negative eugenics," or a set of restraints on population growth and reproduction, in order to prevent a range of ills—from criminality to "imbecility"—from spreading across the United States generally.[40] By the late 1920s, nearly half of the states had passed eugenic sterilization laws. In the 1927 case *Buck v. Bell*, the Supreme Court ruled that such laws were constitutional, Justice Oliver Wendell Holmes punctuating the decision by noting that "three generations of imbeciles were enough."

American eugenic legislation paved the way for the German programs, that began in the early 1920s. In 1923, the Kaiser Wilhelm Institute for Research in Psychiatry established a chair for race hygiene. Other institutes would follow suit, including the Institute for Anthropology, Human Heredity, and Eugenics and the Society for Racial Hygiene, also in Germany, as well as the Galton Laboratory for National Eugenics in London, headed by population biologist Karl Pearson. Eugenics in Germany took up many of the Americans' racial policies.[41] Together, the American and German movements helped to introduce Mendelian heredity (then recently rediscovered by biologists) into the field of eugenics and social policy. Involuntary sterilization laws led to thousands of sterilized individuals in the United States, not to mention the extremes to which the eugenics movement would go in the Nazi regime. In 1933, Hitler decreed the Heredity Health Law, directly inspired by eugenics. At the same time, U.S. societies, such as the Genetics Society of America debated about whether or not to condemn the Nazi policies. According to some accounts, they were never able to reach a decision on the topic; in addition, following the war, many Nazi scientists and physicians were never prosecuted and in fact returned to university posts within Germany.

As Daniel Kevles notes, there is a strong continuity between the American eugenics movement and the emergence of modern genetics in the 1940s and 1950s in the United States and Great Britain.[42] Following the atrocities to which the Nazi program led, so-called reform eugenicists such as Ronald Fisher and J. B. S. Haldane aimed to bring a more scientific view to eugenics study, purged of its racism and doctrine of racial hygiene. To do so, molecular biologists began focusing on early techniques in genetic mapping and linkage analysis. One result was a wave of innovations in the use of this more "scientific" eugenics in the diagnosis and prognosis of a range of illnesses. This emphasis on the medical aspect of genetics—without the rhetoric of social degeneration—led the way to the late-twentieth-century emphasis on genetic testing and hereditary study of the transmission of disease. Although quite different from the negative eugenics of the early part of the century, this "new eugenics" was instead characterized by a consumer model for health care, high-tech testing, and an emphasis on prevention.[43]

The context of eugenics helps to frame this layer of genetic warfare, in which largely defensive measures are taken to protect either the military body of the soldier or the social body of a population. The level of genetic warfare is both *preventive* and *preemptive* at the same time. Several real-world examples give further credence to this third level: first, the Gulf War demonstrated that biological warfare was continuing to make its way steadily into the standard armament of modern war, as revealed by Gulf War Syndrome and the experimental vaccines given to soldiers prior to battle.[44] Second, examples of intranational genocide—in Cambodia, Yugoslavia, and Rwanda—suggest that the possibility of targeting ethnic groups through genetics could offer a potentially powerful tool in the hands of regimes bent on ethnic cleansing or racial war.

Fourth Level: Biocolonial Mission

A more directed use of biowar as a tool of ethnic and political conflict occurred during the eighteenth century, in which we find documented instances of biowar used within a colonial context. One example is British Soldiers' intentional use of smallpox to infect Native American tribes. In 1763, Jeffery Amherst, the British commander in chief in North America, gave an order for the presentation of smallpox-infected blankets to Native American tribes in the Delaware region.[45] The blankets were to be taken from infected patients in the infirmary and given to the Indians as a peacemaking gesture. As General Amherst emphasized, the aim was "to try every other method that can serve to extirpate this execrable race."[46]

It can be argued that colonialism is unthinkable without medicine. Without an ability to ensure the health of a colonial army or the health of colonizing populations, the colonial project is compromised from the start. As David Arnold notes in his analysis of British colonial medicine in India, there is "a sense in which all modern medicine is engaged in a colonizing process."[47] Yet, as Arnold points out, this notion of "medicine in the service of empire" is also two sided. On the one hand, there are instances in which the spread of a disease has worked to the advantage of the colonizer or explorer. On the other hand, there are also instances in which disease—"native disease"—has served to obstruct the colonialist or expansionist enterprise.[48] Malaria, yellow fever, sleeping sickness, and a host of other "native diseases" often served to impede European expansionism as much as other illnesses indirectly furthered its cause. As medical historian Roy Porter notes, "without disease, European intruders would not have met with such success or found indigenes so feeble in their resistance. Yet endemic diseases also held back European expansion into Africa."[49]

Recent efforts to provide assistance in the fight against AIDS in Africa—most notably by the Gates Foundation as well as by the U.S. government—is undoubtedly a positive sign of an awareness of global health issues.[50] But it is also important to assess how such financial aid is spent and whether financing alone is enough in a situation where education, communication, and the complexities of the physician-patient relationship are still primary issues. Furthermore, it is also important to ask whether the global health-care industry or the pharmaceutical industry stands to gain from such relief efforts. Although it is clear that AIDS and malaria in countries such as Africa do constitute serious health crises, it is also important to recall the tangled history of colonialism and medicine, as well as the often one-sided narrative of British "medical missionaries" in India and Africa during the nineteenth century.[51]

Today the logic of this level of biocolonial war is, strictly speaking, not war at all, but rather the establishment of a naturalized, permanent link between "developed nations" and a Western health-care paradigm based on costly prescription drugs. Although such treatments are often quite effective and life saving, their benefits are always abetted by what Frantz Fanon describes as a structure of indebtedness.[52] A number of pharmaceutical companies have noted the potential market for generics in developing nations, and controversies still ensue over the corporate patenting of genetic material and cell lines from diverse regions around the world. A multifactorial health-care approach—including environment, diet, cultural context, poverty, education, and drugs—is clearly what such health crises demand.

Of course, the limit of this biocolonial level is when it is turned inward, within the United States itself. This is what Paul Virilio and Sylvère Lotringer call "endocolonization," in which the social body is invaded internally through genetic screening, in vitro fertilization, medical prostheses, and so forth.[53] If it is true that the newest biotechnologies will be field tested in the United States—DNA chips, tissue engineered skin or organs, stem cell therapies—then this testing will be preceded by efforts by the "medical missionaries" of the biotech industry to establish biotechnology as safe, desirable, beneficial, and, above all, natural.

Fifth Level: Bioinfowar

Thus far I have covered four levels, each existing simultaneously, but to varying degrees depending on historical, social, and political context: a first level of biological sabotage, a second level of biological weapons, a third level of genetic warfare, and a fourth level of biocolonial mission. A fifth and final level is that represented by the integration of molecular genetics and computer science in the biotech industry: bioinfowar.

Bioinfowar is not yet a reality, but it is, arguably, quickly becoming one. It includes what has for some time been the practice of "infowar," or the military conflict played out on the level of computer codes, databases, Internet servers, electronic wiretapping, computer viruses, firewalls, and physical communications infrastructures.[54] The development of infowar does not occur as a technological feat, but takes place in the development of military use of information technologies, most explicitly demonstrated in the Gulf War and the Kosovo conflict.

Recent discussions on the intersections of war, global politics, and technology have raised the issue of how the increasing importance of computer and information technologies have transformed the field of combat into a logistical, screenal Sega System (or PS2).[55] This entrance of both spectacle-based technologies (media-based infowar) and information technologies (communications and hacking) into the domain of war has meant, in part, that the enemy ceases to be a body or mass of bodies, but rather coordinates among other coordinates on a pixel plane. These "wars which did not happen," as Jean Baudrillard states, show two fundamental changes occurring in postmodern war. First, the physical encounter of hand-to-hand combat is increasingly being replaced by the mediated encounter of vision machines. The model here is Orson Scott Card's novel *Ender's Game*, in which a young video game wiz unsuspectingly becomes the futuristic military's top combat pilot. Second, war is increasingly coming to be seen as so much more than actual battlefield combat; during every modern war, there are several other levels of combat: media war, encryption and decryption, finances, the business of production for war, the opportunities for revitalizing nationalism, the dark opportunities for genocide and ethnic cleansing, and the use of new media such as networks, computers, and databases of automated war machines. At the most extreme end of this war business, we enter a condition that Paul Virilio and Sylvère Lotringer call "pure war," or the situation of infinite preparedness for an always deferred war. [56]

Juxtapose this scenario of infowar with current developments in biotechnology: the automation of genome sequencing, the rise of bioinformatics and gene discovery software, DNA microarrays and microfluidic "labs on a chip," data mining software, DNA encryption, and other developments show that biotech is becoming thoroughly computerized, and that the biotech patient of the future will be less an anatomical, individuated body than a computerized profile of gene patterns and statistical predispositions analyzed by bioinformatic expert systems. Yet, for all this, biotechnology remains resolutely material in the drugs, therapies, and diagnostics that regularly rub up against the patient's body.

Biotechnology currently plays a number of roles in biological warfare. One recent area of application has been in portable hazardous bioagent detection systems. Nanogen, for example, has a hand-held biochip device for the detection of aerosolized agents such as anthrax. Another area is in the use of genetic engineering for the design of vaccines to potential pathogens such as anthrax, ricin, or smallpox. As noted previously, the U.S. Project BioShield has as one of its priorities the development of "next-generation medical countermeasures"—that is, new drugs produced by the American-based global pharmaceutical industry. Finally, a third area of application has been in medical surveillance systems for the monitoring of potential outbreaks of a naturally occurring or terrorist type. The WHO and the CDC have such networks currently in place.[57]

What would a merger between infowar and the new computerzied biotech look like? Is the answer here nanotechnology? The use of nanomedical particle systems? The use of robotic drones to disperse engineered pathogens to ethnically targeted regions and populations? Will we see the horrific hybrid of the biological suicide bomber? Bioinfowar seems at once less material than the catapulting of diseased cadavers and more material than the targeted military release of computer viruses on an enemy subnetwork.

To recap: a history of biowar cannot be told from one perspective, be it technological development, scientific progress, or the culture of fear and paranoia. A critical account of biowar would have to take into account the social and political dynamics that enframe the transition from military application to civilian use. In the case of biowar, we can see (at least) five coexistent levels at play in any given event, each of which raises fundamental issues concerning the way in which biological "life itself" is instrumentalized in political, military, and ideological conflict.

Targeting the Body

In any consideration of these different but coexisting levels of biowar, it is important to note also how the concept and the practice of biowar has historically changed. We might ask: How does biowar "target" the body? In biowar, biology is both the weapon and the target, a form of "life itself" that targets "death itself" through the use of a range of pathogens, epidemic infections, and, in some cases, engineered life forms.

As discussed in other chapters in this book, one key historical transition in the concept of "life itself" involved a "taking charge of life, more than the threat of death" in the development of a wide range of medicopolitical practices during the eighteenth and nineteenth centuries: the application of statistics and demographics to account for the "health" of populations, the attempts to reform hospitals in terms of management and infections, urban hygiene programs, the establishment of professional societies dedicated to maintaining and monitoring health standards for a population, and the notion of a "medical police" or a managerial apparatus for ensuring the health of the body politic. Michel Foucault refers to such practices as a form of "biopolitics," a form of power in which the health of the population is also the health of the nation, and vice versa. In these and other instances, "biological existence was reflected in political existence," and the medical often dovetailed into the governmental.[58] Biopolitics "tends to treat the 'population' as a mass of living and coexisting beings who present particular biological and pathological traits and who thus come under specific knowledge and technologies."[59] At the center of biopolitics is a concern over the "population," defined in terms that are both biological and informatic—an attempt "to rationalize the problems presented to governmental practice by the phenomena characteristic of a group of living human beings constituted as a population: health, sanitation, birthrate, longevity, race."[60]

In the context of biowar, the health of the population takes on a distinctive character in light of concerns over national security. The population must be protected or secured in two ways: against the threat of an attack from nature, in the form of disease and epidemics, and against the threat of an attack from a political entity, in the form of biological weapons or sabotage. The first type of security metaphorizes disease as a war, whereas the second treats war as an attack on the biology of the population, as an attack on "life itself." The population, then, is doubly vulnerable: it is vulnerable as a biological entity, subject to disease, famine, and overpopulation; and it is vulnerable as a political entity, related to national and international regulatory policies, military research programs, and a range of social anxieties concerning the level of threat. In the context of biowar, the population is defined by *vulnerable biologies*.

The role of "population" moves Foucault to ask a broader historical question: "How, when, and in what way did people begin to imagine that it is war that functions in power relations, that an uninterrupted conflict undermines peace, and that the civil order is basically an order of battle?"[61] In a series of lectures given at the Collège de France in 1976, Foucault addressed this question within a biopolitical context. His analyses distinguish between three general ways of "targeting the body": a politics of sovereignty, an anatomo-politics (or discipline), and a biopolitics (or governmentality). Each of these approaches defines three types of targeted bodies: the body of the sovereign, the docile body of the individual subject, and the regulated social body of the population. The context of war—war that is the continuation of politics by other means—frames these three bodies as vulnerable biologies in need of an "apparatus of security" to provide protection from threatening forces of nature or artifice, epidemics or weaponry. *The body politic is defined by its vulnerable biologies*. It will be helpful, then, to situate Foucault's lectures within our own context of biowar.

As is well known, Foucault argues that a shift in the historical character of power occurred during the eighteenth century, in which structures of power based on sovereignty gradually receded in the face of a disciplinary power. The doctrine of absolute sovereignty, which Foucault analyzes in the formation of nation-states in fifteenth- and sixteenth-century Europe, is a notion of power that is legitimized through the sovereign's right to rule. As Giorgio Agamben and others have noted, this notion of sovereignty is also a right to the exception from the rule: the sovereign is that power that exists above, yet is included within its domain of rule.[62] As Foucault notes, "the essential function of the technique and discourse of right is to dissolve the element of domination in power and to replace that domination, which has to be reduced or masked, with two things: the legitimate rights of the sovereign on the one hand, and the legal obligation to obey on the other."[63]

However, Foucault also suggests that sovereignty is not simply a "top-down" model, for the sovereign's effectiveness is measured only by the effectiveness of its "truth" in the daily lives of the citizens who are ruled. Here he distinguishes between *sovereignty* and *domination*: "by domination I do not mean the brute fact of the domination of the one over the many, or of one group over another, but the multiple forms of domination that can be exercised in society."[64] Sovereignty, in this early modern sense, not only is exercised from on high, but takes effect throughout the social body. This shift in emphasis helps Foucault lead into his discussion of the emergence of a disciplinary power in the eighteenth century. Focusing on the development of social institutions—military, medical, educational, penal, and governmental—he suggests that a new type of power emerges, in which the "operators of domination" that had been implicit in sovereignty take on a more explicit form. Rather than sovereignty's focus on territory and land, discipline focused on individuals' bodies and on their limited "rights," especially their political and property rights. Rather than extracting commodities and wealth, it extracted time and labor, shifting the state's emphasis from the wealth of territory to the wealth derived from the individual subject. Rather than exercise itself through discontinuous taxation and obligation, it exercised itself through continuous surveillance, even self-surveillance. Finally, rather than relying on the self-legitimizing presence of the sovereign ruler, it established a set of coercions that materialized in a range of social institutions (hospitals, schools, mills, prisons, military barracks, and so forth). Rather than a focus on the divine, "living body of sovereignty," we have a focus on the citizen's, anatomical, physiological body, a focus that Foucault calls an *anatomo-politics*.

What is of interest in this broad transition is that sovereignty did not simply disappear. Rather, sovereign power, formerly embodied in the king, was further incorporated into the workings of discipline. The combination of an emphasis on political subjects' individual rights and a set of institutions for rendering subjects docile as subjects produced a different, more attenuated form of sovereignty—the sovereignty of the social body, of the political subject. Foucault refers to this process as a "democratization of sovereignty": "In other words, juridical systems . . . allowed the democratization of sovereignty, and the establishment of a public right articulated with collective sovereignty, at the very time when, to the extent that, and because the democratization of sovereignty was heavily ballasted by the mechanisms of disciplinary coercion."[65]

This newer form of the soereignty of political subjects, however, produces a certain tension, for, inasmuch as sovereignty is based on an exception to the rule (formerly embodied in the monarch) and is exercised in a blanket way over the citizenry, the democratization of sovereignty claims no exception—or, it claims only the exception not to be ruled by an all-powerful, absolute sovereign. Foucault, in his later work, analyzes this relation as the core of "liberalism"—the suspicion of too much government. In the shift into disciplinary power, then, we see a sort of sovereignty that has been distributed throughout the social body and which is a kind of negative sovereignty, the right of political subjects to exist as subjects.[66] But, as Foucault notes, resolving this tension requires a mediation between sovereignty and disciplinary modes of power:

I think that normalization, that disciplinary normalizations, are increasingly in conflict with the juridical system of sovereignty; the incompatibility of the two is increasingly apparent; there is a greater and greater need for a sort of arbitrating discourse, for a sort of power and knowledge that has been rendered neutral because its scientificity has become sacred. And it is precisely in the expansion of medicine that we are seeing . . . a perpetual exchange or confrontation between the mechanics of discipline and the principle of right. The development of medicine, the general medicalization of behavior, modes of conduct, discourses, desires, and so on, is taking place on the front where the heterogeneous layers of discipline and sovereignty meet.[67]

It is here that Foucault begins to outline the emergence of what he calls a "non-disciplinary" power, a *biopolitics*, in which the role of medicine plays a central role. As we have seen, this biopolitical power is predicated on treating the biological traits of the population as opposed to the individualized body of the political subject. What is the difference between discipline, or an anatomo-politics, and biopolitics?

One technique is disciplinary; it centers on the body, produces individualizing effects, and manipulates the body as a source of forces that have to be rendered both useful and docile. . . . And we also have a second technology which is centered not upon the body but upon life: a technology which brings together the mass effects characteristic of a population, which tries to control the series of random events that can occur in a living mass, a technology which tries to predict the probability of those events (by modifying it, if necessary), or at least to compensate for those effects. . . . Both technologies are obviously technologies of the body, but one is a technology in which the body is individualized as an organism endowed with capacities, while the other is a technology in which bodies are replaced by general biological processes.[68]

Thus, to summarize, Foucault makes a distinction between an anatomo-politics, or disciplinary power, and a biopolitics, or a form of governmentality. Anatomo-politics is a form of power applied to the individualized human body of the subject. Its means include discipline, drilling, testing, routines, habit, and measures that instill a degree of docility in the subject. To this end, it employs an anatomical, physiological, and organismic approach to the body of the subject to be disciplined. The sites of anatomo-politics are often institutions such as the prison, the hospital, the school, the military, and so forth. The aims of anatomo-politics are docility, coercion, and self-surveillance—the internalization of disciplinary measures, an "anatomo-politics of the human body."

Biopolitics, by contrast, is a form of power applied to the massified, human population as biological species. Its means include the use of statistical averaging, demographics, forecasting, and a general quantification of the social and political body. To this end, it employs the biological sciences of species and a medicalization of the social. The sites of biopolitics are often medical fields or instances in which medicine is used in the service of other interests. The aims of biopolitics are regulation, modulation, and control—the homeostatic regulation of the population species as a whole, a "biopolitics of the human race." (See table 6.1.)

Thus, we have three broad historical shifts in the way power is materialized within society: a politics of sovereignty, an anatomo-politics, and a biopolitics. Now, it is precisely in this last shift toward a biopolitics that we

Table 6.1 Targeting the Body

	Biopolitics ("A biopolitics of the human race")	Anatomo-politics ("An anatomo-politics of the human body")
Mode of control	Governmentality, regulation, massification	Discipline, docility, subjectification
Effect	Collectivizing, universalizing	Individualizing, singularizing
Object of control	Population, group, body politic	Person, individual, subject
Interpretive lens	Biology, evolution, medicine	Anatomy, physiology, physics
Unit of control	Species	Body
Tools	Statistics, demographics, informatics	Mechanics, mechanism, engineering
Examples	Hygiene, criminality, heredity, eugenics, neurasthenia, hysteria, degeneracy, birth control, sexually transmitted disease, welfare, insurance, social security	Prison, school, military barracks, hospital, sanatorium, workplace
Result	Life, war, security	Organism, institutions, surveillance

see the role of war significantly change. I noted previously how Foucault saw the role of medicine as crucial to balancing the tension in the "democratization of sovereignty." The issue can be put another way. *The rise of biopolitics implies a kind of biologization of politics*. As Foucault notes, the emergence of biopolitics indicates "the acquisition of power over man insofar as man is a living being, [where in] the biological came under State control, [and] there was at least a certain tendency that leads to what might be termed State control of the biological."[69] This relation between power and life is, then, to be distinguished from an earlier relation, that of sovereignty. Sovereignty claims the right to take life or let live (die and let live), whereas biopower claims the right to foster life or let die (live and let die).[70] In sovereignty, power is exercised by the right to kill, to take life, to deprive the subject of his or her most basic attribute. The subject is therefore neither alive nor dead, unless qualified by the sovereign. By contrast, biopolitics, a new type of power, poses the question of life in a different way.[71]

If sovereignty is predicated on an absolute power, even the power to command the death of another, then biopolitics takes the opposite track: it undertakes the ongoing monitoring and fostering of "life itself." So, again, Foucault's question: "How can a power such as this kill, if it is true that its basic function is to improve life, to prolong its duration, to improve its chances, to avoid accidents, and to compensate for failings? . . . How can the power of death, the function of death, be exercised in a political system centered on biopower?" Foucault offers a response that is in many ways unexpected: "It is at this moment that racism is inscribed as the basic mechanism of power, as it is exercised in modern States."[72]

In this context, Foucault uses the term *racism* in a different sense than it is often used. *Racism* is here an *ism* in that it is, in Foucault's usage, a way in which relations within a single species—a population—are thought of in terms of conflict. Racism in this sense is a biologically inflected political relation in which *war is rendered as fundamentally biological*:

The war that is going on beneath order and peace, the war that undermines our society and divides it in a binary mode is, basically, a race war. At a very early stage, we find the basic elements that make the war possible, and then ensure its continuation, pursuit, and development: ethnic differences, differences between languages, different degrees of force, vigor, energy, and violence; the differences between savagery and barbarism; the conquest and subjugation of one race by another. The social body is basically articulated around two races. It is this idea that this clash between two races runs through society from top to bottom which we see being formulated as early as the seventeenth century.[73]

What is needed, then, is a set of measures for ensuring the security of the population: not only welfare programs, health insurance, health-care policies, and an industry for the development and distribution of medicines, but, in addition, new means of monitoring health, both at the macrolevel (disease surveillance networks) and at the microlevel (individual monitoring of cholesterol, blood pressure, red blood cell count). "In a word, security mechanisms have to be installed around the random element inherent in the population of living beings so as to optimize a state of life."[74] Although such practices have many beneficial aspects, they also introduce new levels of discipline and regulation that become increasingly normalized and naturalized. New scientific paradigms—such as genetics—play an important role is the redesignation of medical normativity and health as being predicated on information, sequence, and code.

In fact, the role of informatics is a key part of biopolitics. As Foucault and others have noted, the mathematical and informatic approaches of statistics, demographics, the census, and the field of classical political economy constitute approaches in which the biology of the population—birth, death, marriage, migration, disease—is accounted for through information. There is a sense that this accounting has only magnified in the twentieth and twenty-first centuries: the increasing quantification of medical diagnosis (X rays, urinalysis, magnetic resonance imaging [MRI], blood pressure), the introduction of computers into medical and health-care management, the development of molecular genetics and genetic engineering, and, most recently, nascent fields in the biotech industry: bioinformatics, genetic testing with microarrays, and computer-aided drug discovery.[75] As Scott Montgomery suggests, "bio-informationalism, to no small degree, provides the discourse of combat with a new, inner layer of action and intent."[76] In molecular genetics, then, health and illness become a matter of conflicting, combating genetic codes in which it follows that the key for successful medical therapy is code breaking or cryptography. "Here, health or disease would be defined as one or another state of control over the body's informational systems."[77]

In addition, the combination of anatomo-politics and biopolitics diversifies the domain of control, encompassing both the individual subject as patient and the collective population as a body politic. According to Foucault, "We are, then, in a power that has taken control of both the body and life or that has, if you like, taken control of life in general—with the body as one pole and the population as the other." It is in this nexus that medicine and biology come to play a diversifying function, exhaustively accounting for the patient and the population. "Medicine is a power-knowledge that can be applied to both the body and the population, both the organism and biological processes, and it will therefore have both disciplinary effects and regulatory effects."[78]

Foucault's comments on the race war should be understood in this context. Not only is health increasingly mediated by and understood through the terminology of informatics, but this informatic view operates in a nonreductive manner, actually proliferating and diversifying the types of medical subjects and collective biological entities that can be studied, analyzed, and treated. "Wars are no longer waged in the name of a sovereign who must be defended; they are waged on behalf of the existence of everyone; entire populations are mobilized for the purpose of wholesale slaughter in the name of life necessity: massacres have become vital . . . the existence in question is no longer the juridical existence of sovereignty; at stake is the biological existence of a population."[79]

Despite the new forms that power takes—anatomo-politics or biopolitics—sovereignty persists, though in a modified form. What Foucault's work suggests to us is that, in a sense, *all war is biowar*, not only in the obvious confrontation it elicits with death, but also in the sense that war conditions the ongoing efforts to establish the population's security. In the current biopolitical context, *it is war that mediates between biology and politics*, but a war thoroughly informed by medicine and biology.

Genome Bomb or Genome Message? or, Affection by Infection

If war mediates between biology and politics, it is important to ask, What kind of war? If all war is in some sense biowar, then how are we to understand the specificity of the different levels of biowar I referred to earlier, or how are we to understand the specific bioterrorist attacks in the United States in the early twenty-first century?

In his analysis of sovereignty in the work of Hobbes, Foucault makes a distinction between the "ideal war" and "real battles," the former being the largely fictional state of nature without government, and the latter being the actual conflicts that ensue between governments.[80] We can take Foucault's distinction between the ideal war and real battles and look at a particular case study of biowar: the 2001 bioterrorist anthrax attacks in the United States. Again, this chapter considers both biological warfare and bioterrorism as instances of a more general "biowar," or the biologically driven mode of political conflict that may include, but is not exclusive to nation-state entities. At the same time, the 2001 anthrax attacks demonstrate the difference between biological warfare and bioterrorism, as seen through the lens of Foucault's distinction between the ideal war and real battles. The legitimation of sovereignty is readily apparent in both instances, but in different ways.

The 2001 anthrax attacks are still, as of this writing, in recent memory for many in the United States. The general facts of the event are well known, though with some gaps.[81] On September 18, 2001, letters containing a "weaponized," powdered form of anthrax were mailed to the offices of NBC and the *New York Post*, and possibly to the *National Enquirer*. The letters contained a hand-written note expressing anti-U.S. sentiments and Islamic fundamentalist views. The letter arriving at NBC was opened by Erin O'Conner, an assistant to Tom Brokaw (to whom the letter was addressed). Some ten days later, after falling ill, O'Conner was diagnosed with cutaneous anthrax. Other similar cases were reported by employees at CBS and ABC. By early October, the first reports of anthrax inhalation cases in Florida, and on October 5 the first death from anthrax occurred. On October 9, letters were mailed to government officials, including Senator Tom Daschle. In addition, prior to and after the first mailed anthrax letters, several "hoax" letters containing a harmless powder were sent to the *New York Times*, NBC, Fox News, and the *St. Petersberg Times*. On October 12, the media released the first reports of the anthrax letters.

The response in some situations was to close down offices (as happened in Washington, D.C.), whereas in other instances investigations into the Postal Service revealed the letters' origin. In addition, investigative reporting over the next year revealed other, disturbing conclusions. A *Los Angeles Times* column noted that "the strain and properties of the weaponized anthrax found in the letters show that it originated within the U.S. biodefense program."[82] The particular strain of anthrax found in the letters was identified as the same strain in the Northern Arizona University database (the AMES strain). After two years, the Federal Bureau of Investigation (FBI) was still not able to provide any suspects of the anthrax attacks, though it is estimated that of the 200 or so scientists involved in the U.S. biodefense program, only 50 would have had the knowledge to produce weaponized anthrax.[83] The discovery of secret bioweapons programs in the United States and the relative disappearance of the anthrax investigation led many—scientists, politicians, and journalists—to criticize the government's handling of the situation, which is ongoing.[84]

On October 25, 2001, House Democrats introduced a $7 billion "bioterrorism bill" (later to become the 2002 Bioterrorism Act), which would call for increased spending on a national health-care surveillance effort, as well as an increase in the national stockpile of vaccines and antibiotics. The day prior to this, health officials announced that a "deal" had been struck with Bayer to purchase large amounts of Cipro (Ciprofloxacin Hydrochloride), the anthrax antibiotic of choice.[85] During this time, several government buildings, including post offices, Senate offices, and business offices, were either temporarily or indefinitely closed off for testing and decontamination. The week of October 8, 2001, *Newsweek* ran a cover story on bioterrorism, the title reading "How Scared Should You Be?" in front of a close-up of a gas mask. It appeared on newsstands the same time that a mechanical fault in a New York subway car inadvertently released smoke into a subway station, causing the panic-led rush of crowds toward the exits.

By 2003, one biotech company, Human Genome Sciences, announced the promising preclinical results of its anthrax drug ABthrax, a engineered, human monoclonal antibody intended for the "prevention and treatment" of anthrax infections. Although a number of Human Genome Science's drug candidates have been in clinical trials for a much longer time, the FDA granted ABthrax a "fast-track" designation because of its promising use in bioterrorist response programs such as those outlined in the 2002 Bioterrorism Act. But ABthrax was not the only drug to receive widespread attention; immediately following the anthrax attacks, there were reports of hundreds of individuals in the United States purchasing gas masks, decontamination suits, and Cipro—online. The overwhelming demand for Cipro has also had a backlash. There is at least one class-action suit against Bayer for its ambiguous marketing of Cipro for anthrax; accusations are that Bayer knew its drug was not successful in treating anthrax, but owing to demand did not discourage its use in anthrax treatment.[86]

If, for a moment, we take the two waves of terrorist "attacks" launched within the United States—the events of September 11 and the anthrax-tainted letters sent through the mail—we are presented with a troubling juxtaposition. As we know, the first wave of attacks resulted in a large-scale tragedy that encompassed thousands of human lives, New York's urban infrastructure, the airline industry, and the nation's economy. The second wave of bioterrorism included, most prominently, a letter sent to the U.S. Senate majority leader Tom Daschle in Washington, D.C. On the one hand, we have an event with no warning and no precautions, which resulted in a tragedy that claimed many lives and brought the airline industry and economy to a standstill. On the other hand, we have an "event," if that is the right word, which has been framed by an excess of warnings, precautions, and concern, and which—thus far—has resulted in several deaths in localized buildings. Both events have been treated with the exhaustive, even obsessive reportage we have come to expect from (American) global media networks. But one cannot help but sense the strange incongruities between these two types of "attacks."

And that is perhaps the point: that they are two different types of terrorist actions, each defining the "event" in a different way. It would be wrong simply to conclude that the September 11 attacks were effective, whereas the anthrax attacks were not effective. Understanding what *effective* or *noneffective* means is the key to understanding the strategy of bioterrorist attacks such as the anthrax scare. Again, the question: How do we assess "effectiveness" in such contexts? From a purely militaristic-scientific perspective, bioterrorism's effectiveness is in the number of people in a targeted area who come into contact with, are infected by, and either become ill or die owing to a pathogenic biological agent (most often a bacteria or virus). However, if we consider the cultural, social, and political environment into which a pathogen is introduced, the effectiveness of bioterrorism has to do with much more than biological contamination; it also has to do with a political contamination of national security. This is the moment in which the biological body becomes inescapably political; politics becomes biopolitics.

To the perspective that says the bioterrorist activities involving anthrax were not successful, we can reply that, on the contrary, they were very successful in generating a state of immanent preparedness on the governmental level, accompanied by a state of "biohorror" on the cultural and social level. This accompanying state of biohorror may or may not have anything to do with the "reality" of biowarfare, and that is its primary quality. That anthrax is not a "contagious" disease matters less than the very fact that an engineered biological agent infiltrated those components of the social fabric we take for granted—the workplace, the mail system, even subways and city streets. If there is one way in which the bioterrorist anthrax attacks "targeted" the body, it was in its very presence in proximity to our bodies, which triggered a heightened cultural and social anxiety about the threat of an immanent contagion. This anxiety is related closely to a certain horror of the body, or, more specifically, to a horror of what biological warfare (whether for defense or for acts of bioterrorism) is able to do to the body.

Events such as the 2001 anthrax attacks elicit fearful images of a new type of war, one in which conventional weaponry will be augmented by more precise, more long-lasting, and more comprehensive genetic weapons or bombs. In fact, this image of the "genetic bomb" is what Paul Virilio points to as the likely outcome of the current "militarization of science" in biotechnology and genetics.[87] For Virilio, the "genetic bomb" is both a literal and metaphorical configuration. It is, quite literally, a new type of weapon, but, as noted previously, a weapon that makes use of biology as its payload (be it viruses, bacteria, or an engineered biological agent). Like all bombs, this genetic bomb will attain its effectiveness through explosion, dispersal, and destruction. But the genetic bomb is also a bomb in a more metaphorical sense in that it serves to proliferate an awareness of the possibility of exterminating the species in a new way. Paraphrasing Einstein, Virilio distinguishes between three types of bombs: the atomic bomb, the information bomb, and, finally, the genetic bomb. Whereas the atomic bomb "sets off the question of a possible end of the human species through extinction of a way of life," the information bomb, produced from research in military mainframe computing, not only makes possible the atomic bomb, but also "allow[s] one to decode the encoding of the human genome map."[88]

The idea of a genetic bomb is not new—indeed, it was concretely imagined by postwar field tests conducted by the United States and the United Kingdom, if not foreshadowed by the deployment of American eugenics policies, which later inspired Nazi medicine. But the doomsday scenario is also a powerful image, having been applied not only to genetics, but also to information technologies (infowar, cyberwar, and the information bomb). Indeed, if the atomic bomb makes possible the idea of total specieswide extermination, then the genetic bomb would appear to follow upon that, but with more refinement: the extermination of genetically targeted populations within a given region. As Virilio notes, the genetic bomb is a more updated term for the population bomb, the demographic explosion that was seen to occur in the United States during the 1950s, noted by Einstein and many sociologists of the postwar era. But Virilio discounts the population bomb thesis and suggest that, now, the genetic bomb is in the process of engineering new divisions within the species, a "super-humanity that has been 'improved,' a eugenic humanity, by virtue of the decoding of the genome."[89]

Indeed, the 2001 anthrax attacks would seem to illustrate Virilio's comments, if in a more attenuated form. A literal genetic bomb—a package of weaponized anthrax—triggered a wave of genetic and medical interventions in the body of the population, from new antibiotics and vaccines to biohazard equipment and clothing, to immunization-boosting for soldiers. According to Virilio's scenario, a kind of preemptive, neoliberal eugenics would soon follow: gene therapy, customized drugs, consumer devices for monitoring an environment. This attitude of genetic preemption is the result of a new neoliberal eugenics that is combined with the national security concerns of the militarization of science.

With such a doomsday scenario, we should also theoretically situate Virilio's comments in relation to their interest in the accident: "the accident is the new form of warfare." Virilio's unique take on the history of technology is that it is a history of technological accidents: "Each time we invent a new technology, whether electronic or biogenetic, we program a new catastrophe and an accident that we cannot imagine." With each innovation in technology comes an unintended innovation in the form of the accident. "When we invented electricity, we didn't imagine Chernobyl. So, in the research on the living organism, on the 'book of life,' we cannot imagine the nature of the catastrophe."[90]

But we can ask: Is what we are dealing with in biowar generally really the apocalypse of the "bomb," be it atomic, informational, or genetic? Virilio's reliance on the tropes of World War II imply that each new militarized science happens as a catastrophic event: Hiroshima (nuclear war), the Gulf War (infowar), and so on. But it is difficult to find such an event, a singularity, for biowar. Should we consider the 2001 anthrax attacks as such an event, a singularity that demonstrates the genetic bomb? We might do so, except that the temporal nature of biowar is quite a bit different from that of nuclear war or infowar. Those living in the United States are, arguably, still feeling the effects of the anthrax attacks. Not, of course, from direct infection, but in terms of national security and Homeland Security initiatives, new restrictions on the movement of biological samples and knowledge within the scientific community, new "fast-track" drugs and other therapies designed for preemptive uses, and the presence of epidemic motifs in popular film and television. It is clear that biowar builds no bombs—or it builds not only bombs—but that it feeds on a series of microevents, brief appearances, failed weapons inspections, and botched attempts. These are not events, but more happenings, situations, even forecasts and speculations. But from them issue significant changes that extend throughout the governmental, health-care, and cultural sectors.

The fact that the anthrax attacks utilized the mail system—perhaps the earliest "Internet"—is not without meaningful implications. Certainly there are other means of conducting such as an attack—germ bombs, crop-dusting planes—but they are less technically feasible than a technology that connects groups of individuals in a network for distributing information. The folding of biological pathogens onto an information network such as the postal system shows us what molecular biologists have been suggesting for years—that DNA is information and that genomes (bacterial and viral included) are computers. The layering of one network—a biological one—onto another network—an informatic one—gives us an uncanny example of the pathogenic qualities of information.

In short, Virilio's rhetoric is constrained by his reference point, which is modern warfare between nation-states. The bomb, of whatever type, is the key negotiator in such conflicts, either in the way it settles conflicts (Hiroshima) or in the way it serves to dictate the terms of diplomacy (in the Cold War). But the example of biowar—and, arguably, the example of emerging infectious disease—illustrates the network properties of biology, *as it affects many by infecting a few*. In this sense, it would more appropriate to refer to biowar as utilizing not a grandiose, genetic bomb, but rather as deploying a number of genome "messages." It is no accident that anthrax, ricin, and other toxins are most often spread via the mail, accompanied by a letter. *It is the message, not the bomb, that is the guarantee of the continuing effectiveness of the threat of biowar.* The message—in a letter, a vial, a package, even a computer file—attempts to have the best of both worlds. It is able to create microevents in which the reality of the threat is substantiated, and, in doing so, it creates a condition of permanent threat and an ongoing "state of exception."

Furthermore, the network properties of biowar illustrate the degree to which it is able to affect so many by infecting the few. And here the distinction between biowar and emerging infectious diseases begins to collapse.

Whether it is the 2001 anthrax attacks or the 2003 SARS epidemics, the network properties of the genome message propagates itself via three layers of the network: the transportation networks of air and road travel, the communications networks of medical databases and disease surveillance, and, of course, the biological networks of infection and mutation. Therefore, we can say that biowar generally and the genome message specifically emerge from the intersection between militarization and the accident. Indeed, in the case of bioterrorism, in which the weapon is an uncontrollable and uncontrolled bacteria or virus, the aim is precisely to effect *an implosion between war and accident*. This is why the language of post-2001 U.S. legislation often makes no distinction between naturally occurring, emerging infectious diseases and intentional acts of bioterrorism. It is disease *or* war; the results amount to the same difference from a medical and political point of view.

Coda: "Life Is Just a Shadow of Death"

I began this chapter by contemplating a political shift, from the BWC in 1972 to the Bioterrorism Act of 2002.[91] I also noted how a permanent "state of exception" has become the rule, with regard to the way biowar targets the body. Add to these factors another disturbing element. In July 2001, the Bush administration, after some failed negotiations with other member nations, pulled the United States out of the BWC altogether.[92] In a statement to the Ad Hoc Group of Biological Weapons Convention States Parties, Ambassador Donald Mahley, the United States special negotiator, stated that the United States would be unable to continue to support the BWC on three grounds: first, the lack of any protocol for enforcing the BWC meant that it could not adequately detect covert proliferation (e.g., terrorist activities). Second, the monitoring and inspection procedures outlined in the BWC could be a financial risk to U.S. pharmaceutical companies in that their proprietary knowledge may be compromised by inspections. Finally, it was argued that the BWC would negatively impact the U.S. biodefense program and its classified defensive research. The decision by the Bush administration came as a surprise to U.S. allies, such as the United Kingdom. With more than 140 countries signed to the BWC and more than 50 ready to sign in favor of a prompt completion of the BWC protocols, the U.S. decision brought the process to a standstill.

The curious move by the United States to pull out of the BWC was, possibly, answered in a *New York Times* article just days before the September 11 attacks. The article, which others have corroborated, reported that the United States had under way three top-secret bioweapons programs, programs that, for all intents and purposes, could be classified as offensive and not defensive research. They included field-testing anthrax bombs in the Nevada desert (conducted by the CIA), the assembly of bioweapons labs using commercially available products (conducted by the Pentagon's Defense Threat Reduction Agency), and the intent to engineer genetically an antibiotic-resistant strain of anthrax (also by the Pentagon, under the Defense Intelligence Agency). In 2003, President Bush secured almost $8 billion for biodefense alone. The next round of BWC negotiations is not set to take place until 2006.

Within such a context, in what ways are the biopolitical concerns expressed by government agencies (such as the BWC) over a "genetic bomb" themselves weapons, under the guise of "national security"? A more cynical rephrasing of this question might suggest that we will never, in fact, see the kind of biological weapons of which the BWC speaks. Instead, what we will see is the use of this rhetoric of crisis, the political structure of the exception, and a biopolitical targeting of the body to gain an unprecedented control over the nation's population on the medical and genetic levels.

What might such a near future look like? I conclude this chapter with three theses, followed by three scenarios.

Three Theses

Thesis One The discourse of biowar is one in which war is biology, and biology is war. War is biology in the sense that the constitution of the body politic is as much a concern of national security as the attack on the body politic from outside forces. The militaristic function of eugenics is to be found here, but, arguably, it is present in Plato's *Republic*, which authorizes selective breeding for the ruling classes.[93] But biology is also war, and there is an equally long tradition of regarding the medical fight against disease as a war carried out on the level of cells, germs, and microbes. In modern times, anthropologists such as Emily Martin have analyzed modern immunology's predilection for war metaphors in describing the antibody-antigen response, a metaphor that, interestingly enough, breaks down in the face of autoimmune diseases such as those caused by HIV.[94] The questions that remain open is how to do away with the hegemony of war metaphors in relation to biology and medicine, and whether alternative models—symbiosis, autopoiesis, network science—can offer a way of doing this.

Thesis Two In the twenty-first century, national security is increasingly expressed as the implosion of emerging infectious disease and bioterrorism. As noted several times in this chapter, a defining characteristic of the twenty-first-century response to biowar has been the nondistinction in policy and legislation between naturally occurring and artificially occurring instances of biowar. The operative term in the plans for Project BioShield in the United States is *or*: emerging infectious diseases *or* bioterrorism. It matters not which, for the end results are seen to be the same: the infection and degeneration of the body politic. Of course, from the perspective of their causes, they are very different: whereas emerging infectious diseases ask us to consider our actions within a network of relations with the environment globally, bioterrorism challenges us with a fundamentalist view of biology.

Thesis Three The integration of biotechnology and informatics in national security concerns culminates in a more general, pervasive *biological security*. The 2003 RAND report *The Global Threat of New and Re-emerging Infectious Diseases* specifically notes how globalization has transformed both biological boundaries and national boundaries, rendering them vulnerable in terms of infectious disease.[95] This transformation has prompted the authors to suggest a shift in policy outlook, from national security to what they call "human security." Whereas national security places its emphasis on the national population, human security would place its emphasis on the individual (but the individual as a participant in an ideal global citizenship). What seems to be replacing national security in such examples is a form of *biological security*, or a security so pervasive that there is no outside; it is simply a security against biology, against biological "death itself" (the conceptual inverse of biological "life itself"). This biological security is the use of biotechnology to defend against biology itself, in a kind of surreal war against biology. Genetic screening, preemptive vaccination, new medical countermeasures, Web sites posting disease alerts: the body is under attack on all fronts, and in the confrontation of biology against biology we see the spectrum of responses, from antibiotics to cosmetic surgery.

These three points—the discourse of war and biology, the implosion of epidemic and war, and the notion of biological security—can serve as conceptual tools for the further analysis of biowar as it exists alongside governmental national security, the biotech and pharmaceutical industries, and the presence of nonstate terrorist actors.

It is thus not difficult to imagine several possible scenarios—not doomsday scenarios, but rather scenarios involving an increasing naturalization of the state of emergency that biowar elicits.

Three Scenarios

SimSARS The popular *SimCity* games allow players to oversee, manage, and regulate a virtual city. Using unique algorithms, the player sets certain parameters and then watches as the complex of events—urban development, financial transactions, health, crime, education, and so forth—unfold before the player's eyes. *SimCity* is predicated on the idea of a complex system: multiple agencies, multiple factors, and multiple demands, all acting simultaneously over time (a user can "speed up" time, in which a *SimCity* day equals a real-time second). The aim of *SimCity* is to know how to set the best starting conditions for the growth of a city, and how and when to intervene in times of crisis (financial crisis, natural disasters, rising crime rates, poverty in neighborhoods). Although the player is in the position of sovereign with respect to the city (a "God mode" allows a player to inflict natural disaster, thereby destroying the city), what *SimCity* teaches, above all, is a mode of regulation, monitoring, and surveillance. In fact, formally, *SimCity* is very much like the field of epidemiology, which involves the tracking of, monitoring of, and intervening in the spread of a particular infectious disease. It is not difficult to imagine a version of *SimCity* based on the control of an outbreak: *SimSARS* (*SimAIDS* would, presumably, be too complex for most computers). The player acts as a "virus hunter," or, better, as the head of a team at the CDC attempting to handle a potential outbreak of anthrax or Ebola. In the tradition of military-training simulations, such a game might be useful for training personnel at agencies such as the CDC or WHO. But in its civilian use it would serve to normalize further the heightened level of monitoring and regulation that the current "disease surveillance networks" operated by U.S. government agencies already demonstrate.

PGP-DNA If DNA is a code, and if the encryption and decryption of codes are a key aspect of any military conflict, then it follows that the most perfect combination of soldier, weapon, and secret message would be encryption using DNA—in the living body. Most modern encryption systems involve three basic elements: a message text (or plaintext), a method of encrypting the plaintext (cipher), and a means of decrypting the ciphertext (the key). Might it be possible to encrypt a message into an actual DNA sequence? DNA, being both remarkably simple (a mere four base pairs) and admirably complex (endless combinatorics), would serve as an ideal medium for encryption. With the basic tools of restriction enzymes, plasmids, and gene therapy, this is not outside the realm of possibility. When the message exists in the living body, an in vivo ciphertext, it will be indistinguishable from any other sequence of DNA in the body. A horrific scenario comes to mind: the ideal suicide bomber is one who is not recognized as such, but who is carrying a lethal, highly infectious virus during its incubation period. In the 2001 anthrax attacks, the message and weapon were delivered together; here, the weapon and the message collapse into one.

"Good" Virus In 2003, the infamous "Blaster" virus worked its way through the Internet, infecting an estimated 400,000 operating systems running Microsoft Windows.[96] During the attack, an attempt was made to construct a countermeasure, a "good" virus, which would, like a computer virus, propagate itself through the Internet, but instead of bringing down the computer, it would automatically download a "Blaster fix" from Microsoft's Web site. Dubbed "Naachi" or "Welchia," this good virus itself ended up causing a fair amount of damage, affecting the systems of the Air Canada offices and the U.S. Navy's computer cluster. But the logic of Naachi is interesting: fight networks with networks, viruses with viruses. In a sense, this may be the analogue to the development of inoculation and vaccination in the nineteenth century. Or, perhaps, we have not yet seen the biomedical equivalent of Naachi. But it is not difficult to imagine. During an outbreak of an infectious disease, what would be the quickest, most efficient way to administer a vaccine or a treatment? Hospitals are jammed, doctor's offices overbooked with appointments, the local drug store out of their supplies. A vaccine can be inserted into a "disabled" airborne pathogen, thereby spreading its anecdote against the epidemic. This is, basically, the logic behind gene therapy. Perhaps a "good" virus is in the making here, using the networks of infection to spread a vaccine rather than a virus. A kind of invisible, molecular war would take place in the very air we breathe. A decidedly yet ambiguously nonhuman form of medicine under the aegis of national security.

We are placed in a very challenging situation. On the one hand, it is painfully clear that the new "enemy" called bioterrorism is not completely a product of military discourse. The "invisibility" of both the network of perpetrators and the pathogens themselves pose an infrastructural threat that manifests itself in very material-biological ways. But, on the other hand, it is difficult not to see how this event of bioterror has triggered in the United States an ever tighter vigilance, surveillance, and monitoring of its own citizens as biological subjects, as a "population" in Foucault's terms. At times, separating the external threat of bioterror from its internalized anxiety (at the cultural, social, and political levels) can be exceedingly difficult. It seems that, for the time being, the best we can do, alongside the preventive measures being implemented, is to make sure that we understand how bioterrorism functions in a way that is more than simply biological. In short, we must ensure that we understand bioterrorism as a set of activities (scientific, political, and cultural) that is able to trigger the "state" of biopolitics.

III

Decoding/Consumption

Consumption is also immediately production, just as in nature the consumption of the elements and chemical substances is the production of the plant.

—KARL MARX, *GRUNDRISSE*

7

The Thickness of Tissue Engineering

Skin Jobs

In May 1998, the U.S. FDA approved a product called Apligraf, an organic, artificially grown skin product developed by the biotech company Organogenesis.[1] Apligraf is the first "off-the-shelf" engineered body part to have been granted FDA approval and is now being selectively implemented in medical centers for the treatment of leg ulcers and general skin burns. Such products are based primarily on research in the field of cellular biochemistry and stem cell research—those cells that contain the capacity to turn into and differentiate into particular cell types (blood cells, nerve cells, muscle cells, bone cells, skin cells). Put simply, researchers are looking into ways in which the body's cells can be coaxed into growing and developing as they did during the first stages of embryonic development.[2] By harvesting cell samples, cloning those cells, and "seeding" them into lattice structures immersed in growth medium, researchers can "cook" the cells, which will, potentially at least, yield a new patch of skin (figure 7.1), or a new organ.[3] Commonly known as *tissue engineering*, this field not only promises the ability to generate entire organs and even limbs, but in projects framed by university, governmental, and corporate biotech organizations also emphasizes the practical necessity of such research for transplantation, immunology, and medicine generally.[4] Tissue engineering is not a field of speculation; it is a set of practices currently being implemented in health care and medicine, as well as in a range of clinical trials and laboratory experiments.

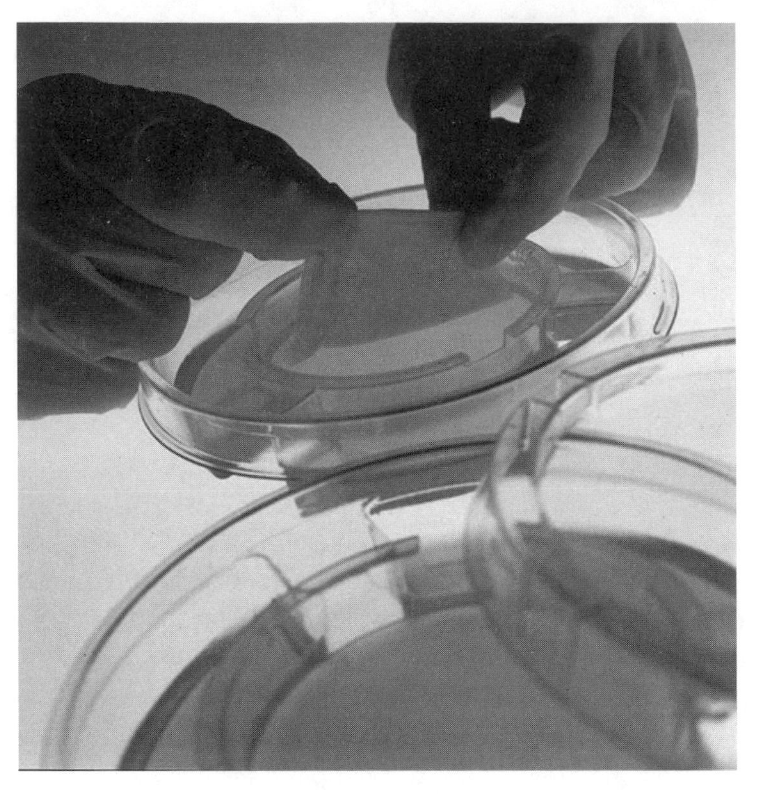

Figure 7.1 Promotional image for Apligraf, a bioartificial wound therapeutics product manufactured by Organogenesis.

Although advances in tissue engineering research have shown great promise in the technical mastery and understanding of tissue and organ formation, how might we assess the larger medical and philosophical effects of tissue engineering in biomedicine's view of "the body"? This chapter suggests that a close examination of tissue engineering practices shows that the "regenerative body" imagined by tissue engineering presents us with a series of aporias, tensions, and contradictions with regard to the status of the biomedical body. Through its techniques, tissue engineering separates the body from the technological domain at the same time that it constantly transgresses that boundary, forming unique biomedical hybrids composed of stem cells, biopolymers, and lab-grown tissues. Research in tissue engineering is transforming conventional biomedical approaches to the body, health, and what Georges Canguilhem has identified as the dynamic between the "normal" and the "pathological."[5] It is from within these transformations that the assumed relationships between bodies and technologies and between the natural and the technical will likewise be affected.

Tissue engineering's interest in this notion of a regenerative body is also an instance of the modern investiture in power relationships in the articulation of the biological population, Michel Foucault's biopolitics. Tissue engineering is not a liquidation of the body or an incorporation of the natural by the technological, if by "natural" we mean an essential, prediscursive notion of "the body itself." Similarly, tissue engineering is not about the devaluation of the body, whether regarded as mechanistic object operated by the self or as the unpleasant conditionality of the "meat."[6] Tissue engineering is, however, very much involved in a series of practical, technical renegotiations with biological materiality and biological function, and in this sense it follows our traditional definition of biotechnology as a technical enframing of "naturally occurring" biological processes. It is in this sense that tissue engineering is involved in the medical, philosophical, and political production of what biomedical science will define as a body. Through a look at tissue engineering's main technical approaches and at the extrapolations of those approaches through Canguilhem and Foucault, I highlight the tensions and aporias in the kinds of bodies imagined by tissue engineering practices.

Regenerative Medicine

On a technical level, tissue engineering involves an integrative, multidisciplinary array of techniques whose primary character is that at each stage it is the "natural" biochemical processes and properties of the body that form the central point of focus. As researchers Lawrence Bonassar and Joseph Vacanti state, tissue engineering is based on the "concept that the repair and regeneration of biological tissues can be guided through application and control of cells, materials, and chemoactive proteins."[7] Although the development of artificial skin substitutes had been explored as early as the 1960s, these temporary, nonorganic substitutes were designed to provide short-term tissue maintenance.[8] By contrast, tissue engineering has developed a set of novel techniques for enframing the living biological body, for providing the right biochemical conditions through which cellular growth and development will take place. Thus, the first thing to note about tissue engineering—in its techniques as well as in its conceptual approach—is that it is exemplary of the way in which I have been discussing biotechnology: as the technically configured mobilization of biological components and processes toward novel ends.

It will be helpful to begin with a brief description of tissue engineering's main techniques. Put simply, the logic of tissue engineering is to combine "cells and materials" to grow tissues and organs, and then to implant those tissues and organs into the patient's body.[9] Therefore, current tissue engineering can be broken down into a three-step process: obtaining cell samples ("cell sourcing"), cultivating those cells (often in combination with biomaterials), and then reimplanting those cells and tissues into the patient's body.[10]

In the first step of cell sourcing, a cell sample, or biopsy, is taken from the patient—that is, the "cells" part of the "cells and materials" recipe is acquired. The trick in this initial step is to isolate a significant cell sample without doing damage or causing complications to the donor site. This procedure is less of a problem for some instances (e.g., epithelial cells from a burn victim taken from any part of the skin), but for more complex organs (such as heart or liver) it can be delicate.[11] For this reason, some researchers have advocated using therapeutic cloning techniques for replicating a cell source, or developing "universal donor cells" whose membrane receptors have been "silenced,"thereby making them applicable to any patient and thus circumventing histocompatibility issues.[12] Once healthy, intact cells are isolated, they can be cultured to produce a batch of what will basically be source material for regeneration.

The second step involves biomaterials integration. Once the desired cells have been isolated and cultured, they are then implanted or seeded into a kind of biomaterial skeleton that will provide structural support—what researchers variously call a "matrix,"a "lattice architecture,"or a "polymer scaffold."[13] This three-dimensional structure ensures that cellular growth also has a form, so that, for example, regenerative cartilage cells will grow in the shape of an ear or a nose (both of which have been accomplished in the lab).[14] Recent developments in the field of biomaterials science have privileged polymer and polypeptide structures because they can actively help the process of cell and tissue development and are biodegradable—that is, they simply dissolve away as the regenerative cells grow into the damaged site.[15]

Finally, the third step takes this combination of cells and materials and places it in an environment intended to induce regeneration. Once the scaffold or matrix has been seeded with cells, the entire apparatus is set in a "bioreactor,"which gives the cells an initial jump-start toward regeneration.[16] Often a biochemical "cocktail" of proteins or growth factors is added to aid this process of regeneration.[17] When the cells are stable and set within the scaffold or matrix structure, the whole complex is then surgically implanted (back) into the damaged target site in the patient's body, where, ideally, the cells will continue to grow, meshing themselves with cells in the surrounding environment in the body, and vascularizing the new tissues (growing the necessary network of tubular structures—capillaries, vessels, veins, arteries).

This process might be referred to as the standard model for tissue engineering technique, and it has been referred to as a standard model both in scientific writing about tissue engineering and in tissue engineering research.[18] However, a number of developments have also expanded and diversified this standard model, some of which deserve mention here for the way in which they conceptually transform the body at the cellular and molecular level into a kind of storehouse of regenerative potential.

One alternative development for the regeneration of cells and tissues has been the concentration on stem cells as a potentially reliable cell source.[19] Researchers working in the field of stem cell research are attempting to understand the ways in which the body's embryonic stem cells begin the process of cellular differentiation as the embryo develops.[20] Likewise, the recent discovery of adult stem cells has suggested that such capacities for cell morphology continue to exist in the adult body, offering another, less controversial means of cell sourcing.[21] Such stem cells are often referred to as "pluripotent" because in their undifferentiated state they contain the potential to turn into a constricted range of the body's cell types (muscle, bone, nerve cells, etc.).[22] The aim of stem cell research is, through the subtle manipulations of the cellular environment, to control and direct the development of stem cells so that, for example, muscle cells may be grown for the regeneration of muscle tissue or mesenchymal stem cells found in the bone marrow may be induced to differentiate into cell types such as bone or neurons.[23]

Although the main therapeutic goal of stem cell research is cell therapy (which most often translates into cell injections at the target site), tissue engineers have also been interested in the ways that stem cells can provide a novel source for tissue regeneration using tissue engineering's standard structural model ("cells and materials"). Unlike traditional protocols for cell sourcing, which sample cells from the damaged site, stem cells might provide a reliable, long-term source of diversifiable raw material from which a range of tissues can be grown. Thus, for instance, mesenchymal stem cells might become the main cell source for the regeneration of bone, cartilage, tendon, or muscle, and the cells would not need to be taken from the damaged tissue site. In addition, the ability to control cellular function in the lab has in some cases proved difficult; some cell types (such as cartilage) taken by a biopsy will display irregular functionality outside the body or will in some cases dedifferentiate altogether.[24] The use of stem cells offers an alternative in which the careful balance of stem cell colonies and polypeptide growth factors provides a means of directing cellular behavior.

The convergence of stem cell research and the standard, structural approach in tissue engineering has been an area of ongoing attention. Stem cell research itself is a distinct field, but the main question has been how to incorporate the findings and techniques of stem cell research into the generation of structurally complex tissues and organs. Writing about tissue engineering in 1993 (five years prior to the discovery of stem cells), Robert Langer and Joseph Vacanti identified cellular differentiation and growth as a primary area of future research for tissue engineering. In 1998, the same year that James Thomson and colleages published their findings on human embryonic stem cells, Bonassar and Vacanti identified the use of stem cells in tissue engineering as a possible means of overcoming the problems of in vitro cellular control (namely, enabling sustainable function and preventing dedifferentiation): "The issue of phenotype expression and dedifferentiation has led to the investigation of pluripotent stem cells as a source for engineered tissues."[25] A year later, researcher Roger Pedersen cautiously echoed this possibility in *Scientific American* by differentiating stem cell research, which is primarily aimed at controlling differentiation, and the use of stem cell research in tissue engineering, which aims not only to direct differentiation, but to coordinate that differentiation into complex, functioning structures: "All the differentiated cells discussed so far would probably be useful in medicine as isolated cells, or as suspensions; they do not have to organize themselves into precisely structured, multicellular tissues to serve a valuable function in the body."[26] In the same issue, Langer and Vacanti took up this challenge by suggesting three possible ways of using stem cells as a cell source for tissue engineering: (1) by using human therapeutic cloning to create embryronic stem cells specific to a particular patient (both costly and controversial, they state), (2) by creating "universal donor cells" from stem cells by "masking" the outer membrane's highly complex structural ability to distinguish self from nonself, and (3) by concentrating on an intermediary type of cell called a "progenitor" cell, which is partially differentiated and still maintains enough flexibility for its process of differentiation to be controlled.[27] Finally, researcher David Stocum has summarized these challenges by proposing that the most viable convergence between stem cell research and tissue engineering is either stem cell therapy (direct injection to target site) or cell sourcing (using stem cells as initial raw material in tissue engineering's structural approach). As he states, "the idea is to transplant stem/progenitor cells, or their differentiated products, into a lesion site where they will form new tissue, or to use them to construct a bioartificial tissue in vitro to replace the original tissue or organ. . . . The use of stem cells is preferable to the use of differentiated cells harvested directly from a donor because stem cells have the potential for unlimited growth and thus supply."[28]

Another alternative development in tissue engineering research has been to sidestep the laboratory altogether. Several leading researchers have suggested that although current tissue engineering techniques have provided a means of generating tissue and organ structures in vitro, the long-term goal of tissue engineering is to be able to do the same in vivo, in the living body of the patient.[29] Such a strategy requires not only a more thorough knowledge of how biomaterial polymers behave over time in the body, but also an extensive knowledge of the genetics of cell morphology and differentiation. One proposed technique involves the surgical implantation of biopolymer structures seeded with stem cells and growth factors (an undeveloped apparatus, whereas the standard tissue engineering model first fully regenerates the tissue in the scaffold and then implants it into the body).[30] Another, similar strategy involves "spiking" a surgically implanted biopolymer scaffold with genetically engineered plasmids, which then insert themselves into the neighboring cells at the target site.[31] The plasmids contain inhibitors, promoters, and other genes for the production of proteins needed for cellular regeneration. As the surrounding cells take up these genetic cues, they begin replicating and filling in the area created by the biodegradable scaffold.

However, at the time of this writing, many of these developments are still highly experimental, some even hypothetical.[32] For our purposes here, we can include them as extensions of tissue engineering's standard protocol (cell sourcing, biomaterials integration, and apparatus culturing), principally because they reinforce tissue engineering's central concern, which is how effectively to enframe and modulate the "natural" processes of cell and tissue growth, development, and regeneration. Therefore, although significant developments have extended the study of biomaterial polymers or cell sourcing in tissue engineering, this basic model has persisted since the 1980s, when the first experiments into the synthesis of living tissue equivalents were carried out.

An important point here is that at no point in these processes are mechanical or nonbiological, nonorganic artificial materials or components incorporated as part of the core process of cellular and tissue regeneration. All "natural" biochemical processes are maintained in their normal functioning, and (according to researchers) only the extracellular, environmental context has shifted, from the body's interior to the lab and back again. The regenerated tissue thus ultimately derives from the patient-subject's own biological resources.

The Normal, the Pathological, and the Regenerative

Tissue engineering was originally a medical response to the problems of tissue and organ failure and the shortage of organ donors, whereas previous and current alternatives were and are mostly reduced to organ transplantation.[33] Tissue engineering thus already emerges from a biopolitics of the population as a medical-biological problem, where health statistics and the distribution of biological materials (e.g., organs in transplantations) are translated into a problem of the regulation of biological materials produced directly in the laboratory ("off-the-shelf" tissues). An economy of body parts (transplantation; xenotransplantation or from animal to human; surgical intervention) is thus replaced by an economy of autogeneration (the generation of tissues from one's own cells) that is both circular and self-making.

This replacement points to an important distinction between the type of medicine practiced in tissue engineering and a range of modern medical perspectives on the biomedical body and disease. In *The Normal and the Pathological*, Georges Canguilhem outlines several basic medical approaches to disease, each of which emerged from a larger complex of a philosophy of life. For instance, with the work of the physiologist F. J. V. Broussais during the early nineteenth century, pathology or disease began to be more clearly articulated in relation to a normal state of health.[34] Broussais, in his work on excitation, irritation, and physiology, suggested that pathology and physiology form one continuum and are really part of the same state, differing only in degree (excesses or deprivations of an equilibrium state). Canguilhem goes on to show how Claude Bernard, in his work on the experimental elaborations of a pathophysiology, clarified Broussais's principle by suggesting that the pathological state differs only quantitatively from the normal state; thus, every pathological state corresponds to a normal or equilibrium state, and such differences can be measured within the context of the experimental laboratory.[35] As Canguilhem states, "every conception of pathology must be based on prior knowledge of the corresponding normal state, but conversely, the scientific study of pathological cases becomes an indispensable phase in the overall search for the laws of the normal state."[36]

This concept of disease—what we might call a "homeostatic" medicine—is primarily restorative in its therapeutics. Its central aim is to "read" the pathological state through a variety of measurable and visible signs and to return the organism to its normative state of health and equilibrium through a counteracting of the excesses or deprivations in the organism. Central to this view of the homeostatic organism is the role of measuring technologies and the laboratory in the assessment of excess or deficiency.

Canguilhem also cites a second conception of disease, one markedly different from the equation between pathology and physiology elucidated by Broussais and Bernard. René Leriche's early-twentieth-century work on the physiology of reflex and the role of physical pain in disease marks, for Canguilhem, an important point in the notion of disease as a qualitative rather than merely quantitative phenomenon.[37] Leriche concurred with Bernard that there is essentially no sharp dividing line between pathology and physiology on a purely quantitative level, but Leriche did point out that on a qualitative level—that is, on the quality of life of the organism—pathology or disease is a distinct state quite different from a normative, healthy state. As Canguilhem notes, the "state of health is a state of unawareness where the subject and his body are one."[38] This "silence,"as Leriche termed it, is broken in disease, which is less a deviance from a norm and, as Canguilhem puts it, "more really another way of life."[39] Although not exactly proposing a phenomenological medicine, Leriche suggested that physical pain and the anatomical lesions and physiological dysfuntionalities that are an index of that pain, constitute a whole field of structural and systemic elements that are qualitatively different from and run against the organism's normal operations. As Canguilhem states, according to Leriche, pain is not a symptom of disease, but a disease itself; disease not only founds normal pathology, but constitutes a qualitatively different way of life or mode of operation for the organism.

These two conceptions of biological normativity, health, and disease—a homeostatic model (disease as essentially indistinguishable from health, as in Bernard) and an externalist model (disease as "other" to health, as in Leriche)—are founded on the notion that the healthy, normal state is the one that is in some way returned to via the intervention of medicine. This "therapeutic" notion of medicine—of health and the normal as the state that forms the foundation for the derivations (be they quantitative or qualitative) that constitute disease or pathology—still informs a majority of mainstream Western medical practice, as evidenced by contemporary clinical trials and applications in gene therapy and genetic drugs.[40]

If, following Canguilhem's suggestions, the homeostatic ("quantitative") and externalist ("qualitative") perspectives form two primary philosophies of modern therapeutic medicine, tissue engineering presents us with another model, based not on the quantitative measurement of disease or on the qualitative differentiation of the effects of disease. Instead, I suggest that tissue engineering is in the process of constituting a unique biomedical normativity based on a notion of the body as "regenerative" and as self-healing. However, as my description of tissue engineering's techniques illustrates, this capacity for self-healing is also a fully technologized capacity. The body of the tissue engineering model does not simply spontaneously heal, but requires an elaborate apparatus for properly enframing the regenerative potential of cells and tissues.

In tissue engineering, the biomedical body returns to itself in a spiral that simultaneously moves upward (an infinitely reproducible body) and downward (an expendable body). With tissue engineering, the body is not treated by medicine, is not improved or repaired by medicine; rather, tissue engineering is part of a new type of medicine that instead facilitates the generation of the body's own materials from itself. The difference here is between a range of technologies intended to supplement the body (from drugs to surgical repair to prosthetics in modern, therapeutic medicine) and a biotechnology whose aim is to be able literally to synthesize or materialize the body (from the engineering of genes to the growing of organs). There is no return to a previous, healthy state, though this is often how the medical application of tissue engineering is presented. Rather, there is a proliferation of the body's cells outside the body, the production of the body's materials and processes in the lab. The model here is not restorative or therapeutic, but rather generative and synthetic. Tissue engineering extends its biological and biochemical processes beyond the corporeal boundaries of the patient-body itself, although at the same time maintaining a direct connection to the patient-body.

What kind of externalizing process is this? The way in which tissue engineering externalizes the body can be contrasted with the ways in which media technologies similarly affect the body. When Arthur Kroker speaks of the "outering of the body," he is primarily focusing on the capacity of media technologies to distribute our sensorium externally (e.g., the Walkman externalizes hearing, the cell phone speech, camcorders sight).[41] When Marshall McLuhan speaks of the "extensions of the self" in media technologies, his basis is specifically neurological (the communicative and connectionist dimensions of electronic media that create the fertile ground for a complex, interconnected global village).[42] Both Kroker and McLuhan are talking about technological externalizations that function by forming a physiological or functional trajectory from the biological, human body. That trajectory has as its destination some technology or media that is taken as ontologically distinct from but nevertheless not disconnected from the body.

By contrast, tissue engineering is made possible by an approach in which biological *structure* and biological *process* are separated and isolated. The biological process of growing kidney cells in a lab is taken as independent of the structure of the renal system and the bladder inside the living body. Tissue engineering creates a space for the externalization of the very operational and material elements of the biological body, in conjunction with a series of techniques and technologies in which *certain aspects of the body biologically function outside the body* from which they have been extracted. According to tissue engineering research, the cells sampled from the patient, which regenerate themselves in a structural matrix in the laboratory, operate exactly as they do inside the body (for example, during embryonic development, when cells begin to grow and differentiate). Researchers' suggestion that the natural, biological body operates the same outside (in the laboratory) as it does inside implies a kind of biological externalization. The specific context and biological milieu of the living system is fractured by an approach in which biological function is extracted from biological structure.

In this sense, tissue engineering implicitly instantiates a shift from a modern medicine based on homeostatic and externalist principles to a *regenerative medicine* based on the principles of biological renewal. If a modern medicine intervenes from without, either aiming to restore the organism to its normative state (the homeostatic approach) or aiming to identify disease as a qualitative state (the externalist approach), then a regenerative medicine recontextualizes the body from within, isolating and abstracting the functionality of the organism. If a modern medicine treats the body as marked by disease and therefore a deviation from a biomedical norm, then a regenerative medicine purports to treat the body as marked by operational deficiencies open to improvement in design (e.g., lag in recovery times, vascularization capabilities, diversity in stem cell differentiation, in vivo injections).

Our Tissues, Our Selves

The externalization of the body also has implications in how novel types of biomedical subjects may be formed in tissue engineering–related practices. This phenomenon can be analyzed by extending the "biopolitical" perspective from previous chapters. In much of his later work, Foucault began to refine his theories of the relationships between power, knowledge, and their manifold points of contact with the subject's body. He increasingly reserved the term *biopolitics* for a particular strategy of power relations, exemplified by the "art of government" in the modern state, in which what was of primary concern was a collective body, or the "population," viewed in biological terms. Biopolitics is thus a political management and regulation of the population, but at the biological level or at the "species" level. With the emergence of political economy, demography, and population studies, modern subjects not only were situated within a range of disciplinary and institutional contexts, but also were considered collective biological bodies with their own dynamics that could be articulated and organized as sources of instrumental knowledge. To reiterate, biopolitics was thus "the endeavor, begun in the eighteenth century, to rationalize the problems presented to governmental practice by the phenomena characteristic of a group of living human beings constituted as a population: health, sanitation, birthrate, longevity, race."[43]

As in the discussion of genomics and biocolonialism in chapter 4, one of the key strategies of biopolitical power identified by Foucault is its ability simultaneously to universalize and individualize through its biological-species perspective.[44] A population can be a source of knowledge production as a collective biological species, but this is possible only if the parameters of individualized biological subjects are also taken into account and differentiated.

The universal category that enables a conceptualization of the species-population to take place is the notion of the continuity and universality of the individual human body itself. In this sense, biopolitics makes the individualized biological body the foundation for a flexible system of informational knowledge production (e.g., health statistics, governmental health policy, health records, hospital management). The population is biopolitically accountable in each and every one of its units, as collectivities of individuals and as individuals constituting a collectivity.

Tissue engineering also universalizes the biomedical body in this way, but it subtly shifts the modes of application of the universal category of the human biological body. As previously mentioned, tissue engineering emerged as a response to problems in organ transplantation, which are primarily problems of the immune system's rejection of foreign tissue. Thus, although medical science can posit the universality of the biological body (implicit in the very idea of organ transplantation), the problems of immunocompatibility frustrate such categories: bodies, though the same, can nevertheless identify each other as foreign and as other. The central claim of tissue engineering approaches is that they bypass tissue-matching problems and long organ-transplant waiting lists by beginning with the patient's own cells. No biological materials are therefore introduced into the body that do not originate from that body to begin with. Therefore, the return of one's own cells to one's own body is simply an extended process of supplementing the body's own self-healing mechanisms—it is "as if" the body has never left itself. Although tissue engineering takes each individual body as unique, it also implies that one technique will be valid across a population (that each instance of immunocompatibility is unique in the same way; that vascularization is a problem in the same way for all instances).

In tissue engineering, biomedical bodies are universal instances valid across a population in terms of the biological processes occurring within themselves. The biopolitics of tissue engineering works through a highly compartmentalized universality of closed, self-regenerating biological processes, which are nevertheless consistent across a population. The practical models of health care relating to this field—cell banks, off-the-shelf organs, donations of discarded embryos from fertility clinics—form bioeconomies of body parts that illustrate the apparatus by which tissue engineering is able to produce a vision of the regenerative body (a body always capable of regulating itself with the assistance of a range of biotechnologies).

However, the body as seen by tissue engineering is not just a unidirectional process of externalizing biochemical and physiological properties. Although Kroker's and McLuhan's externalizations have as their specific destination points either technologies or mediations to other subjects, tissue engineering's trajectory has as its destination the biomedical patient-body itself. The process can be technically configured as a loop, but functions bioethically in its effects more like a spiral. Cells are isolated, extracted, and multiplied in a lab culture, then generated in a bioreactor, then seeded into a polymer structure with other growth factors, where eventually the desired portion of tissue grows and then is implanted back onto the patient's target site, where the cells grow, differentiate, connect, and vascularize the structure and its immediate environment, while the polymer gradually degrades. Tissue engineering research claims that regenerated tissues and organs will be "just like new" or the same as before, but it is precisely this language of proximity that suggests that tissue engineering forms a kind of self-differentiating, self-exceeding spiral. The regenerated tissue or organ is recognized as self because the cells used to generate the tissue come from the patient's own body. Yet the very act of biologically producing and introducing a tissue mass or organ into the patient-body also suggests a certain qualitative difference in the very act of artificially (re)introducing biological materials into the body. Biologically speaking, how can the self be synthesized and introduced into the self? This reflexive relationship of the body to itself raises several open-ended questions: Does the involvement of a range of biotechnologies fundamentally change the ontological status of the particular body part regenerated? Is the regenerated organ or tissue mass exactly the same as the "original"? What are the phenomenological and psychological dimensions of this process of autoalterity? If the biomedical body of tissue engineering is dispersed throughout these techniques and technologies, how and where do we situate the body that is supposedly "proper" to the patient-subject?

Looking at tissue engineering through the lens of Foucault's biopolitics, we can see a move toward a medicine that is both biologically individualized (in terms of compatibility issues) and universalized (in the vision of "off-the-shelf" organs or "universal donor cells" described earlier).[45] The "material" of the body becomes at once intensely proper to one's self (only my flesh can be transplanted onto my body) and totally open to a generalized therapy (through tissue engineering, my body or anyone else's body can be grown on demand). Situated between modularity and custom design, tissue engineering is a dual statement about the biomedical subject: my body is specific to me, and my body is your body—and both can be generated in the lab for medical therapy.

"We Are Born Without Armor"

In analyzing tissue engineering through Canguilhem's and Foucault's lenses, I have highlighted the ways in which tissue engineering separates biological structure from function and in doing so creates the conditions for a simultaneous individualizing and universalizing of the body of the biomedical subject.[46] These conditions—a new paradigm of regenerative medicine that approaches the body through a biopolitical perspective—set the stage for a series of tensions, contradictions, and aporias in the kinds of bodies imagined by tissue engineering research. Although I use the terms *contradiction, tension, and aporia* interchangeably, *aporia* is central to my analysis in that it defines a situation of two equally valid but incompatible situations, which therefore necessitates a new formulation of the problem. In speaking of "the aporias of the regenerative body," we can begin to think about how the biomedical body is made to accommodate incompatible notions of "bodies" and "technologies." Each of these aporias is some variation of this boundary between bodies and technology, the natural and technical orders.

Ideal Biologies, Real Bodies

The first type of aporia is that the regenerative body finds itself caught between idealism and empiricism. The idealism is embedded in the very logic of tissue engineering: that the body's natural components can be effectively retrained or reprogrammed to operate in novel, "nonnatural" contexts. When combined with research into stem cells, where pluripotency means that stem cells can differentiate into a range of mature, adult cell types, the idealism in tissue engineering looks forward to a highly diversified biological resource—"universal cells." This view elicits an image of a body that is totally malleable and open to the interventions of tissue engineering techniques. Idealism is countered by an empiricism that places a great deal of emphasis on the body as a natural, organic resource—a resource that can be harnessed for medical ends. The point of tension is that one can grow any part of the body, but that one is also limited to the biological parameters of a particular body as source material.

As previously stated, tissue engineering is establishing—through its research, clinical trials, and product development—a unique biomedical norm, encapsulated by the phrase *regenerative medicine*. To posit a regenerative body is antithetical to postmodernist claims for the body's disappearance, as, for example, when Jean Baudrillard suggests that the body and sex have unmoored themselves as referents and are thus always preceded by the "hyper-reality" of the body as a "model," as a sign, or as a media image.[47] By contrast, tissue engineering is deeply committed to the materiality of the body and to the notion of the (biological) body itself as the very foundation of its practice. But this foundation is not fixed, for this commitment to the materiality of the biological body is coextensive with a commitment to a technology of the biological body, a biotechnology that always operates at a distance from the integrity of the natural-biological body itself.

Without an approach to the body that begins from the body's natural capacities for cellular differentiation and regeneration, tissue engineering is no different from other medical approaches (e.g., artificial limbs) that more explicitly rely on technological aids. Tissue engineering therefore assumes the body as preexisting in a natural state in order that its techniques of recontextualization will be understood as not essentially altering that pretechnical body.

Immediate Mediation

Tissue engineering's balancing act between idealism and empiricism leads to a second aporia, which is a kind of fantasy of transparent technology, of media that do not mediate, of "biomedia" that are fully present to themselves. This fantasy is to conceive of a technology that is indirect, invisible, and transparent to the natural and biological orders—a technology that simply helps things along, creating the right context or conditions for the facilitation of certain medical outcomes. In traditional uses of biotechnology in agriculture and animal breeding, the development of this technology meant first gaining a knowledge of natural processes and then utilizing them as processes geared toward slightly modified ends (greater crop diversity or homogenization of livestock). A similar logic follows in biomedicine. "Technology" in tissue engineering seems to disappear because no prosthetics, mechanical parts, foreign DNA, or even major surgical interventions are required. With the lack of a readily identifiable technological apparatus, it appears that the body produced in tissue engineering is a fully "natural" body—something akin to a biotech vitamin supplement. Technology is thus invisible yet immanent. This is an example of *biomedia* described in earlier chapters, where the biological body becomes its own technical resource, through practices of biomolecular recontextualization. Technology in tissue engineering is indirect, facilitative, and transparent—not a mechanical apparatus, but a series of treatments, actions of enframing, and shifts in contexts and environments.

The more indirect and invisible the technology becomes, the more the body projected by tissue engineering becomes an autonomous, enclosed, proliferative sort of biomodule. The more the discourse of the natural body is asserted within tissue engineering and medicine, the more this vision of a regenerative body is instituted as a normative constraint defining the normal, healthy, biomedical body. This version of the biomedical body is simultaneously natural (that is, according to researchers, essentially unmodified by the techniques utilized) and unable to do without a technology separate from it (that is, a body whose definition implies the necessity of a set of regulative biotechnologies). The result is that the body of medicine once again is objectified, but in a particularly unique way, such that the body can be seen, in the right conditions, as a self-regenerating, self-curing "black box."

Tissue engineering deeply invests the materiality of the body with a force that attempts to break out of the constraints of materiality and corporeality. For example, if we return to the opening description of the standard techniques in tissue engineering, we can see a set of practices that harbor within themselves some implication about the body surpassing itself. In the first step, the very idea of cell sourcing looks forward to the capacity to replicate an unlimited supply of raw material for tissue and organ regeneration (especially when stem cells become the cell source). In the second step, the integration of cells and materials (cells seeded in a biopolymer scaffold) evokes a process of spontaneous order in which, given the right combination of materials, the body simply self-assembles with minimal intervention. And in the third step, the cultivation of the apparatus and its surgical implantation reinforce the modular, highly objectified approach to the body seen in anatomical and medical science.[48] From this perspective, tissue engineering can be regarded as an instance in which the body gradually arises out of what Arthur Kroker calls a "torture chamber" of mortality[49] and rapidly heads toward what can be viewed as a body-without-death.

Transcend Aside

A third aporia to consider is that the body in tissue engineering is poised between medical healing and biotechnical redesign, marked by a lateral transcendence in which the body remains biological and material, but is also impelled technically to surpass itself. This body is simultaneously expendable (because, theoretically, tissues can always be grown back) and infinitely reproducible (as in the idea of "off-the-shelf organs"). Such duality puts forth a kind of revaluation or a reinvestiture of biological materiality, implying that the natural may persist through the artificial, that the organic may persist through the nonorganic, that the biological body may persist through technology. Tissue engineering uses its assumptions concerning the transparency of technology in effect to upgrade what biological materiality can become. Matter is not fated to the effects of microbes, injury, and time; rather, matter is amenable to techniques of supplementation, biomolecular modification, and biological design. Current research on stem cells, telomeres, and biomaterials is concerned principally with these questions of efficiency, sustainability, and effective design of therapeutic techniques. The body thus remains fully biological, but also enabled with attributes that surpass its prior functionality (e.g., assisted in vitro or even in vivo cellular regeneration). Tissue engineering thus conceives of a strange body that is constantly surpassing itself, a body-more-than-a-body.

This regenerative body is doubly marked by a movement of production and a movement of excess—the model here is not incorporation or recuperation, but production and (re)generation. To posit a regenerative body implies that we are presented with bodies that are, theoretically speaking, infinitely reproducible. Furthermore, this notion of a body that can be regenerated coincides with the anatomical logic of body parts that form assemblages (components of cells, tissues, organs). Thus, it is not the whole body that is regenerative, but rather body parts or those sections articulated through anatomical and physiological science as being open to laboratory isolation and experiment. But along with this infinitely reproducible body, regeneration also implies a region of excess of that which is displaced by regeneration. Thus, along with the infinitely reproducible body, there is also the expendable body, the "spare parts" that researchers speak of replacing ("I can always grow it back").

The significance of these aporias of the regenerative body is that they collectively illustrate the degree to which biotechnology fields such as tissue engineering are attempting to negotiate the radical changes they bring to conventional medical notions of what counts as "natural" and a "body." These versions of the regenerative body illuminate a deep-running anxiety over the maintenance, at all costs, of the divide between the natural and technical orders, while at the same time conceiving of nature as something that can be manipulated. That the aporias mentioned here—idealism and empiricism, transparent technology, the body-more-than-a-body—remain unresolved tensions suggests that biomedicine's philosophical perspectives on biological life need to be reconsidered in a new light. At the basis of these contradictions is the specter of technological dehumanization that has haunted industrial biotech since its beginnings in genetic engineering. It is a fear of any number of posthuman futures that mistakenly align technology with instrumentality. The very introduction of technology in proximity to the body—the material foundation of what it means to be "human"—brings with it the anxieties of technical instrumentality and the fears of the loss of "the human."

Tissue engineering is exemplary in this instance because it obsessively grows the biological body while it masks the fact that its techniques fundamentally alter traditional biomedical and biological views of the body. What we can distill from these aporias of the regenerative body is *a will to preserve the biology/technology divide, at the same time that this division is constantly transgressed*. Implied in tissue engineering's practices are several notions: first, that the divide between body and technology is absolute; second, that a body without technology is more healthy, more natural; and third, that the body-more-than-a-body can be achieved using techniques that simply recontextualize preexisting biological processes. In order to achieve this body-more-than-a-body, tissue engineering must mix up all sorts of categories. In Bruno Latour's terms, tissue engineering must first undergo processes of "hybridization" in order to arrive at a process of "purification," in which the biomedical body may be transformed into a experimentally viable response to the medical problems of organ and tissue failure.[50] It is to a consideration of the ways in which tissue engineering establishes norms and facilitates hybrids that I now turn.

Tissues Are Actants

Thus far, my analysis of tissue engineering has tried to show how it produces a range of characteristics of the ways in which the biomedical body is being transformed. Continuing this line of inquiry, I now turn to Bruno Latour's mapping of the modern scientific episteme as a means of understanding tissue engineering's doubled division and transgression of the natural/technical boundary.

Latour suggests that the modern perspective is founded on a two-part, asymmetrical epistemological process. Visible and self-evident, there is the process he calls "purification," whereby elements (in our case, stem cells, biotech companies, public policies on cloning, DNA sequencing machines, media reportage, individual medical patients, etc.) are organized and arranged into a set of dichotomies, each distinct—human/machine, nature/technology, science/politics, self/other, and so forth. Concurrent with this process of purification, but invisible or interstitial, is the process Latour refers to as "hybridization." Whereas purification operates through qualitative differentiations, hybridization operates by establishing connections and networks among elements; indeed, it can be said that hybridization is that act of connecting.[51]

These processes—purification and hybridization—and the elements they produce—subjects/objects and hybrids—together form the modern worldview. That is, they are interdependent, and they form a complex interplay of mutually constituting elements that explicitly (purification) and implicitly (hybridization) form the contours of the modern episteme. Latour's overarching suggestion is that the unrepresentable, unthinkable quality of hybridization has begun to saturate the visible zone of purification, resulting in both a symptomatic relativist pluralism and a need to reconfigure critically the modern worldview into a "nonmodern" one that places equal emphasis on the processes of hybridization.

Hybrids, as Latour describes them, are a number of things; they are, above all, heterogeneous collectivities interstitially constituted by mediation; the process of mediation or hybridization is not transparent, but transformative. If hybridization and hybrids are considered as transformative, productive, and affective forces, we are then invited to think about the kinds of networks that a field and practice such as tissue engineering weaves. As Latour states, "they [hybrids] become mediators—that is, actors endowed with the capacity to translate what they transport, to redefine it, redeploy it, and also to betray it."[52]

On the one hand, tissue engineering appears to be a very literal example of the constructionist argument (the tissues of the biological body are actually synthesized or produced in the context of the laboratory); on the other hand, the "product" of tissue engineering—the tissues of the human body itself—are, of course, always presented and represented as fully organic, natural, and biologically pure biomaterials. That is, although on one level tissue engineering appears to be perhaps the quintessential fusion of biology and technology, on another level—that of its rhetoric in textbooks and journals, that of the way it enframes its practices and research—it rarely if ever acknowledges its own intimate linkages between the biological and the technological.

For Latour, one of the primary characteristics of hybrids is that the interstitiality they occupy resists a complete reduction to the poles of subject or object; they form what Latour calls "actants."[53] For instance, tissue engineering is a technology that is certainly the result of active, intentional human intervention, which, from one perspective, is to be regarded as merely a tool used by researchers or medical doctors. Tissue engineering is an object, then, in the sense that it is the means—that is, the technology—by which a particular biomedical field may regulate, produce, regenerate, and render docile a biopolitical body of biotechnology.

As a life science, with a certain claim to the "truth" of the human biological body, however, tissue engineering and its technologies (cellular analysis technologies, advanced microscopy, cellular cloning, biomaterials synthesizing, bioreactor technologies, surgical tissue grafting) are also very active participants in the process of production or (re)generation of the body and the formation of new biomedical approaches and models for generating body knowledges. That is, on the part of the hybrid technologies of tissue engineering there is an active intervention that is not the agential intentionality of the subjects who are assumed unproblematically to "use" the technology. Put simply, tissue engineering is impossible without technologies such as cellular cloning, and such technologies fundamentally and actively constitute the ways in which scientific, medical questions may be asked about the body, from the level of technique and technology.

Tissue engineering forms a unique configuration between the body (specifically, the biomedical body) and technology (specifically, biotechnology and molecular genetics) such that the technological is fundamental, implicit, and constitutive of the biological. Such a relationship depends, nevertheless, on an absolute, original division between biology and technology. This ontological division happens in spite of the fact that the techniques of tissue engineering consistently transgress that division, forming hybrids of all kinds. On the one hand, the practices of tissue engineering attempt to harness the naturally occurring processes of cellular regeneration, in which skin cells, for example, are placed in the right environment for growth to occur. On the other hand, tissue engineering is filled with an array of techniques and technologies combined from different specializations such as genomics, biomaterials, cloning, and stem cell engineering—a complex apparatus for enframing the body. Technology is, in tissue engineering, not opposed to the body (that is, technology does not consume or totally incorporate the body, as it does in many science fiction narratives), nor is technology external to the body (that is, it does not exist totally independent from the body). *The "technology" of tissue engineering is instead internal to the very biological processes of the body, but it does not simply supplant or displace those processes.*

If tissue engineering begins from an assumption concerning the absolute division between body and technology and then proceeds through its practices to construct a notion of a natural, biological body defined by the body's regenerative capacity, then it is centrally concerned with a twofold movement: simultaneously to preserve the referent of the natural, prediscursive body-in-itself and to redefine the qualities of that body through the use of biotechnologies that minimally (that is, transparently) intervene in this process of redefinition. This natural-biological body is made possible only through a particularized, indirect, and facilitative use of biotechnologies. These technological practices and techniques are the nonvisible interstices of the body in tissue engineering. They can be regarded as the media of transparency that form the network of connective tissue that constitutes this vision of the regenerative body.

Connective Tissues

As we have seen, tissue engineering relies on an ontological split between biology and technology, the natural and the technical. Its overall theoretical approach is one in which a more "natural" body is treated through new therapies that harness the body's own resources. Yet in its practice tissue engineering constantly transgresses this split, combining adult or embryonic stem cells with a biopolymer scaffolding set in an engineered growth medium to await transplantation back into the patient's own body.

In utilizing biotechnologies in ways they are not manifest—and thus appear only at the interstices—tissue engineering seems to be gearing itself toward a standard of the biomedical body that strategically eliminates one entire sector of the biological body's contingencies (chromosome degradation, tissue aging and decay, and the markers of the body's mortality).[54] Again, although this vision may include the best intentions, what needs to be asked is whether it is more valuable for what it excludes than for what it promises.

This chapter has argued for a medical-philosophical reconsideration of the possible transformations brought about in fields such as tissue engineering to our broader notions of what a body is, within a framework that is at once scientific, technological, and social. Instead of assuming the biology/technology divide (which, as seen earlier, only engenders aporias), tissue engineering may consider the ways in which its techniques make meaningful statements about how human bodies may be conceived within medical practice and health care. It would be a mistake to assume that a natural body is innocently treated by technologies, just as it would be misleading to try to conceive of a more natural, more "healthy" body produced through advanced biotechnologies. Instead, tissue engineering can consider the specific contexts it produces (e.g., tissue-engineered, lab-grown organs for single-patient retransplantation owing to organ failure of the kidney) as "tissue networks" that encompass a range of heterogeneous factors: the body of the individual patient, enframed by biomedicine; the tissue engineering apparatuses (tools and techniques for biopsies, therapeutic cloning, PCR, bioreactors, biopolymer scaffold, growth media, genomics stem cell resources, surgical techniques); institutional context (research, clinical trial, medical practice, product development); disciplinary relationships (with surgery, genomics, cloning, stem cells); biomedical challenges (compatibility, sustainability in lab-grown organs, in vivo regeneration); and bioethical challenges (policy decisions, products and services, stem cell resources, cloning).

By a thorough situating of the instances of biotechnical regeneration, tissue engineering can take into account the polyvalent ways in which bodies are reconfigured, renegotiated, and radically recontextualized. Challenges that appear to be technical (e.g., how to enable the regeneration of a functioning vascular system in a tissue-engineered liver) are also philosophical, social, and ethical challenges (e.g., if such a process can be triggered in vivo, has medicine switched from healing to "upgrading"?). Without ever losing sight of the material practices that constitute a particular biotechnology such as tissue engineering, we equally need to consider the ways in which traditional relationships between bodies and technology, natural and technical orders, may be critically rethought, with an eye toward the medical-cultural effects of such "technical issues."

Tissue engineering presents a version of the body in which all capacities for regeneration, replacement, and supplement come from within the same body of the subject itself through a range of techniques, technologies, and engineered biomolecules. What the biotech industry values in the body projected by tissue engineering is the value of the biotech body to reproduce its own materiality internal to its purportedly original, natural conditions. Again, tissue engineering, in implying the potential physical and biological manipulability of the human body, is not suggesting that the body is somehow "less real." But without a consideration of the dynamics of hybridization whereby scientific practices such as tissue engineering are always already situated by discourses, knowledge systems, and technologies, it becomes all too easy to desire habitually, at any cost, the vision of a body transcending itself, sublimating itself, curing itself, and yet still "a body."

8

Regenerative Medicine: We Can Grow It for You Wholesale

Biologies Without Technologies

Our skin is in constant contact with our surroundings: our pores constrict and loosen in response to temperature, dead skin cells fall off our bodies like ashes, and, when cut, punctured, or scraped, our skin will repair itself, sometimes regenerating new skin cells, at other times forming scar tissue in response to more serious injuries.

According to modern cell biology, the human body is composed of some 100 trillion cells, with more than 200 different kinds of cells in the body, ranging from bone to muscle to skin to blood cells.[1] Each type of cell, though they all contain the same genetic information, performs a specific set of functions in the body, and each type responds differently to the onset of disease, injury, or the regular wear and tear the body undergoes. In short, the body contains within itself an incredible set of biological self-healing and regulatory systems that are at all times in communication with our environment. Now imagine that this potential might be harnessed, even "reprogrammed," so that cells, tissues, and even entire organs might be "grown" on demand, in vitro or in vivo, and all from the patient's own cells.

These are some of the questions raised by the field of *regenerative medicine*. The idea of regenerative medicine is simple yet powerful: study the body's own natural mechanisms of healing, and then develop a set of techniques and technologies for harnessing those mechanisms. The term "regenerative medicine" has been used frequently in research publications, journals, magazines, and even textbooks on cell biology and tissue engineering. For instance, William Haseltine, CEO of Human Genome Sciences, has repeatedly defined regenerative medicine as the ability to harness the naturally occurring regenerative properties of human biology. As he states, "We are a self-assembling organism. . . . That information is there to be captured and used. . . . It's a fundamental principle of regenerative medicine that we only have to trigger the body to do what it needs to do."[2] This view has been echoed in the academic sector by researchers in tissue engineering. For instance, tissue engineering pioneers Lawrence Bonassar and Joseph Vacanti have defined their field as focusing on "the concept that the repair and regeneration of biological tissues can be guided through application and control of cells, materials, and chemoactive proteins."[3]

Medically speaking, regenerative medicine arises out of a range of specific concerns, including histocompatability in organ transplantation, treatment of degenerative disorders such as Parkinson's disease, and even the possibility of moderately extending the natural human lifespan. Research into embryonic and adult stem cells as well as research into the genetic mechanisms of cellular aging have opened a whole new field of potential medical intervention. Leading tissue engineers David Mooney and Antonios Mikos have described the promises of regenerative medicine in the following way: "Ten millennia ago the development of agriculture freed humanity from a reliance on whatever sustenance nature was kind enough to provide. The development of tissue engineering should provide an analogous freedom from the limitations of the human body."[4] Such visions have also prompted some to voice questions concerning the ethical implications of research on embryonic stem cells or the long-term impacts of extended life span. The late 1990s saw the emergence of a series of national and international debates over the regulation of research using stem cells derived from human embryos. At the same time, a number of biotech companies specializing in regenerative medicine, such as Geron Corporation and Advanced Cell Technologies, have either advanced products into clinical trials or, in some cases, gained U.S. FDA approval for selected products and treatments.[5]

Thus, although there is definitely a need for effective therapies for debilitating diseases such as Parkinson's, the emerging field of regenerative medicine also brings with it a host of questions. How would such a control over biological processes change our sense of what medicine does and what "the body" is? What would be the scientific and the medical implications of such a capacity? What would be the social and cultural implications?

Perhaps we can look to science fiction for a different perspective on such questions. Science fiction is worth considering because it is one area in which science, society, and speculation come together. The main purpose of much science fiction is, of course, entertainment (and merchandising), but it is exactly this "pop" mentality that makes it interesting. For instance, a number of science fiction films in the late twentieth and early twenty-first centuries have taken up controversial topics in biotechnology, from genetic engineering (*The Fly*; *Mimic*; *Spider-Man*) to human cloning (*The Sixth Day*; *Star Wars Episode II*), to biological warfare (*Resident Evil*; *28 Days Later*), to genetic screening (*Gattaca*), to regenerative medicine (*X-Men*; *The Hulk*). It is this last example—that of regenerative medicine research involving stem cells—that is the point of focus in this chapter. The science fiction versions of scientific and social issues are, without a doubt, exaggerated and fantastical. Embedded in their exaggeration and hyperbole, however, we can also see the hopes and fears of emerging biotechnology fields. The promise of regenerative medicine generally—indeed, the very concept of "stem cells"—has already raised a host of ethical, political, and philosophical questions in policy considerations by the President's Council on Bioethics. Many of these same concerns are dramatized—and allegorized—in science fiction.

For example, the recent science fiction films of the *X-Men* series provide us with two biological "icons" for the field of regenerative medicine. There is, on the one hand, the character Wolverine, who, aside from having surgically grafted, extractable metal blades, also has the ability to recover rapidly from any injury through tissue regeneration. As the *X-Men* mythos goes, Wolverine is thought to be the product of military experimentation on humans, using biotechnologies to craft the ultimate fighting machine, with an anatomical structure fortified by a near-indestructible metal and a biology designed to regenerate efficiently and repair itself in response to injury. Wolverine bears the markers of both industrial technology (his cyborglike metallic claws) and contemporary biotechnology (the flesh wounds that we see healing right before our eyes).

In contrast to Wolverine's militaristic robustness, there is the character Mystique, a kind of mutant femme fatale who is defined by a single ability: molecular biomorphing. Though the computer graphics artists working on the *X-Men* film did, of course, use graphical morphing software for imaging Mystique, within the narrative world of the film Mystique actually morphs her body at the cellular level. We usually think of morphing as two-dimensional, image-based process in which an animation is created by extrapolating between two different images. Mystique does not use morphing as an illusion, but rather, we might guess, she actually controls her body on the cellular and molecular level. This is the capacity to shape-shift on a biological, three-dimensional level, something that speaks to her gender assignation (the stereotype of fluid female identity) and to her personality traits (mute, impersonal, cold, without any personal history).

Between these two characters—Wolverine's fortified, regenerative body and Mystique's flexible, biomorphic body—are two slightly different cultural expressions of the scientific field known as regenerative medicine. Put simply, regenerative medicine is a field within biotechnology that harnesses the power of the body's own healing and regenerative properties toward medical application.[6] The basic idea of regenerative medicine is to take a biological process such as cellular regeneration and technically redirect that process toward a more "natural" approach to handling tissue and organ damage as well as transplantation. As discussed further on, regenerative medicine not only is a field of science research, but also presents us with radically new views of what "health" is and how the body may be defined and redefined. The primary claim I would like to make is that practices in regenerative medicine demonstrate a complex regulation of the division between biology and technology, and in doing so it demonstrates the anxieties surrounding the contingencies of the biological body.

As a methodological note, it is worth reiterating that leading-edge research fields in biotechnology such as regenerative medicine are not only scientific and technical practices, but also social and cultural practices. The kinds of meanings they produce through their research also have implications for thinking about the future of medicine and biology, implications that extend beyond the domain of medical research, from policy decisions to technology development to popular culture. This also means that the ways in which fields such as regenerative medicine view the body are never simple. The body is complexly sculpted in manifold ways by science, technology, institutions, economics, bioethics, education, and media. It follows that in attempting to come to an understanding of fields such as regenerative medicine, we need an approach that is neither an "anti-biotech" position (which might claim a "return" to the natural body) nor a kind of "technophilic" position (in which science and technology are the answers to social concerns). In between these positions we need to locate a site of negotiation, to understand the potentials that inhere in regenerative medicine, but to do so through a critical perspective.

Ready to Wear

As might be guessed, the incentive for regenerative medicine is both medical and economic. According to a recent survey, more than $400 billion per year is spent in the United States alone on patients suffering from organ failure or tissue loss, amounting to some 8 million surgical procedures involving organ or tissue transplantation.[7] But the real problems are the shortage of organ donors for such procedures. It is estimated that more than 100,000 people die in the United States from organ or tissue failures, many while they are on an organ-donor waiting list. Regenerative medicine is claiming to provide a solution to such medical challenges. William Haseltine, CEO of Human Genome Sciences, has heavily promoted the use of the term, and the field itself combines the practices of gene therapy, tissue engineering, stem cell research, biomaterials, and bioengineering. Researchers in different labs have already shown that many body parts—skin, cartilage, bone, kidneys, heart valves, and bladders—can be grown in the lab, prompting some of them to envision a future of "off-the-shelf organs."[8] In 1998, the U.S. FDA granted approval for a tissue engineered, bioartificial skin product called Apligraf, made by Organogenesis.[9] Around the same time, Joseph Vacanti's lab at the University of Massachusetts shocked the public by growing a tissue engineered human ear on the back of a mouse.[10]

A good way to talk about the techniques of regenerative medicine is to differentiate it from current biomedical approaches that similarly address the body's incapacitation through organ and tissue damage. Within medical practice, there are currently two basic approaches to organ and tissue failure.[11] The first involves the use of hardware to take the place of the malfunctioning organ. Bionic devices, artificial organs, and prosthetics are examples. The paradigm here is that of displacement. The mechanical organ—be it a pacemaker or a dialysis machine—functions in place of the biological organ, but because such hardware is often not physically amenable to the body's daily functioning, it also often requires a radical change in the patient's physiology and lifestyle.

The drawback of the medical hardware approach is that by imperfectly miming biology, such devices can require extensive maintenance, even when surgically implanted. Thus, their longevity and sustainability are put into question because they function mostly as short-term solutions.

A second biomedical approach involves organ and tissue transplantation. It can be from one human donor to another; it can be donated biomaterials from tissue banks; or it can be an animal-to-human organ transplant (xenotransplantation). Transplantation requires a view of the body as modular—an automobile with equally interchangeable parts. This paradigm of replacement takes one kidney to be like another, simply necessitating an exchange. Unfortunately, two major problems have recurred with transplantation since its beginnings: organ donor shortages and immunocompatibility. The first is simply a case where the demand far exceeds the supply, and even when organ donors are available, the problem of rejection by the immune system can often negate all possibilities except a donor who is a blood relative.

The problems inherent in both of these approaches have to do with the body's unwillingness to incorporate these foreign objects—medical hardware and foreign tissue. On both the technological level and the biological level, the body thus responds through a breakdown, a material and functional incommensurability. The alternative offered by regenerative medicine is to take the body itself as its own resource. This is not so much a technical paradigm of displacement or replacement as it is one of regeneration. Regeneration is not the search for an adequate substitute for the body, but rather the use of the very processes of development and maintenance to let an organ, tissue, or cell exist again—a kind of biological reiteration within the body.

Regenerative medicine researchers use several techniques and one primary biological resource.[12] That resource is the stem cell, discovered in humans in 1998 by Geron Corporation (human embryonic stem cells) and named the "discovery of the year" by *Science* magazine in 1999.[13] Briefly, stem cells are those undifferentiated cells that contain the capacity for specialization within the body. Researchers have been concentrating on human embryonic stem cells, which are often acquired from infertility clinics, as well as on human adult stem cells, which have recently been isolated in the brain.[14] At the very first stages of embryonic development, just after fertilization, the embryo in its single-cell state can be described as *totipotent*; that is, it contains the potential to turn into any type of cell (its potential is total). Just after this, the embryo undergoes a series of cellular divisions, and as it does this, the cells gradually begin to specialize (cellular differentiation). The cells at this early stage are now further limited in their range of differentiation, and are known as *pluripotent*. Each cell is eventually taken down one line of development, specializing itself through a tight correspondence of form and function (e.g., muscle cells tend to be long, spindlelike, and flexible, whereas epithelial skin cells are tightly packed in layers). Though the exact genetic and molecular mechanisms that control this process are largely unknown, researchers have found that, when inserted into different cultures with growth medium, stem cells will spontaneously begin differentiating according to their new cellular milieu (figure 8.1). Furthermore, the controversies surrounding the use of stem cells from discarded embryos has recently been sidestepped by research into adult stem cells and engineered stem cells (or "embryoid derived cells"), both of which make use of the stem cell's lessened but still promising "multipotent" capacity.[15]

Current regenerative medicine research is exploring two general approaches.[16] The first borrows techniques from gene therapy and involves the site-specific transplantation of prepared stem cells into the region in which regeneration is to take place. As with gene therapy, the challenge is effective delivery, and researchers often use a viral vector. The idea here is to prompt the cells to replicate, either through the isolation of the genes or genetic trig-

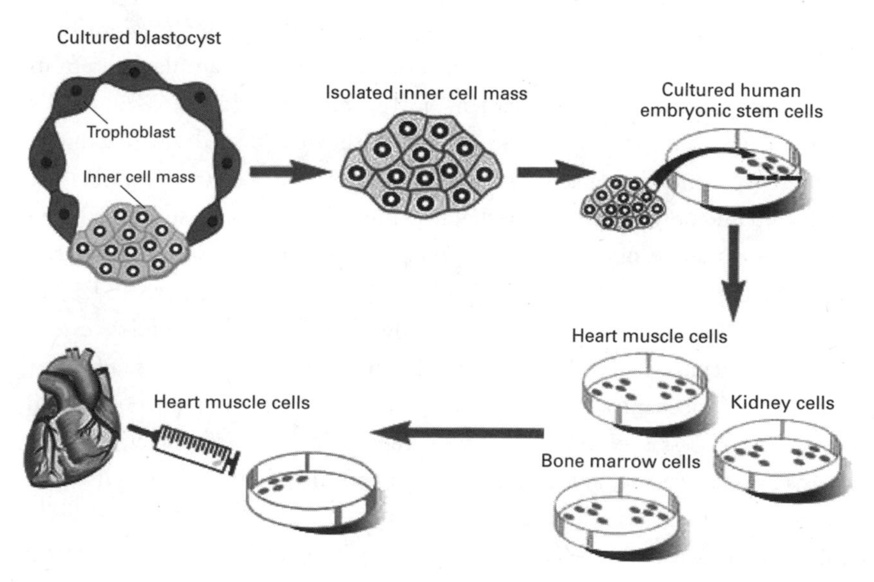

Figure 8.1 Culturing stem cells. Image courtesy of the U.S. NIH.

gers in adult somatic cells, or through the culturing of embryonic stem cells that would then be injected into the patient's body. This approach is both the most promising and the most difficult. Finding the genetic triggers that take a cell down one line of specialization or another is an extremely ambitious task, given that in all probability multiple genes and signal pathways are involved. However, researchers do know that certain stem cells can be directed along certain developmental paths: adult hematopoietic stem cells can develop into blood and neuronal cells; adult stem cells from the brain can develop into neurons, bone marrow, and blood cells; and adult mesenchymal stem cells can develop into bone, muscle, cartilage, and tendon cells.[17]

The second approach in regenerative medicine has focused less on stem cells and more on the ways in which organs, tissues, and limbs form during embryonic development—the field known as tissue engineering, discussed in the previous chapter.[18] Tissue engineering proposes to make use of the body's own cells for the regeneration of tissues and entire organs, with the assistance of biochemical growth factors, bioreactor technology, and biomaterials science. In animal experiments and human clinical trials, tissue engineering researchers have demonstrated that it is possible to grow and reimplant entire organs into the organism. To date, bone and cartilage structures, skin, the liver, pancreas, kidney, bladder, and neural cells have been grown using this approach.

The most common procedure thus far—pioneered by Joseph Vacanti and Robert Langer—involves the treatment of a biopsy or cell sample taken from the patient. Let us say it is a patient with a diseased kidney.[19] Those cells are then replicated in culture and infused with growth medium to prompt cellular division. The cells are then seeded into a lattice structure made of biodegradable polymers to guide the growing cells and make sure they grow in the proper shape of a kidney. The whole complex is then surgically placed back into the patient, where the patient's body takes up the new cell and organ structure, allowing vascularization (the growth of a network of blood vessels) to occur. In contrast to other biomedical approaches—be it the use of drugs or the use of mechanical devices—regenerative medicine uses the body's own biological processes of cellular growth, replication, and regeneration. This approach is, of course, aided by an array of techniques, including cell delivery techniques, isolation of cell lines of stem cells, development of biochemical growth factors for cells, and the construction of biodegradable lattice structures for the seeding of cells.

Thus, in response to modern medicine's "technological" approach to the body (either through medical hardware or the mechanistic view of transplantation), regenerative medicine attempts to facilitate a view of the biomedical body that is essentially self-healing.[20] On the one hand, it would seem that this more "natural" approach to medical therapeutics requires only a minimum of technical enframing, simply the right type of environment in which the self-healing body may emerge; the question here is less technology, but more context. On the other hand, we might also question this surreptitious disappearance of technology, in which it appears that cells and tissues will automatically regenerate themselves in a fashion analogous to the repair of minor skin cuts, for example. Where is the technology in regenerative medicine? Is it to be found in various laboratory tools? Is it to be located not in "gadgets" or hardware, but in techniques, processes, or actions? It is this transformation of what we mean by technology with which regenerative medicine is, on one level, engaged. The issue is important because some approach to the boundary (the separation?) between bodies and technologies is imperative for medical practice to differentiate between patient status and medical technique, between an autonomous biological body, and a technical apparatus for analyzing that body.

Is the Body a Pathogen?

A key to understanding the perspective of regenerative medicine is to consider how "biology" or "body" is defined through its research. For instance, a stem cell "discovered" in the body might be taken as naturally occurring, whereas stem cells derived through therapeutic cloning might be taken as technological or artificial. Although never totally questioning the boundary between bodies and technologies, regenerative medicine does create a zone of ambiguity by defining the biological body as both fully natural (preexisting medicine and science, operating independently, self-regulating, and self-producing) and fully amenable to biomedical augmentation, enhancement, improvement, and design. As an organic system that is both natural and open to technical improvement, the body as described by regenerative medicine invites a certain type of intervention in the body's "natural" operations—interventions that supposedly do not alter its ontological status as natural.

To put it another way, the body in regenerative medicine is both the resource and object of technical action. In a kind of feedback loop, regenerative medicine takes "technology" to reside in the biological itself; the "tools" that regenerative medicine uses are not external prosthetic devices or mechanical organs, but rather the very processes of cellular regeneration and differentiation themselves. By focusing on certain cellular processes that occur to a limited degree in the body, regenerative medicine creates novel conditions in which these natural, biological processes are foregrounded. Such an assumption at once takes the body as something that is lived (individual subjects are always embodied and specific) and as an object that is composed of a universal set of parts ("off-the-shelf organs"). This body is still totally "biological," fully organic, "vital" through and through; but in its new, technically optimized context, it also acquires a new range of capacities. This is not (only) a body that has been medically healed, but a body that has been upgraded using its "original" design principles.

A great deal has been added to this body, but a great deal has also been glossed over in favor of the more ambitious promises of regenerative medicine.[21] The body in regenerative medicine is in many ways a highly insular body, both in its technical configurations and in its ideology. Technically, this body is a dream of the autonomous, totally self-sufficient biological system, which, by using its own regenerative properties, is able to regulate itself regardless of any environment—a biological monad, the body existing in a kind of vacuum. Ideologically, this notion is supported by the far-reaching desires embedded in regenerative medicine research. From the claims for a future of increased longevity (by Geron Corporation) to the promises of a niche-marketed medical future (by Human Genome Sciences), regenerative medicine is largely concerned with ensuring the body's protection from the dangers of certain contingencies that threaten its well-being. "Bodily contingency" here refers to a set of complex factors that both negatively and positively define what it means for us to exist as bodies: environmental specificities, disease, injury, aging, lifestyle, and mortality itself. Those contingencies, which often enframe the body as a set of limitations, are ironically the very things that define the biological body as a body. Put simply, the body can be thought of as a kind of membrane, constituted by sets of constraints, limitations, and specificities. But these contingencies, these elements that define the body by constraining it, also define its flexibility and mobility—anatomically, physiologically, immunologically, neurologically, and so forth. Bodily contingencies are difficult to deal with for obvious reasons, but at the same time it is equally difficult to imagine where the line between the eradication of disease and the engineering of cellular immortality should be drawn.

Regenerative medicine may thus be seen to remove two modes of contingency from the body. One might be referred to as "external," which involves a deemphasis on the roles that environment and context play in forming dynamic relationships with bodies. For regenerative medicine, as for modern molecular genetics, the body is first and foremost the product of the causal dynamics of its genome, genes, or DNA. Despite shifts away from reductionist approaches in molecular biology, this "central dogma" still holds, as is clear from the investment (both figurative and economic) by biotech companies in gene targets, gene therapeutics, and drug development.[22] Curiously, regenerative medicine has chosen not to study the relationships between body and context, but to focus solely on the body's capacity to "regrow" itself on the cellular level to operate in a preventive, regenerative way (capable of responding to any external influence, be it disease or injury). The dangers in the general deemphasis on environment and context is that the body is viewed in a cellular vacuum, which takes all external context as homogeneous and which, as a result, offers one universal response (the genetic triggering of cellular specialization and replication).

In a related vein, regenerative medicine also seeks to remove an "internal" contingency. Regenerative medicine not only seeks to insulate the body from any environmental influence, but also seeks to extricate the contingencies that mark the biological body as internally unstable, dynamic, and marked by temporality. Geron Corporation's focus on biological aging, Curis's focus on medical therapeutics, and tissue engineering's focus on renewable organs point to a more general desire effectively to engineer biological mortality out of the body.[23] In many ways, an immortal biology is somewhat of an oxymoron, just as a vision of our bodies in bubbles denies the very thing that makes bodies what they are: a constant negotiation between an interior and an exterior. Although no one will counter the efforts to understand and treat diseases such as Parkinson's or Alzheimer's, regenerative medicine treads a difficult bioethical line in setting its sights on the biological basis of aging, for instance. That line becomes more blurred as research shifts from working through the body (where a certain concept of medical therapy and healing is still at work) to a kind of lateral transcendence (in which biotechnology is used to escape from the body). In this sense, to posit a body "cured" of all its contingencies is something like wanting to privilege the senses through massive sensory isolation or sensory deprivation.

What these techniques of regenerative medicine seem to demonstrate is that biological processes such as cellular regeneration can occur independent of the context within which those processes take place. The techniques of regenerative medicine point to an abstraction of biological processes—be it in the lab, in the patient's own body, or through gene therapies. This would seem to imply that the body, although remaining fully biological, can be redesigned so that it becomes more efficient and technically optimized. But how might these bodies in regenerative medicine be different as living, acting bodies?

The Concept of Regeneration

The very idea of regeneration brings to mind science fiction scenarios of self-repairing wounds and even immortality. Indeed, the pop cultural transmutation of regeneration may stem in part from the biological sciences. In biology, regeneration has two primary meanings: the replacement of lost tissues and organs lost in injury or the healing of wounds. The former is demonstrated by many plants (the ability to generate a plant from a shoot or leaf) and certain crustaceans (crabs) and eichnoderms (starfish). Some reptiles and amphibians have the ability to regenerate a lost limb—say, a tail—if it has been trapped by a predator (a process known as autotomy), and the planaria flatworm is a favorite subject of high school biology classes for its ability to regenerate when cut and even for its ability to grow two heads when split.

However, in most animals, certainly humans, regeneration is limited to the second definition—the healing of wounds. The distinction is noteworthy, for, although a wound may heal, its healing is not the replacement of lost tissues, an organ, or limb. Nevertheless, the biological healing process has been a source of fascination for biologists for some time. We may not have the capacity to regenerate entire organs or limbs, but at the cellular level our bodies do have the capacity, in most cases, to heal with a great deal of efficiency. A simple cut on a finger initiates a set of responses that include clotting of the blood, deploying the proteins for the clotting of the blood, and initiating rapid cellular replication to cover the wounded area, resulting in tissue that may not be as flexible (because it is scarred), but still functions as normal tissue.

In a specifically biological sense, however, *regeneration* is also different from *development*, the latter a primary interest in morphogenesis, or the study of biological form. In biological development, cell division and differentiation play key roles in the growth of an organism. In vertebrate animals, this growth is accomplished by three basic stages: a cleavage stage, in which the zygote begins dividing, resulting in a ball of cells known as the blastula; a gastrulation stage, in which the dividing cells become arranged in three "germ layers," or a gastrula; and an organogenesis stage, in which further differentiation in cells eventually results in the formation of a particular type of tissue (skin, muscle), organs, or limbs.

Even in morphogenesis, the line between development and regeneration is not decisive. Although development happens "once" in the life of the organism, the processes of cellular differentiation can occur more than once in the life span of an organism. The recent discovery of adult stem cells has led to further inquiry into the ways that development is, in a sense, a lifelong process. The main question is whether or not there is something in the regenerative properties of living cells that also is present in the development of the organism itself. This is, arguably, the wager of stem cell research: that there exists in stem cells a *biological potentiality* of pluripotency or even totipotency in cellular differentiation.

This basic process of regeneration—whether in terms of healing wounds or regenerating whole organs or limbs—serves as the source of inspiration for biotechnology research in regenerative medicine. Indeed, at some basic level, cells in our bodies are dying and regenerating each day, and it may be said that regeneration is in some sense isomorphic with the very cellular processes of "life itself." Aristotle had already noted this in his distinction between "coming-to-be" and mere "alteration" in animals.

Aristotle's treatise *On Generation and Corruption* begins by asking a question: Is "coming-to-be" different from mere "alteration"? That is, is the emergence of a body, of some thing, a fundamentally different process from a change in that body or that thing? In the context of regenerative medicine, Aristotle's question can be rephrased as follows: Is cellular differentiation from stem cells a "coming-to-be" of fully differentiated, functional cells, or is it a mere alteration within a cell that remains, in some basic way, the same?

For Aristotle, both monists such as Empedocles and atomists such as Democritus, begin from the assumption that the change of any thing, living or nonliving, is dictated by its capacity to be divided. Whereas Empedocles argues that there is a basic, singular magnitude ("the One") from which different things emerge, the atomists see everything as infinitely divided. However, this premise concerning divisibility leads to a number of paradoxes: a singular magnitude is actually constructed by the particular things that emerge from it (because it already exists in them in potential), and, likewise, the atomistic body is built on nothing because it is divisible through and through.

The assumptions that both monists and atomists have concerning change in general are that coming-to-be is defined by "association and dissociation," whereas the change in what is continuous is defined by alteration. In other words, coming-to-be is the assembly or disassembly of parts, whereas alteration is a mere change or shifting of the same parts.

Aristotle's response is to suggest that coming-to-be is in fact distinct from alteration, but not based on the divisibility of bodies. Coming-to-be is defined by a change in the *relationship* between matter and form, but alteration is a change in *either* matter or form, but not in their relation:

For unqualified coming-to-be and passing-away are not effected by "association" and "dissociation." They take place when a thing changes, from this to that, as a whole. But the philosophers we are criticizing suppose that all such change is "alteration": whereas in fact there is a difference. For in that which underlies the change there is a factor corresponding to the definition (i.e. a "formal" factor) and there is a material factor. When, then, the change is in these constitutive factors, there will be coming-to-be or passing-away: but when it is in the thing's qualities, i.e. a change of the thing *per accidens*, there will be "alteration."[24]

Coming-to-be is predicated, therefore, not only on some sort of prime matter, but on a dynamic process of potentiality and actuality: "In one sense things come-to-be out of that which has no 'being' without qualification: yet in another sense they come-to-be always out of 'what is.' For coming-to-be necessarily implies the pre-existence of something which *potentially* 'is,' but *actually* 'is not'; and this something is spoken of both as 'being' and as 'not-being.'"[25]

Aristotle's distinction between coming-to-be and alteration leads us to ask if the same distinction holds for stem cells. Are stem cells a form of *coming-to-be* or of *alteration* in Aristotle's sense of the terms? On the one hand, cellular differentiation is a coming-to-be of a fully differentiated, functional cell where before there was not one, but instead a stem cell, something like "the One" of Empedocles or a prime matter. On the other hand, cellular differentiation is not a genesis, but rather a change within the cell, triggered by gene expression and the environment of the cell. The same cell exists, but simply changes. The stem cell and the fully differentiated cell are thus isomorphic. The question here is whether this process is reversible, whether a fully differentiated cell can be transformed back into a stem cell. According to Aristotle, this transformation would make the process not a coming-to-be but rather an alteration.

This interest in the naturally occurring properties of regeneration in the body has led to the questions of how regeneration occurs, and what are its mechanisms and its logic. For Aristotle, regeneration broadly speaking required some foundation. Whether it is a singular magnitude, a resource from which all things emerge (as the monists would have it), or whether it is simply the collisions and interactions among atomic elements (as the atomists would have it), the process of regeneration requires, in Aristotle's view, two things: a prime matter or "stuff" out of which organization might occur and an organizing principle or prime mover actively to make order from nonorder. Aristotle's hylomorphism is a response to this question concerning regeneration. Matter cannot be separated from form, but the function of form is as the organizing principle, that which sculpts or molds matter into a body. Regeneration—as the fundamental biological property of all living things—is thus the result of a transformation of matter by form, of a basic resource by an organizing principle.

Aristotle's assumption is that in the process of regeneration there has to be some "stuff" that exists prior to the coming-to-be of a body, to serve as the source material, as it were. In addition, the active principle of form necessitates a force from the outside and a principle of organization externally applied to the source material. Although the view from molecular genetics would place Aristotle's principle of form in the DNA or genes, current research in biocomplexity has brought new perspectives to the questions first raised by Aristotle. Brian Goodwin and Ricard Solé, for instance, have shown how cellular differentiation and morphogenesis do not require a prime matter or prime mover. In their discussion of studies on somite formation in the chick, amphibian, and mouse embryos, Goodwin and Solé suggest that genes play only a minor role in cellular differentiation.[26] Rather, morphogenesis is seen to emerge from "self-organizing" processes that are the collective result of internal networks of interactions (gene expression, protein-protein interactions) as well as of cellular environmental conditions. For Goodwin and Solé, genes "do not occupy a privileged position in making decisions about alternative pathways of differentiation. Yet they clearly constrain the possibilities open to cells."[27] There is no external, controlling prime mover, as Aristotle would have it. Nor is there any primeval matter from which a body is constructed. As Goodwin and Solé state, "it is not genes that generate this coherence, for they can only function within the living cell, where their activities are highly sensitive to context. The answer has to lie in the principles of dynamic organization that . . . involve emergent properties that resolve the extreme complexity of gene and cellular activities into robust patterns of coherent order."[28]

Thus, the concept of regeneration in regenerative medicine seems poised between an Aristotelian emphasis on a prime matter and external formal agency, on the one hand, and a view of the self-organizing, emergent properties of the cellular milieu, on the other. As a biotechnology research field, regenerative medicine can be understood as a particular interpretation of the Aristotelian distinction between coming-to-be (or development) and alteration (or regeneration). Regenerative medicine reiterates this distinction through a wide range of projected applications that go from life-extension biotechnologies (development) to medical therapy for degenerative diseases (regeneration). Of course, the line between medical therapy and biotechnical enhancement is not as clear as limit cases seem to suggest. Attempts to augment the human life span may seem mere folly, but research into neurodegenerative disorders such as Parkinson's disease has gained much support and shown a significant level of progress. But what about examples that are not so clear-cut, such as the use of stem cells for gene therapy or the research into telomeres as a possible cancer therapy? It is, certainly, too soon to tell whether such projects will show significant promise.

We have, then, a series of terms that collectively define the field of tension within regenerative medicine (table 8.1): Aristotelian *coming-to-be* and *alteration*; *development* and *regeneration* in biology; *life-extension* biotechnologies and *medical therapy*; *enhancement* and *medicine*.

Certainly, however, the line between these terms is far from decisive. In fact, the examples mentioned are only heuristic. With powerful biotechnologies such as stem cell research, tissue engineering, and other fields of regenerative medicine, each application will confront in some way this tension

Table 8.1 The Conceptual Field of Tension in Regenerative Medicine

"Coming-to-be" in Aristotle	"Alteration" in Aristotle
Development in biology (morphogenesis)	Regeneration in biology (autotomy in certain organisms)
Life-extension biotechnologies	Medical application of stem cells
Extramedical enhancement	Medical therapy

between extramedical enhancement and medical therapy. What is important to note about regenerative medicine broadly is that *the concept of "regeneration" in regenerative medicine considers coming-to-be as a mode of alteration*. In other words, regenerative medicine is a way of understanding development within the context of regeneration. Recall that Aristotelian coming-to-be is analogous to the process of morphogenesis in the biological development of the organism. Key in this process are the processes of cellular division and cellular differentiation. Yet in most organisms this process happens only once; or rather, it is a continuous process that in many ways defines the organism. By contrast, regeneration, as the replacement of injured tissues, presumes the prior development of the organism and is thus analogous to Aristotelian alteration.

This field of tension impacts what Susan Squier calls "the tissue culture point of view," or the ways in which various discourses on the biomedical body impact the meanings that science research has within and outside of the specifically medical context. As Squier notes, the tissue sciences such as xenotransplantation, life-extension therapies, or stem cell research are positioned between "the boundary of the species and the boundary of the human life span."[29] The regenerative properties of cells are explored in two ways that reflect table 8.1: "Two forms of growth are typically investigated in tissue grown in vitro: organized growth and unorganized growth. Researchers have accessed them—with iconic overdetermination—by culturing the embryo and the cancer cell, potent images of life and death."[30] If the concept of regeneration is poised between Aristotelian coming-to-be and alteration, then we can also say that regenerative medicine articulates two types of regenerative cells: the cell defined by organization (driven, indeed, by Aristotle's four causes) and the cell driven by proliferation and production—the cell that differentiates and the cell that produces.

The cell that produces, then, not only produces for a biological use value, but in the context of the health-care industries it can also be made to produce for a biological use value that is also an economic exchange value (e.g., a medical therapy or bioartificial tissue product). Furthermore, this relation between biological and economic value is largely mediated by a social network of exchange constituted by donors, volunteers, and structures of social indebtedness. Stem cells are thus carriers of what Catherine Waldby calls "biovalue." For Waldby, biovalue "refers to the yield of vitality produced by the biotechnical reformulation of living processes. Biotechnology tries to gain traction in living processes, to induce them to increase or change their productivity along specified lines, intensify their self-reproducing and self-maintaining capacities."[31] Aristotle expresses this correlation between medical and economic valuation in ambiguous terms: "As in the art of medicine there is no limit to the pursuit of health . . . so too, in this art of wealth-getting there is no limit of the end."[32] Again, regenerative medicine is in many ways a paradigmatic biotechnology precisely because it identifies a mode of biology that is at once "natural" and yet *potentially* (in Aristotle's sense of the term) technical. It will thus be worth our while to consider the concept of regeneration with respect to the "time of life" of the cell and the organism.

Biotechnical Speed

As we have seen, the approach of regenerative medicine is different from current biomedical approaches to tissue and organ failure in that it uses the body's own cellular processes toward novel ends. In addition, the very concept of "regeneration" is set within a field of tensions between extramedical enhancement and medical therapy. Yet the bodies in regenerative medicine are never static, fixed bodies, but always embedded in different biotemporal cycles and processes; regenerative medicine is, in this sense, constantly concerned with the body's sustainability through biological time. One way to address this unique relationship between bodies and technologies is to consider media theorist Paul Virilio's views concerning the connections between speed and technology: "Being alive means to be lively, quick. Being lively means being-speed, being-quickness. . . . All these terms challenge us. There is a struggle, which I tried to bring to light, between metabolic speed, the speed of the living, and technological speed, the speed of death which already exists in cars, telephones, the media, missiles. . . . Politics should try to analyze this interface, because without this analysis a fatal coupling will be established."[33] The tension Virilio highlights here is exactly the fact that although technologies of speed have quite literally been progressing at the speed of light, the techniques of the body have largely been left to obsolescence, wired into the "polar inertia" of interactivity. Elsewhere, Virilio outlines four modalities of speed in the current, dromological order: metabolic speed (or "animal speed"), which includes physiology but also psychology and emotions; and technological speed, which can be either mechanical speed (or "automotive speed") or the speed of spectacle (or "audiovisual speed").[34] The former has come to replace the use of the body and animals as a means of transportation (Virilio cites the mass production of automobiles and "automobility"), and the latter has come to replace the exclusive reliance on the human sensorium (the gradual ubiquity of a tele-vision).[35] In addition to metabolic and technological speed, there are, Virilio states, "couplings" between the two, the metabolic-automotive coupling and the metabolic-audiovisual coupling.[36]

But what of these tensions Virilio evokes between metabolic and technological speed? On the one hand, it is clear that new transportation and communications technologies are advancing the science speed to new levels, new units of temporality, new definitions of presence. On the other, the human body, the supposed bearer of such advances in speed technologies, has been left behind, still defined by antiquated notions of the long duration of the living, the slow and meandering pace of the "natural" operations of the organism.

Or has it? Virilio's dichotomy between metabolic and technological speed is helpful because it asks not how the body itself is changing as a result of technology, but rather how the "couplings" of the dynamics of body-technology relationships are being transformed by new, technical notions of development and morphogenesis. From such a perspective, it is clear that the body is also a generator of a unique type of dromologistics, a generator of speed in such a way that more traditional notions of the organism, biology, and health are challenged.

For this reason, we would do well to take up Virilio's notion of metabolic and technological speed if only to question the very ontological division this coupling presupposes. As we have seen in a variety of other contexts (tissue engineering, genomics, bioinformatics), if there is one thing that biotechnology demonstrates, it is that the body itself can be deployed as a technology, that it can accelerate its own internal metabolic speed beyond the domain of what it is "naturally" capable. The very techniques that these approaches use also show us the extent to which the body itself comes to inhabit a unique type of temporality, partially defined by the biological processes of growth, development, and decay, and partially defined by the technical enhancements of a punctuated, triggered regeneration.

Within the tension Virilio highlights between metabolic speed and technological speed, we might locate a *biotechnical speed* in which the biological body is constantly surpassing itself. Biotechnical speed involves a technical approach to the body, whose goal is to maintain the integrity of the biological body as a "natural" entity, while also improving it's capacities. This is a body that is not just responsive to environmental signals, but that seeks to optimize itself independently of its environment. The body of biotechnical speed emerges within a set of conditions in which it can biologically self-engineer, culminating in a newfound optimization (increased efficiency and effectiveness) and a new found cellular versatility (regeneration at any location).

Taking these comments into consideration, we might, like biotech research itself, peer into the near future to speculate on the kinds of bodies that regenerative medicine might produce through biotechnical speed. Again, as pointed out at the beginning of this chapter, we need to see these bodies in all their complexity; our near-future scenario is not simply a utopia or a dystopia, but a network of cells, technologies, therapeutics, politics, and economics. Thinking about the body on a philosophical and political level, we can extrapolate three possibilities, based on current research.

Immune System 2.0

Although regenerative medicine research has been able to stimulate singular instances of cellular regeneration, its dream is to be able to reprogram the body on the cellular-molecular level to accomplish this regeneration on its own. Reprogramming the body requires thinking of the body's immunity operations as not only offensive (attacking designated molecules) but also defensive (capable of triggering regeneration at diseased, weakened sites). The goal of the regenerative medicine research into diseases such as Parkinson's or Alzheimer's is not just the replenishment of, for example, neuronal cells, but the programming of the body's immune system to respond to disease by triggering targeted regeneration. If we extend this basic process—not just the replacement of cells, but the design of a biological replacement system in the body—then regenerative medicine becomes a kind of directed, controlled metastasis.

Gerotech

As with all biomedical innovations, the bioethical line between necessity and vanity, disease and design, is ever present in regenerative medicine. Biotech companies focusing on cellular aging, in which, the telomeres, or the protective caps of chromosomes, get progressively shorter each time a cell divides, are pointing to telomere research as the fountain of youth. If regenerative technologies can be utilized to counteract the telomere-shortening process, then cellular aging would turn into cellular longevity, and biological aging would turn into biological immortality. The age-old concept of defeating the aging process and cheating death has found its newest exemplar in the very idea of regeneration. Once cells, tissues, and organs can be grown on demand, outside or inside the body, then the line dividing medical necessity and metaphysical vanity becomes more ambivalent.

Off-the-Shelf Organs

We should never forget that biotechnology is first and foremost an industry. Although medical application provides an impetus for the research, its true test will be whether viable products and services can emerge out of this research. The existence of bioartificial skin products on the market is testimony to this bioeconomic drive. For regenerative medicine, this test means patenting and marketing the techniques used to trigger cellular regeneration for medical purposes; the development of novel, genetically designed drugs, aimed to trigger regeneration or stem cell development biochemically, is also another area of focus. What regenerative medicine shows us is that the body can be approached as a biological storehouse, a living cell bank located inside the body that is constantly replenished. This body defines itself not according to cycles of growth and decay, but according to biological surplus, encouraging a view of the body that is defined by disposability.

As these examples indicate, biotechnical speed transforms the biological body into a technology, a continued, controlled deployment and enhancement of the body's cellular capacities. These bodies defined by biotechnical speed are hyperresponsive (the upgraded immune system), biologically interminable (cellular immortality), and marked by disposable bioeconomies (off-the-shelf organs).

Both technical optimization and cellular versatility also imply a general improvement in the body's performance. A body that can, within the right type of context, self-initiate the regeneration of damaged tissues and cells is also a body that is constantly pushing the limits of its performance. This is what Virilio refers to as "the race to achieve limit-performances" within the technosciences.[37] For Virilio, this emphasis on the "limit-performance" moves technoscience away from discourses of medical truth and closer to technical paradigms measuring functionality, an ambiguity between medical healing and technical effectiveness.

What is left of the body when both external and internal contingencies have been biotechnically removed? Different research agendas present us with different visions of the future of the biomedical body. In terms of aging, the body becomes self-recyclable, harnessing its stem cells—the biomolecular fountains of youth—to replenish lost cells continually. In terms of disease, the body becomes a biological war machine constantly undergoing autovaccination by targeting and recognizing possible pathogens; indeed, the environment itself becomes one great pathogen for regenerative medicine. Finally, in terms of tissue and organ replacement, the body becomes a "wet" storehouse of resource materials, which may be freely bought and sold, exchanged and engineered, in a sudden materialization of the biotech economy. The point of contention here is not the benefits of developing treatments for disease. Rather, the issue raised by fields such as regenerative medicine is how different ways of approaching the biological body articulate certain views of how disease, health, normativity, and the body are understood.

The Labor of Regeneration

These temporal aspects of regenerative medicine lead to a consideration of the unique type of "labor" performed by the cells, enzymes, and genes that constitute the regenerative body. Recall that, in Marx's terms, "labor-power or capacity for labor is to be understood as the aggregate of those mental and physical capabilities existing in a human being, which he exercises whenever he produces a use-value of any description."[38] Marx comments that labor power requires that three conditions be met. The first is that the worker must offer his or her labor power on the market as a commodity. That is, in order for labor power to exist, the worker must be able to conceive of labor as being amenable to quantification. This openness to the quantification of labor enables labor or productive activity in general to become specifically labor power. Once this occurs, labor power can be sold as a commodity.

But this openness to quantification actually means that the worker is able to separate labor from the body and do so in a way that renders labor as property—that is, property to be sold. Labor cannot become labor power, and labor power cannot become a commodity unless the worker establishes the connection between body, labor, and property. Therefore, the second condition is that the worker must be willing to offer his or her labor power (separated property of quantification) for sale. That is, the worker must be able not only to quantify, but to exchange, circulate, and sell that labor power. Implied is the willingness to extricate the body's productive activity from the body of the self. Self is therefore disengaged from labor (or only to the extent that labor will produce returns for the body of the self).

Also implied is that the worker has nothing else to sell except his or her labor power. That there are those who own money and those who sell their labor is not a natural condition, Marx notes, but a product of historical conflict. The form that that conflict historically takes is an exchange of quantitative values. What kind of a quantity is labor power? Marx states that "the value of labor-power is determined, as in the case of every other commodity, by the labor-time necessary for the production, and consequently also the reproduction, of this special article." Furthermore, "it represents no more than a definite quantity of the average labor of society incorporated in it. Labor-power exists only as a capacity, the power of the living individual."[39]

As a quantity, labor power is determined by a time measurement—how long it takes to produce a commodity. Note that this time is not necessarily how long it takes to produce an object with use value for the worker, but the time it takes to produce an object of value for others. This value is determined not only by external social customs, but principally by the amount of labor power necessary to sustain the life (body) of the worker. Here Marx edges toward biological terms again in speaking of labor power as a quantity that is defined by biological sustenance of the body doing the labor. The vicious circle is that the body must work a certain amount in order to sustain itself, but it can sustain itself only if it is healthy to begin with. Biological processes get translated directly into economic terms: "thereby a definite quantity of human muscle, nerve, brain, &c., is wasted, and these require to be restored. This increased expenditure demands a larger income."[40]

The role of time—of many kinds of time—is a key factor in the broader issues raised by regenerative medicine. Antonio Negri, in his discussion of the pervasive expansion of capitalist modes of production over many aspects of social life, notes not only that time becomes a major point of contestation between laborer and capitalist, but that this contestation takes the form of an ontological and material struggle: "If social labor covers all the time of life, and invests all of its regions, how can time measure the substantive totality in which it is implicit?"[41] If a quantitative valuation of time—of labor time—becomes the basis for labor power, the wage, and the guarantee of surplus value, then the primary challenge of a capitalist view of time is to render qualitative social activity into quantitative labor power. What results is the "impossibility of distinguishing the totality of life (of the social relations of production and reproduction) from the totality of time from which this life is woven."[42] Time, in Negri's analysis, is not a mediation of a laboring body, nor is it simply a representation of the product of labor. The very concept of labor power involves a notion of time that is constitutive, that in fact produces the very laboring bodies it describes. The nexus of labor power, body, and capital "develops materially, in a tendential and necessary sense, toward the *tautology of time and life*."[43]

This development leads to a situation in which, literally, life is put to work. For Marx, this occurred when the individual sold his or her capacity for work or production as labor power. Michael Hardt and Negri—expanding on work done within the Italian autonomist tradition—have recently suggested that in the informational, global economy, labor power becomes "immaterial labor" in which no material product is produced, but rather acts of communication, calculation, and affect.[44] But in the biotech industry, labor power is transformed in a different way. Regenerative medicine is unique in that it works toward being able to harness naturally occurring biological processes. In this sense, the ideal biotechnology is one in which there is no technology at all, in which biology does all the work itself. Thus, the species-being is not just molecular-genetic; it is that which is made amenable to the technical paradigm of regenerative medicine. Hardt and Negri's immaterial labor becomes a kind of *biomaterial labor*, enabling biologies to produce themselves in novel contexts. *Labor power in regenerative medicine is cellular, enzymatic, genetic—a labor power defined by a biological potentiality.*

Marx, writing in the mid–nineteenth century, was of course talking about industrial modes of production. But his comments concerning production are relevant to us for the way they reconfigure the biotech industry. In fact, I suggest that regenerative medicine, inasmuch as it participates in the biotech industry and health-care economy, constitutes a novel mode of production, one that is distinct from industrial modes of production or more recent modes of flexible accumulation or informatization. In Marx's later work, production is defined as being composed of three elements. Each of these elements has its correlative in the biotech industry, but in ways very different from the mode of industrial capitalism of which Marx was speaking. The first element of production involves the conditions of production, such as the natural resources available for work. As previously noted, for regenerative medicine the condition of production is biological life at the cellular, molecular, and genetic levels. The second element includes the forces of production, or the means by which natural resources are transformed into products. For regenerative medicine, these means include the means by which biological processes are harnessed, appropriated, and recontextualized. The notion of biological "information" is a key component to the forces of production. Biological information acts as both an immaterializing and a materializing force, enabling one to move between an in vitro tissue culture and a computerized genome database. The third element involves the relations of production, or the relations of property and exchange. For regenerative medicine, the relations of production are many, from genetically designed drugs to stem cell–based therapies, to patents on techniques or compounds, to levels of funding and investment in research. The relations of production enable a mobility between information, property, and matter, especially in the case of patents and the potential marketing of regenerative medicine therapies.

It seems, then, that when looked at as a mode of production, regenerative medicine relies a great deal on the ability to define biology in relation to techniques for working on biology and in relation to the economic valuation of such techniques. In this sense, we might describe regenerative medicine not as a mode of industrial production or as a mode of information, but as a *mode of regeneration*. As a mode of production, regenerative medicine begins by approaching biological process as a potential coming-to-be, as a set of known variables with known ends. The labor power of regenerative medicine includes the biological, cellular, and molecular processes of regeneration itself. In regenerative medicine, the conditions of production are constantly regenerated or replenished by the forces and relations of production. But more than this, regenerative medicine produces its own conditions of production; that is, it defines the biological domain as the domain of biological potential, in which a process such as cellular differentiation is viewed as a set of known steps that produces predictable results, which are then amenable to medical forces and economic relations. Hardt and Negri make a similar point when they distinguish between that which is "outside measure" and that which is "beyond measure."[45] That which is outside measure is a spatial exteriority, that which is not yet measurable.

What does this mean for regenerative medicine specifically and biotechnology generally? I should note that the idea here is not to apply Marx point for point from the industrial factory and the human body of the worker to the biotech lab and the nonhuman body of cells, enzymes, and DNA. Rather, we want to see what Marx's concept of labor power can tell us about the biotech industry and its mode of production.

One thing that stands out is that in biotech there is no laboring body. There are, as previously stated, the laboring bodies of researchers, lab technicians, administrators, and so forth. But there is no body that actually does the producing. That job is handled by biology, biology put to work. This is most obvious in the case of tissue and cell cultures, in which the metabolic and developmental capacities of cells are maintained almost indefinitely in laboratory conditions. Their labor is a strange kind, for the very principle of biotechnology is that, ideally, one has to do very little; there is a sense that the biology takes care of itself. A number of characteristics make the labor power of regenerative medicine unique. One is the nature of the labor itself, in which all efforts go toward creating the right conditions for naturally occurring biological processes to take place (e.g., cellular differentiation, division, apoptosis). The labor of biotechnology is, in a sense, purely biological—that is, the labor of biotech aims to be transparently biological.

But genes expressing themselves in an embryo and a culture of stem cells are two different contexts. Labor in one becomes labor power in another, which is made especially evident in the private sector, where stem cell–related patents and the promises of stem cell therapy form a part of several biotech companies' business plan. This labor power is unique for a second reason: no human owner of the labor power volitionally sells that labor power as a commodity. Marx's conditions of labor power have been rewritten by the biotech industry. Instead of a human worker, who views his or her labor power as property to sell, exchange, and circulate, we now have a nonhuman biological network of cell lines, tissue cultures, and genomic databases. Labor is not, then, real-time labor of the physical body; instead, it is the archival labor of cell cultures, databases, and plasmid libraries.

Labor power in biotech is put to work in two kinds of bodies. One body is the nonhuman "molecular" bodies of cells, enzymes, and DNA. These bodies can be derived from one of two sources. One source is human beings or animals by genetic or tissue sampling. This source is controversial, especially in cases where patenting of biological materials is involved, or where genetically and ethnically isolated populations are targeted for study. The second source is much more preferable in that it bypasses the human element altogether and relies on the source code of DNA. Though much regenerative medicine research has ostensibly "black-boxed" the cell, trends in research point to the further integration of cell biology with genomics and proteomics. The data that currently exist are overwhelming, and with the click of a mouse a researcher can download the sequence for a given gene from a given organism and then synthesize that gene in the lab.

The second body in regenerative medicine is human, or rather on the verge of the human (some might say posthuman). This is the body of the patient in a clinical trial, the recipient of a gene therapy, a stem cell therapy, or a bioartificial tissue transplant. The body of the patient is a unique type of laboring body. The introduction of new genes, proteins, or cells impels the body to work in a new way; this is the basic idea behind gene therapy. The modified body is impelled to produce something new—a gene that is not expressed, a protein to block gene expression. This is fine when it works over the long term and can demonstrate both safety and effectiveness. But often this modified, laboring patient-body is not set for the long term. It needs intervention from the outside. This is where the patient's body of the patient exists in a bizarre kind of biological estrangement, when cells taken from that body are used to culture cells that will be modified and put back into the patient. This body is, then, self-sustaining, but only on condition that it be semipermanently wired into the biology of the wet lab.

"Life Runs Painfully Gray down the Walls of Time"

When tissues and organs can be grown on demand, irrespective of bodily contingency or surroundings, our relationships to our bodies and our notions of health and disease are bound to be altered.[46] Regenerative medicine is a unique approach to the biomedical body not only on a scientific level, but, more important, in the way in which it regulates the relationship between biology and technology, the natural and the technical, as both are contextualized by the sciences.

The science fiction expressions of this type of regeneration can be found in the *X-Men* series, the example with which I began this chapter. As genetic mutants, the X-men form ambivalent hybrids caught between genetic discrimination and a kitschlike, heroic posthumanism. If we take regenerative medicine's desire to extricate bodily contingency from the body, we can see two bodily imaginaries at work; the characters of Wolverine and Mystique represent two poles of the regenerative, biotechnical body—one marked by the military-industrial complex, the other manifested through the ability to shape-shift.

As an expression of the autonomous self-referential subject, Wolverine represents the totally fortified, insulated, defensible body. Along with the cyborg enhancements of postmodern warfare, Wolverine's capacity for biological regeneration makes him the autonomous, self-sufficient, biotechnical subject, able to withstand any environment or context of threat. The only way to transform the mortal, biological body into a monad is to make it into biological armor, which is, in effect, to attempt to remove mortality from the body. By contrast, Mystique represents a deeply rooted anxiety surrounding the implications of a total biotechnical control of the body. That is, if Wolverine represents the dream of a totally fortified, autonomous body, Mystique represents the uneasy implications of taking this logic of biotechnics to its logical conclusion. Wolverine's body is primarily defensive, certainly combative; it is industrial, biological, structural. Its entire purpose is to maintain its own self-identity, even in the face of death and even if it means going to the extreme of removing death from the body. Mystique's body is about a biotechnical control of the body to such a degree that "the body" becomes something other than a fixed, stable identity, even something other than the structural, anatomical anthropomorphism that traditionally defines the human body. Mystique's body is biotechnical control to the extent that one may lose control of the body; it is biotechnical control to the extent that the traditional notions of the human body—and the human itself—become regularly transgressed, morphed, and mutated.

If one can technically condition the body at the molecular, genetic, and cellular level, and if one can essentially design all of the body's contingencies out of the body, then it seems that medicine has far surpassed its traditional role as a therapeutic practice and has fully entered the domain of biodesign.

The next logical step here is to move from a pervasive biological regulation of the body to an equally pervasive total redesign of what the body is capable of. Mystique's biomolecular shape-shifting is simply the dark underside of Wolverine's overconfident, hypermasculine dream of a totally fortified body. Although these two characters' gender identifications may be seen as predictable, both Wolverine and Mystique, as a single discourse on the future of the biomedical body, "perform" the limits of an enabling biotechnology. Wolverine thus becomes the biotechnical dream of the self-contained, humanist subject, whereas Mystique represents the scientific and cultural anxieties of total formlessness (or total control).

Regenerative medicine recontextualizes the biological body as capable of being optimized and made more versatile by the creation of certain technically enhanced conditions, through which an upgraded version of the biological may emerge. It provides us with an example of a biotechnology in the literal sense of the term—a technical design of life. The confusions that result are that the body supposedly benefits from technical intervention, while all the time remaining a "natural" biological entity unmodified by technology. It is this ambiguity regarding the relationships between biologies and technologies that inhabits science fiction hybrids, as we see in *X-Men*, revealing both the techno-optimisms and the abject anxieties of biotechnologies.

Thus, the main challenge for any bioethical, political, and philosophical thinking about fields such as regenerative medicine is how to conceive of the role of the subject or patient in the decidedly nonhuman biologies of stem cell research or tissue engineering. The biomaterial labor of regenerative medicine is at once desubjectified, occupied only by the complex biological agencies of gene expression networks, cellular differentiation, and developmental pathways. Yet in the medical application of regenerative medicine, individual subjects are clearly involved in clinical trials, gene therapy experiments, and the business of the field. A central challenge will be the problem of a "collective corporeality," as Negri puts it.[47] There exists no phenomenology of the patient that addresses the uncanny, almost science fictional examples of "immortal" cell lines, totipotent stem cells, and proposals for life-extension therapies. The question is, Can such a view of a biomedical, bioethical, collective corporeality exist while also acknowledging the new forms of biomaterial labor being created in the biotech industry?

the multitude of nonhuman, material-semiotic actors involved in any techno-scientific process (machines, cells, scientific papers, policy regulations, diseases, etc.). This means that it is not only a matter of questioning the subject, but of the "givenness" of the subject—its nature. What gets constructed as nature and as the natural in scientific discourse and practice dictates the kind of ethics and politics that can develop.[4]

Consensus, Dissensus, Post-Media

But what kinds of ethics and politics can develop? In a series of essays dealing with globalism's "planetary computerization," philosopher and activist Félix Guattari asks how subjectivity is currently being articulated in an age of information technologies.[5] These processes of articulating subjects he calls "collective apparatuses of subjectification," and they take a specific form in contexts where technoscientific knowledge legitimizes ways of thinking and acting in the world. Guattari begins answering his initial question by analyzing three main types of apparatuses of subjectification: pathways of power, pathways of knowledge, and pathways of self-reference or self-transformation.[6]

The primary question for Guattari is how to relate the third pathway (self-transformation) to the first two (power and knowledge). In other words, amidst all of these apparatuses of subjectification (technologies for conditioning subjects), how can one conceive of a mode of self-making that will serve as an alternative to either the pathway of power or the pathway of knowledge? "The burning question, then, becomes this: Why have the immense processual potentials brought forth by the revolutions in information processing, telematics, robotics, office automation, biotechnology and so on, up to now only led to a monstrous reinforcement of earlier systems of alienation, an oppressive mass media culture and an infantilizing politics of consensus?"[7] Guattari suggests that before we can think of how to identify sites that are reduced to the pathways of power (such as the state) or knowledge (such as science), we must first think about how to create the conditions for "new existential territories." That is, before thinking about new modes of subjectivity, we must first think about experience, existence, and selfhood in relation to science and technology. The possibility of outlining new ways of thinking and acting in the world are therefore predicated on finding new ways of thinking about technology in relation to the "subject."[8]

9

Conclusion: Tactical Media and Bioart

In her essay "The Promises of Monsters: A Regenerative Politics for Inappropriate/d Others," Donna Haraway begins by making the point that what counts as "nature" is both constructed and necessary.[1] As an instance of "artifactual reproduction," with extended global histories in science, colonialism, and political economies, nature becomes for Haraway less a thing and more a "topos" or field of renegotiation. Haraway is insistent on the constructed aspect of our notions of nature and the natural ("nature for us is made, as both fiction and fact"), as well as on the scientific discourses that articulate and act upon nature ("biology is a discourse, not the living world itself").[2] However, she does not emphasize the constructed aspects of nature in order to destroy it as an efficacious concept. Rather, it is precisely through this artifactual quality of nature that she proposes a refiguration of nature, both as a field of negotiation and as a negotiating or mediating dynamic. The category of nature, with its complex histories and permutations, both structures the field of possible inquiries and investigations concerning life, culture, and technology and mediates the possible connections between various "material-semiotic nodes."[3]

The main force of Haraway's "biopolitics of postmodern bodies" is that it is an inquiry into the ways in which the notion of an agential, intentiona self-aware, and autonomous subject is variously instituted into a range contexts within contemporary technoscience. Haraway does not ever deny political reality of "the subject" as it is incorporated into everyday prac Instead, she proposes a complexification of the subject by offering a v

Guattari poses the possibility of a "post-media era" as one alternative. As its name indicates, the post-media era would be one that both incorporates media (that is, it is not anti-technology), but also transforms media such that they are no longer assumed to be separate from the subject (that is, they are no longer an object, a commodity, a tool, or something opposed to the subject). Guattari does not define what this post-media politics would be, but he suggests that one route is to move from "consensual media" (media homogenization, standardization, and corporate ownership) to "dissensual post-media" (bottom-up media that emphasize polyvocality, that are critical).[9]

How might Guattari's notion of post-media operate in relation to biotech? As we know, continuing developments in the areas of computer animation, the construction of virtual environments, telerobotics and motion capture, and an array of technologies for broadcasting and webcasting innovative work are becoming available not only to scientists, but also to artists, performers, and cultural activists. If biotech research is increasingly becoming "computerized," and if a range of computer technologies are available to the nonspecialist currently, then it makes sense to inquire into the kinds of politically aware, critical, and ethically conscious interventions that are enabled by this nonspecialist engagement with the "medium" of biotech. The term *bioart* is often used to refer to projects that deal with biology as an artistic medium (as opposed to painting, sculpture, photography, or video, one assumes). However, I find it worthwhile to refuse this convenient tag for a number of reasons. It not only marginalizes (or niche markets) art, effectively separating it from the practices of technoscience, but the notion of a "bioart" also positions art practice as reactionary and, at best, reflective of the technosciences. Though the term *bioart* may indeed refer to artists and artworks circulated primarily within the gallery system, we might do better and ask how cultural research dealing with biotechnology can take seriously its interdisciplinary nature. As an example, we may briefly consider four groups specifically working in the intersections between biotechnology, art, and new media. Each employs different strategies of engagement and critique, which also means different technologies and techniques. Each also involves some level of collaboration between artists and scientists, attempting to complicate the redundant circles into which "science wars" discussions can often fall.

SymbioticA and the Tissue Culture & Art Project: The Culture of Flesh

In a text commenting on the way in which technoscience facilitates a "gene fetishism," Haraway has the following to say: "Organisms emerge from a discursive process. Biology is a discourse, not the living world itself. But humans are not the only actors in the construction of the entities of any scientific discourse; machines (delegates that can produce surprises) and other partners (not 'pre- or extra-discursive objects,' but partners) are active constructors of natural scientific objects."[10] The strange, alien artifacts that populate many bioscience labs—from DNA chips to lab-grown tissues—challenge our traditional thinking about the boundary between the living and the nonliving. Bruno Latour often refers to such strange entities as *actants*, neither actors (in the sense of the agency presumed to lie in human subjects) nor objects acted upon (in the sense of instruments or resources). Similarly, Haraway begins to outline a new category for the production of the natural and the technical that is not simply about a dichotomy between nature and artifice, biology and technology, human and machine. Both Haraway and Latour are attempting to move beyond the exclusively ideological critiques of science and toward the creation of a space inhabited instead by a variety of qualitatively differentiated actants, each with its own force, contingency, and context.

However, as both Haraway and Latour note, the actants are also "monstrous," especially when they are placed in a position of radical difference to the more familiar (or naturalized) conditions of the science lab or medical clinic. Perhaps fields such as tissue engineering offer the prime context in which to investigate the extent to which our own bodies—as biologically constituted—are redeployed as actants. Although it is easy to see how a mechanical prosthetic (or laptop or PDA or cellular phone) can become an actant in our everyday activities, the issue of bioartificial tissues is more complex. The tissues are my own, yet they exist outside of and separate from my body. They *live* outside of and separate from my body. Are they indeed "my" tissues? Or are they my tissues once they have been grafted (or regrafted?) onto my body and have assimilated to their new (originary?) environment?

For some years, a small group of artists has been exploring the phenomenological and biomedical paradoxes that tissue engineering elicits. The Tissue Culture & Art Project, begun in the mid-1990s in Perth, Australia, intentionally set out to create a technically feasible laboratory environment in which artists could learn about and investigate biotechnology from a cultural, ethical, and anthropological standpoint.[11] The project is also housed under SymbioticA, a larger project that aims to bring together critical artistic inquiry with the techniques and practices of cellular biology and tissue engineering.[12] In conjunction with the Department of Anatomy and Biology at the University of Western Australia, SymbioticA and the Tissue Culture & Art Project have initiated and presented their work in a variety of contexts, from art festivals to science conferences.

For instance, the Tissue Culture & Art Project artists working at the SymbioticA labs have utilized tissue-culturing techniques to create miniature figurines out of animal stem cells, some—as in their *Semi-living Worry Dolls* project—seeming to fulfill a kind of ritual function (the artists encourage viewers to "tell the dolls your worries"). They have also used architectural and sculptural principles to design novel scaffolding structures upon which to grow tissue, creating "pig's wings" (miniature wings from mesenchymal pig cells) and other such impossible, anomalous anatomies. The discourse of the monster and the monstrous is appropriate here; the types of monstrous bodies SymbioticA designs are biologically and physiologically nonfunctional, and yet still "living." They occupy that ambiguous, intermediary zone between subject and object, a sort of "tissue actant."

If tissues are indeed akin to Latour's notion of actants, then SymbioticA's projects can be understood as inquiries into the issues that arise in the consideration of these strange teratologies—teratologies no stranger than those that exist in tissue engineering labs themselves. Can there be a politics that effectively takes into account these nonhuman actants, entities that are much more than inert objects and yet much less than autonomous organisms? How can we keep from falling into the too easy habit of reducing all actants to agential origins (e.g., the notion that, yes, there are these nonhuman machines, but ultimately humans design and operate them)?[13] The Tissue Culture & Art Project's *Extra Ear Quarter Scale* (figure 9.1), developed in collaboration with the performance artist Stelarc, addresses this question. The *Extra Ear* project aims to grow a bioartificial ear that will, in concept, be transplanted onto Stelarc's head.[14] The ear is fully "biological" in the sense that it utilizes tissue engineering principles and is not a mechanical prosthetic or cybernetic device. But the ear is also excessive, artificial, in the sense that it is not needed by the body to live. The cells of the extra organ are the body's own, and yet the

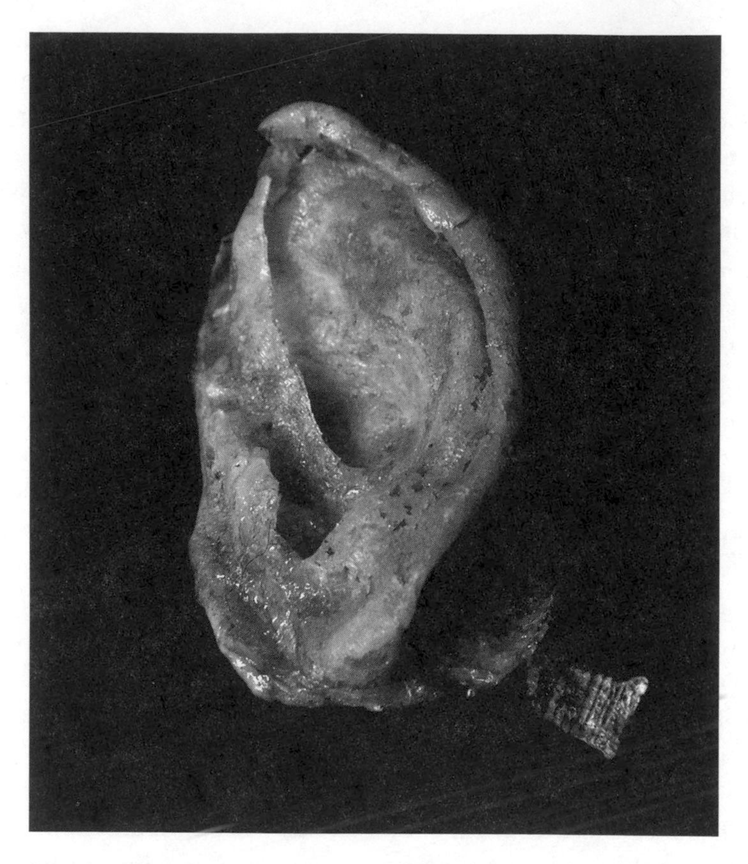

Figure 9.1 The semiliving body: *Extra Ear ¼ Scale*, by the Tissue Culture & Art Project (Oron Catts and Ionat Zurr), in collaboration with Stelarc. Courtesy the Tissue Culture & Art Project.

organ is something that is functionally alien to the body. This and other projects raise the issue of biological property, or, rather, the relationship between biology and property that is at the core of current debates over tissue banking institutions, stem cell economies, and genetic patenting. On the one hand, we assume that our bodies are indeed isomorphic or at least "proper" with ourselves. On the other hand, there is also a booming business of tissue banking (sperm, ova, etc.) that presupposes that the ownership of "a" body is, strangely, predicated on the ownership of "my" body (informed and legal consent, and so on).[15]

SubRosa: The Sex of the Species

As an ideology and social theory, *sociobiology* is, generally speaking, about the application of biological knowledge (mostly from evolutionary biology, hereditary genetics, and population biology) to social phenomena. Heralded as a new synthesis of science and sociology by proponents such as E. O. Wilson, sociobiology attempts to account for social behavior as in some way biological at its basis (such as studies on the evolution of altruism).[16] At its most reductive, sociobiology seeks to isolate, describe, and explain a wide variety of social anomalies (from medical diseases to "social diseases" such as criminality) from the perspective of the biological sciences. At its best, it aims to develop a theory of the coevolution of genes and culture in the understanding of social phenomena. Sociobiological trends have recently gained a fair amount of publicity, largely over what some say are their discriminatory and eugenic implications. Debates over the degree to which certain hereditary or genetic disorders or both are linked to questions of race and ethnicity have sparked a controversy as to how far science should advance on social questions. Similar debates have occurred over the possible genetic variants contributing to other so-called genetic predispositions as wide-ranging as homosexuality, infertility, and mental disability.[17]

Sociobiology today is arguably not so much a school or scientific discipline as it is a way of managing the social as a collectivity. This is most evidenced by the role of biological knowledge in the areas of reproductive technologies, genetic diagnostics, and the increasing reach of the pharmaceutical industry into all aspects of human behavior. At issue with the investigations of sociobiology is this enframing of the social as a biological population (that is, defined by patterns of growth, reproduction, demographics, disease, mortality). The population is regulated through the means of scientifically distinguishing health from disease, the normal from the pathological, and by the intention to locate a biologically rooted cause of a range of social phenomena manifest in the population. This form of management proceeds by many techniques, from surveys to statistics, to health care, to social policy, to scientific experimentation. Its goal is always a search for complex and productive (that is, coercive and nonrepressive) ways to regulate the social body operationally.

Although sociobiological perspectives often seek to isolate a biologically based causality to social and cultural phenomena, the SubRosa collective approaches the social as a field thickly demarcated by the political signifiers of gender, race, and class relationships. As a self-described "cyberfeminist" group, SubRosa takes up a wide range of topics related to biotechnologies.[18] In their performances, projects, publications, and workshops, SubRosa members instigate a series of confrontations between culture, economics, and biology. For example, early SubRosa projects investigated the ways in which gender was being transformed by biotechnology. Faith Wilding's *SmartMom* project extrapolated medical and reproductive technologies into a full-spectrum surveillance system, and the SubRosa *Biotech Sex and Gender Ed Workbook* appropriated the format of a women's health manual, but in the context of the cultural and ideological aspects of new reproductive technologies. Recent performances such as *Expo EmmaGenics* (figure 9.2) and *Biopower Unlimited*! explore the rhetoric and semiotics of corporate biotechnology trade shows, complete with CD-ROM, sales pitch, powerpoint, advertisements, corporate branding, and the use of molecular biology lab techniques for demonstration purposes.[19]

If sociobiology attempts to view the social in terms of the assumptions of the scientific and biological, then anthropologist Paul Rabinow suggests a reversal, *biosociality*, which is about the critical questioning of sociobiology

Figure 9.2 Demonstrating "ART" (Assisted Reproduction Technologies): appropriating the trade-show format in *Expo EmmaGenics,* a project by the SubRosa collective. Courtesy SubRosa.

and is a helpful term for describing SubRosa's activities.[20] Rabinow's biosociality makes two suggestions: that sociobiological trends reduce the social to the biological (and by extension the scientific), and that sociobiology, through its assumptions, discourses, and research, in many ways produces the social as essentially a biological problem.

Although SubRosa's members do not simply take an "antiscience" position in their projects, they do raise contentious issues that demand accountability within the biotech and health-care industries. Their workshops focus on education and awareness building in specific local communities, especially as regards assisted reproductive technologies (ARTs) and the ongoing technical advances in in vitro fertization. Contrary to the technophilia that inhabits much medical technologies, SubRosa asks how the population is monitored, classified, and channeled (in technologies such as ARTs) into a range of viable, culturally homogeneous positions (e.g., the configuration of tissue, sperm, and ova "banks").

Biotech Hobbyist: Will the PC Happen to Biotech?

In most contexts, the notion of the "amateur" or "hobbyist" is looked down upon, signifying as it does one who is not properly trained, who is not an expert or specialist in a field, and who does not take seriously that which he or she practices. In another context, amateurism and hobbyism have been taken in a quite different, opposite manner. For example, prior to the explosion of the PC industry, a number of small zines circulated in the 1970s that were dedicated to the idea that computers could be customized, tweaked, and modified for a range of purposes. These "computer hobbyist" zines (e.g., *Creative Computing*, *Computer Hobbyist*) circulated alongside the first computer kits (e.g., the Altair 8800, the Apple I). In this context, the amateur or hobbyist was someone with a great deal of knowledge and an interest in the nonspecialist use of that specialist knowledge.

However, all of us are also aware of what happened to the so-called liberalism of this geek subculture: the PC, Microsoft, the MacIntosh, an entire industry built around the idea of personal computing. It is this contradiction that the collective Biotech Hobbyist explores. A loose group of individuals from a range of disciplines, Biotech Hobbyist—started by artists Natalie Jeremijenko and Heath Bunting—proposes that what happened to computing may also happen to biotech. That is, the question Biotech Hobbyist (half-ironically, half-seriously) poses is: Will the PC happen to biotech?

As a way of investigating such questions, Biotech Hobbyist projects explore the idea of the "kit."[21] In a do-it-yourself fashion, projects draw on science education to explore biotechnology techniques such as DNA extraction, sequencing, and PCR using readily available materials and tools (e.g., kitchen PCR, DNA extraction with a blender, sequencing with sugar). Other projects focus on specific issues, such as Jeremijenko's project *OneTree*, which involved the cloning and then planting of 100 trees around the San Francisco Bay Area. Among other things, *OneTree* demonstrated the fallacies of the reductionist understanding of genetic cloning: each of the planted cloned trees developed in different ways, in part owing to the different urban and environmental conditions of the Bay Area. Another Biotech Hobbyist project, *PbC (Personal bio-Computing)*, proposes to build a low-tech DNA computer using the standard lab techniques of gel electrophoresis and PCR (figure 9.3). Other projects explore open-source bioinformatics, tissue culturing, biotech video games, genetic horoscopes, and antipatenting software tools.

Biotech Hobbyist projects employ a kind of "tactical media" that supports amateur practice in that the amateur or hobbyist makes use of whichever

Figure 9.3 Garage biotech: gel electrophoresis kit, lo-fi PCR, and laptop bioinformatics in a Biotech Hobbyist lab. Courtesy Biotech Hobbyist.

media are necessary for a given context (the "digital divide" becomes a "biotechnical divide"). Amateurism and hobbyism are important here because they imply both an interest in learning and an interest in reappropriating, repurposing. We can learn from gaming culture: "mods" for biotech. However, this does not mean the freedom to do whatever one wants with one's own mods, nor does it license a total lack of knowledge or responsibility. What is important to understand about this amateurism and hobbyism is the way in which the practitioners construct for themselves customized environments for learning and critiquing (which implies the development of unique communities of differentiated specialists). Tactical media are nonteleological but directed; they support multiple agendas and fully imply all the complexities of ethical action.

Critical Art Ensemble: Delivering on the Promise of Biotechnology

For a number of years, the group known as the Critical Art Ensemble (CAE) has utilized a range of media and practices—video, cultural theory, net.art, activism, guerrilla performance, and research collaborations—to extend its analyses of our contemporary "machinic" culture into new strategies for resistance and "electronic civil disobedience."[22] CAE's book, *Flesh Machine*, lays out three power structures in which the social-political, symbolic-ideological, and biological domains are managed by techniques and technologies aimed at the naturalization of the imperatives of global capitalism. These power structures—the war machine, the vision machine, and the flesh machine—utilize forces of global conflict, representation, and health management effectively to integrate the individual subject into a labyrinth of systematic violence, media overstimulation, and the exhaustive visualization and mapping of the species-body.[23]

One of CAE's projects involves a presentation by representatives from a biotech corporation simply called BioCom. Given at art festivals, academic conferences, and various conventions, the presentation shows BioCom offering to its subscribers an exclusive online sperm donor and fetal-design service in which, for example, female customers can actually "shop" for their preferred genetic mates using online video conference interviews, a Web-based screening application, and an individualized report from BioCom's genetic screening lab pertaining to the genetic information of sperm donors (e.g., disease predispositions, etc.). At BioCom's labs, the latest in ARTs enables the company to fertilize the egg of a customer expeditiously with the DNA from the sperm of her selected donor or to utilize cryopreservation technologies to extend the value of a particular egg or sperm. Combining phenotypic review (person-to-person interviews online between a customer and potential donor), genotypic analysis (by a BioCom research group), and professional fetal-design services (mediated by a BioCom sales representative), BioCom integrates the potentials of computer-mediated communication (CMC) with the reliable advances in genetic engineering and reproductive technologies.

Although not in the "official" spaces of conferences or conventions, another group, the Society of Reproductive Anachronisms (SRA) offers an activist-based countermeasure to the high-tech promises of corporations such as BioCom. Appealing to the general public through adhoc street presentations or at recruiting stations on college campuses, SRA promotes a rigorously Luddite version of biological reproduction and the health of the species. Through pamphlets, posters, discussion, and an independently run Web site, SRA emphasizes the more Malthusian values of natural childbirth, person-to-person sex with reproduction as the goal, and the innate, natural laws of selection dictating the occurrence of disease, mortality rates, and overpopulation.

As can be imagined, the responses by audience members and passers-by to both groups vary considerably. What each presentation does—what it is geared to do—is bring up questions concerning the degree of control held by the flesh machine (that is, the scientific-medical establishment) on the human body and its functions as a dynamic organism. For CAE, the strategies of command and control of the war machine combine with the will to visibility of the vision machine, and they produce a flesh machine that requires absolute, total visualization of all levels of the body. For the flesh machine (again, mostly concentrated in the medical-scientific community), visualization equals production of knowledge, which equals a significant degree of technological, ethical, and economic control.

CAE isolates as a point of critical intervention in the flesh machine the potential rupture between consumerist and social imperatives as they relate to the flesh machine. CAE's example is the controversial "gay gene," which would simultaneously essentialize sexuality while also reducing it to biological determinism. In social terms, given the conservative tendency in the West toward compulsory heterosexuality, the gay gene is dangerously near the domain of pathology and would thus be a candidate for designation as a disease gene. In economic terms, however, greater diversity always extends the domain of the consumer market, especially as biological attributes (from body type to sexual preference) attain cultural value in the media. Such tensions illustrate the current trend toward the need for negotiations between government-medical imperatives (the rhetoric of research to fight disease) and corporate-engineering imperatives (the rhetoric of tailor-made biological products such as pharmaceuticals).

CAE's more recent work has involved a deeper collaboration between art, activism, and science, a combination they term *contestational biology*. For CAE, contestational biology involves the critical and ethical redeployment of biology to enable it to function politically. CAE offers a set of principles for their unique form of resistance that includes the demystification of scientific knowledge, the defusing of public fear, the critique of utopian rhetoric, and the building of respect for conscientious forms of collaboration and amateurism.[24] The ends of contestational biology can be varied, from raising public awareness to posing the question of "how to create models of risk assessment that are accessible to those not trained in biology."[25] In one project, CAE has worked with biologists to engineer a safe, environmentally sound, herbicide-resistant strain of plant (figure 9.4). The herbicide Roundup Ready works

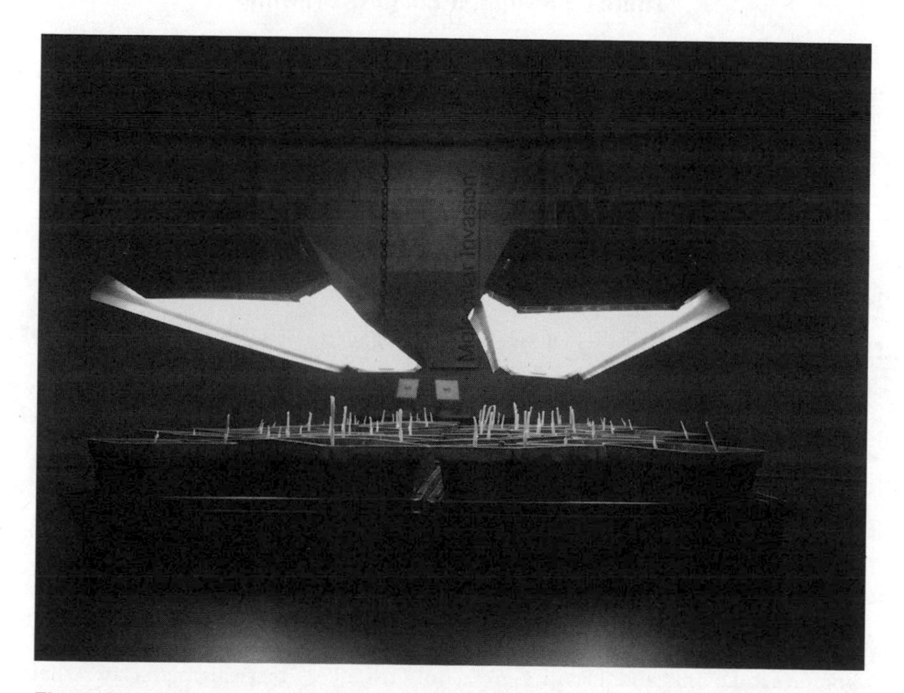

Figure 9.4 Demonstrating "contestational biology" in genetically altered, environmentally safe, antiherbicide crops, from a CAE project. Courtesy CAE.

by inhibiting a particular enzyme (5-enolpyruvylshikimate-3-phosphate [EPSP]) found in most microorganisms and plants, in effect starving the organism. Agricultural biotech companies have then genetically engineered their crops to overexpress a modified form of the enzyme, thereby making only the crop plants resistant to Roundup Ready. CAE proposes a "biochemical intervention" that would inhibit the synthesis of the GM enzyme, thereby making the crop plants vulnerable to the Roundup herbicide. Further, CAE proposes doing this by using—appropriating—a compound manufactured by Monsanto itself, one that, when combined with Roundup, will inhibit the EPSP enzymes in a plant. The aim is not only to demonstrate the vulnerability of GM crops, but also to devise a means of doing so that is environmentally conscious and that in its own way effectively reintegrates plants and organisms into their environmental context. As CAE state, "the gene(s) or biological processes that modify the organism can be targeted, and *turned from a trait of adaptability into one of susceptibility*."[26]

"There—An Animal Shadow Crawling"

Though it will be seen as a truism, it is important to reiterate that biotech is always about bodies.[27] In his work on the history of sexuality, Michel Foucault reserves the term *biopolitics* to speak of the normalizing, coercive strategies of management, or "governmentality," of the social-biological body. As he states, "the disciplines of the body and the regulations of the population constituted two poles around which the organization of power over life was deployed."[28]

In a way, the possibility of any critical engagement with and intervention in biotechnology is articulated in the space between the body (individual or social) and "life itself" (be it biological or informatic). The examples of bioart given earlier are not meant to be representative of the wide range of artworks that deal with genetics and biotechnology today. But each of them arguably points to real fissures in the social, cultural, and material effects of biotechnologies. This would seem to fulfill at least one aspect of Guattari's post-media: the ability to create a site for the exploitation of fissures. In this sense, Guattari's concept looks forward to more recent debates over "tactical media." For Geert Lovink and David Garcia, for instance, "tactical media are what happens when the cheap 'do it yourself' media, made possible by the revolution in consumer electronics and expanded forms of distribution . . . are exploited by groups and individuals who feel aggrieved by or excluded from the wider culture."[29] Although tactical media grew largely out of activities in video activism and later in critical Internet culture, the question also applies to biotechnologies: As both the knowledge and the tools of biotechnology become more and more "consumer grade"—in educational media, kits for classroom use, inexpensive lab equipment—will there also be "tactical media" of biotechnology? As we have seen, CAE argues that there are and should be. The group's term *contestational biology* is meant to address such issues politically and technically. Its unique combination of activism, workshops, performance, and lab biology show that contestational biology can be a new form of praxis.

But there are also many challenges ahead for any kind of post-media, tactical media, and contestational biology. As with the computer and network paradigms, a significant learning curve is associated with biotechnologies. Cell and molecular biology, biochemistry, and plant biology may be involved. In addition, the techniques and technologies are changing often. Although some techniques are more or less standards (e.g., gel electrophoresis), other techniques are constantly changing (e.g., computer-driven gene sequencing). Finally, collaboration raises new sets of challenges in working across disciplines, both eschewing the reductionism of antiscience positions and fostering a critical view of biotechnology. The site in which such activities are carried out—the art gallery, the science fair, the classroom, the lab—says a great deal about the modus operandi of bioart.

In addition, all of these challenges are enframed by a basic distinction between human user and nonhuman tool. The very idea of post-media or tactical media implies a rift between human users and the tools themselves; "do it yourself" and appropriation at once challenge us to think of technology differently and ask us to dissociate "use" from "technology." Alex Galloway suggests a different approach to this problem. Instead of a reliance on a notion of "use" dissociated from technology, "there are certain *tactical effects* that often leave only traces of their successes to be discovered later by the ecologists of media."[30] Galloway is, in part, discussing the Internet, in which the properties that define it (network topologies and connectedness) are also the properties that constitute its vulnerabilities (e.g., hacking and computer viruses). An analogy can be made to biotechnology, which, as I have suggested, places a great deal of emphasis on gene-based research, genetic drugs, and genetic diagnostics. Despite efforts to the contrary, the central dogma or gene-centric approach still holds sway in biotech research, especially in the pharmaceutical sector. But the biology of the cell is much more than the activity of genes or DNA; the "network" properties of gene expression, protein-protein interactions, metabolic pathways, and cell signaling pathways suggest that "command and control" in the living cell is inherently complex. Thus, the very property that defines biological and genetic "life itself"—its network properties—also serves conceptually to undermine the biotech industry's exaggerated emphasis on genes, gene targets, and DNA.[31] In this regard, there is much work to be done on both the science and culture fronts. In so-called post-genomic research, this overwhelming complexity is quickly demonstrating how essential it is that nonreductionist approaches to molecular biology become a part of medical application. And in the cultural sector, the icon of DNA—and its attendant metaphors of command and control—still pervade much science fiction and popular writing on science. Biotechnology is thus never simply "life itself," but rather the continual process of technically articulating something as self-evident as "life itself."

In our contemporary scene, where the intersection of technology development, unimpeded research and application, and economic imperative drive the technosciences in the future of "health" and the "body," all of the strategies presented in this book—science fiction, biopolitics, general economies, biomaterial labor, species beings, actant networks, and post-media dissension—have indeed become not only options but also in many ways necessities. The important point to make is that these practices and necessities need not be specific to the advertisements of "Big Pharma" or the glowing media reports of the human genome. With each of the projects described earlier, "practice" or "research" is simultaneously political and technical. As practiced by the groups considered here, these strategies are also critical gestures, gestures commenting on the contingencies of the future and the necessity of reimagining the present.

Appendix A: Biotechnology Fields and Areas of Application

Agricultural biotechnology Also "ag-bio" for short. The use of biotechnologies such as genetic engineering in the growth, development, and design of plants and seeds. Agricultural biotechnology has a wide range of application, from genetically modified (GM) foods to genetically-modified organisms (GMOs) for industrial use to the production of fabrics such as cotton for clothing.

Biocomputing The use of biological components and processes for research in computer science (making a "computer" out of DNA, for instance). Not to be confused with bioinformatics or computational biology. Subfields include DNA computing, membrane computing, and enzyme computing.

Bioinformatics The use of computer and information technologies for molecular biology research. Software, databases, benchtop computers, and imaging and modeling are specific areas of application. Many fields, such as genomics and genetic screening are aided by bioinformatics, which, as its name indicates, is the integration of biotechnology and information technology. Also sometimes called computational biology.

Biological warfare The use of hazardous biological agents as weapons, either for offensive uses or for defense and national security. Most developed nations have or have had biological warfare programs. Overlaps to some degree with chemical warfare and, increasingly, with bioterrorism, though the latter is distinguished by its being more "low tech" and carried out by non-nation-state actors in low-level conflict.

Biometrics The use of a range of scanning technologies to identify individuals by their unique biological "signature" (iris scan, fingerprint scans, gait recognition, blood tests). Biometrics research is also incorporating genetic technologies such as DNA fingerprinting.

Biomimicry The study of the structural and functional aspects of natural forms for application in industry (materials, chemicals, production). Overlaps with biomaterials science.

Biotechnology In the broadest sense, the use of biological components and processes (occurring in nature) for ends that are medical, industrial, and economic. Although biotechnological practices can be traced back to antiquity (livestock breeding, agriculture, fermentation), the emergence of a "biotech industry" in the late 1970s and early 1980s heralded a new phase in biotech, spurred forth by genetic engineering and the early efforts to map the human genome.

Cloning A set of techniques for artificially replicating the genetic material of a cell or an organism from one generation to the next. Used most frequently in the research lab, but also some limited usage in agriculture and transgenics. There are different types of cloning, such as human therapeutic cloning from stem cells.

DNA chips More often called *microarrays*, these chips are handheld devices with DNA strands attached to a silicon or glass substrate, over which sample DNA may be applied. Used in diagnostic settings for testing the presence or expression of a specific compound. Some overlap with the related fields of bioMEMS (micro-electrical-mechanical systems) and microfluidics.

Gene therapy The use of genetically engineered DNA either to supplant the lack of a gene or to repress the overexpression of a gene. Many gene therapy techniques involve the use of two elements: a *plasmid*, or small circular DNA carrying the desired gene, and a *viral vector*, or a hollowed-out viral shell used for delivery. Human gene therapy clinical trials have met with both success and disastrous side effects.

Genetically modified (GM) foods The use of genetic engineering to create seeds, plants, and entire crops in agriculture, whose genetic makeup will enhance their productivity (growth, pesticide resistance, overall output). GM foods form a part of agricultural biotechnology (or "ag-bio"), which has had a significant and controversial presence in the Western world.

Genetic engineering A set of techniques for intervening in the process by which genetic material is replicated and recombined in living organisms. The first genetic engineering experiments in the early 1970s involved the use of specialized, naturally occurring enzymes to cut and paste DNA sequences in bacteria. This DNA sequence was called *recombinant DNA*. The biotech industry in its modern form started alongside the first genetic engineering experiments, but the implications of the experiments also raised controversial ethical issues.

Genetic screening The use of genomics-related tools to screen a patient's DNA for the presence or expression of certain disease-affiliated genes. New technologies such as microarrays and digital computers have made genetic screening a much more quantifiable, even statistical process. Genetic screening in medicine is not a form of forecasting, but in most cases is used to aid in the diagnosis of a prescription drug or therapy.

Genomics The sequencing, assembly, annotation, and analysis of the entire genetic makeup of an organism. Genome projects have been initiated for a wide range of model organisms, including baker's yeast, the roundworm, the fruit fly, the mouse, bacteria and viruses, and, of course, the human being. Contemporary genomics is made possible largely by new computer and robotics technologies.

Medical genetics The use of biotechnological diagnostics tools in medical and clinical settings. Includes genetic diagnostics, new reproductive technologies (NRTs), and nontechnical services such as genetic counseling. Not all biotechnology research is intended for medical application, though the most significant concentration has been in two areas: testing (e.g., microarrays) and drug design.

Molecular biology The study of biological life at the molecular level. Often used interchangeably with *molecular genetics*, a field that came to maturity during the post–World War II era and that is associated with the names Francis Crick, James Watson, François Jacob, Jacques Monod, and many others. Molecular biology emerged from a complex interchange with the then-current fields of cybernetics and information theory, from which the common tropes of the genetic "code" are derived.

Nanotechnology The study of the atomic and molecular properties of matter, with the aim of being able to control matter at the atomic and molecular level. Subfields include nanomedicine, nanobiotechnology, and environmental uses of nanotechnology.

New eugenics Term used by sociologists to describe the transformation in eugenics outlook since the 1930s and World War II era. Unlike early eugenics, which was primarily a negative regulatory practice, the new eugenics promotes consumer choice and enhancement through biotechnology.

New reproductive technologies (NRTs) The application of biotechnology toward all aspects of biological reproduction. May include in vitro fertilization, surrogate motherhood, research into artificial wombs, and the practice of preimplantation diagnostics.

Pharmacogenomics The use of genomics and bioinformatics toward the development of genetic drugs and therapies. Pharmacogenomics often emphasizes the promise of custom-tailored, individualized drugs that will minimize side effects and adverse drug reactions (ADRs). Pharmacogenomics is differentiated from *pharmacogenetics*, the latter making use of genetic technologies to advance broadly the specificity of drug activity at the genetic level.

Population genomics The use of genomics and related fields to study the genetic makeup of particular populations, often identified according to biological, racial, and ethnic characteristics. Population genomics is intended primarily to aid in medical research and the study of disease, but it has also been utilized in other fields such as archaeology and genealogy.

Proteomics The study of the sequence and structure of proteins, their function, and their interactions with other proteins. Although the number of base pairs in the human genome may be known, the total number of proteins is not. Proteomics is often split between research that emphasizes sequence (amino acid sequence) and research that emphasizes structure (protein-folding problems).

Regenerative medicine The study of the natural regenerative properties of the body, with particular focus on stem cells and their medical-therapeutic potential. Regenerative medicine overlaps a great deal with tissue engineering and is increasingly employing techniques from genomics. Other areas related to regenerative medicine focus not on stem cells but on chromosomal structures such as telomeres.

Structural genomics The study of the relationships between DNA or RNA sequence, on the one hand, and amino acids and proteins, on the other. Structural genomics is positioned between genomics and proteomics, or between the study of sequence (in DNA) and the study of structure (in proteins).

Tissue engineering The study of how cells and tissues grow, differentiate, and regenerate, with the aim of being able to grow tissues and entire organs in the lab for purposes of transplantation. Tissue engineering is often considered to be a part of biomedical engineering broadly, and several bioartificial skin products are currently on the market.

Transgenics The use of genetic engineering to create organisms whose genetic makeup includes genes or DNA from another organism or organisms. Often used in agriculture (pesticide-resistant plants), drug development (transgenic animals producing enzymes), and laboratory testing (human disease genes in laboratory animals).

Appendix B: Techniques and Technologies in Biotechnology Research

Artificial chromosomes The insertion of extra chromosomes into the nucleus of a cell, either for the purposes of storage (in bacteria or microorganisms) or for the purposes of genetic therapy. An advanced version of bacterial and yeast artificial chromosomes (BACs and YACs).

Autoradiography A method by which a radio-labeled strand of DNA or RNA produces an image on photographic film. Can also be performed with entire chromosomes. Useful as a way of visually tagging or identifying molecules or regions on a molecule such as DNA.

Bacterial artificial chromosome (BAC) A linear recombinant DNA molecule that replicates in bacteria. Usually of moderate size and most often used for cloning specified DNA fragments. Also see *yeast artificial chromosome* (YAC).

Batch culturing A method by which a large amount of cells are cultured for a specified period of time after inoculation onto the culture medium. They can then be collected for processing and further study in the lab.

Bioconversion The use of microorganisms to convert a specific compound into a valuable product through a set of enzymatic reactions. The process minimizes the artificial or chemical synthesis steps in the production of a commercially viable compound.

Biodegradation The breakdown of compounds through the use of microorganisms and plants.

Biolistics The method of delivering DNA into animal and plant cells by DNA-coated microprojectiles shot under high pressure and high velocity. Some form of biolistics or pulsed projection is often used in cloning.

Bioreactor A growth chamber (most often stainless steel) for culturing cells and microorganisms, usually used for the production of useful or valuable compounds. Also called a *fermenter*.

Bioremediation The use of microorganisms and sometimes plants to convert contaminates or toxic substances in soil and water into nontoxic by-products.

Complementary DNA (cDNA) A complementary double-stranded DNA synthesized in the lab from a single-stranded messenger RNA (mRNA), using the naturally occurring enzymes reverse transcriptase and DNA polymerase. Such cDNA strands represent the original DNA sequence from which the RNA was derived, minus the excised sections (or *introns*). cDNAs were widely used in the sequencing of the human genome.

Contig A collection of overlapping, shorter, cloned DNA fragments that are subsequently arranged in the appropriate chromosomal order. Contig regions are used extensively in genomics as markers; a *contig map* shows the physical location of contig regions along the chromosomes.

DNA fingerprinting The use of genetic screening technologies to identify an individual by his or her unique genetic "signature." DNA fingerprinting is often used in legal cases, particularly to identify criminals, to decide paternity cases, in forensics, to aid in bioconservation, or to exonerate criminals wrongly accused.

DNA library A collection of in vitro cloned DNA fragments, most often existing within bacterial plasmid DNA and stored for further study. Related to expression libraries.

Electroporation Akin to biolistics and microinjection, but making use of electrical pulses, which temporarily open the cell membrane for the introduction of foreign DNA.

Expression library A collection of DNA fragments inserted into a vector that has a promoter sequence next to the insertion site, enabling the regulatory elements of the vector to control (turn on or off) the expression of the inserted DNA fragment.

Fermentation The breakdown of organic compounds by cells or microorganisms in the absence of oxygen to generate adenosine triphosphate (ATP); in the biotech industry, cells and microorganisms are cultured in fermenters (a bioreactor) to produce desirable compounds.

Gel electrophoresis A commonly used method for the separation of DNA or RNA fragments according to their length. A gel (most often agarose—a thick, viscous sugar) is used as a matrix through which DNA or RNA fragments are passed along an electric current. The smaller fragments will pass through the dense gel matrix more quickly than the longer fragments. DNA or RNA fragments can be measured, analyzed, and sequenced according to this method. Proteins can also be analyzed through electrophoresis by similar methods.

Gene bank In its original use, a gene bank denoted a physical facility where plant material—seeds, tubers, or cultured tissue—was stored. But the term now also refers to computer databases that hold DNA, RNA, and protein information (such as GenBank or the Protein Data Bank).

Gene sequencing machines Computer-based technologies for automating the gene sequencing process, laid out principally in techniques such as restriction mapping and gel electrophoresis. Scanning, visualizing, and analysis software is employed to arrange large amounts of DNA fragments according to their sequence. Gene sequencing machines were extensively used in the sequencing of the human genome.

Gene splicing The insertion of a DNA fragment into a vector and the transfer of the vector (a recombinant molecule) into a host for propagation.

Genetic linkage map A map that indicates the genetic distances (physical distance along a chromosomal region) between pairs of polymorphic (nonidentical) DNA regions. Their position relative to the center of the chromosome affects the degree to which "crossing over" occurs during cell division.

Hybridoma The use of tumor cells to enable indefinitely existing cell lines. An "immortal" cell line produced from the fusion of B lymphocytes (which produce antibodies) to lymphocyte tumor cells.

Karyotype A technique for visualizing chromosomes, in which homologous chromosome pairs are separated during cell division and arranged numerically in a line. Used for the detection of chromosomal abnormalities.

Monocolonal antibody The particular antibody (specific for one antigenic determinant) produced by a hybridoma clone (itself derived from a B lymphocyte cell). Monocolonal antibodies are frequently used in the design and manufacture of drugs and therapies.

Oligonucleotide synthesizer A combination of automated, wet-lab components (solution containing denatured adenine, cytosine, guanine, and thymine molecules) and computer or dry-lab components (computer software) for automating the oligonucleotide synthesis process. Costly but efficient means of generating desired DNA sequences for laboratory research (e.g., polymerase chain reaction [PCR] primers).

Pairwise sequence alignment A common technique in bioinformatics that involves the successive alignment of a target sequence against the sequences in a computer database. Basic scripts, data-mining utilities, and string-matching software suggest the closest matches based on sequence homologies. The Web-based Basic Local Alignment Search Tool (BLAST) performs a variety of alignments against the GenBank database.

Physical map A map that indicates the distance in base pairs between markers or gene fragments along a chromosome. Markers can be restriction enzyme–recognition sites, sequence-tagged sites (STSs), and restriction fragment length polymorphisms (RFLPs).

Plasmid An extrachromosomal, circular, double-stranded DNA molecule that replicates independently of the chromosome. With genetic engineering techniques, foreign DNA can be inserted into a plasmid. Most often found in bacteria, plasmids are frequently used as delivery vehicles for foreign DNA inside vectors.

Plasmid library The use of genetic engineering techniques to store foreign DNA fragments in a plasmid molecule in an in vitro state, most often in bacteria.

Polymerase chain reaction (PCR) A method for the rapid and efficient amplification of DNA fragments through successive stages of heating (denaturing) and cooling (annealing) of DNA strands. The programmable machine that carries out this procedure is a *thermal cycler*.

Primer A short, single-stranded DNA (oligonucleotide) that is hybridized to a complementary DNA template, to which further free-floating nucleotides can be added. Primers are used extensively in polymerase chain reaction (PCR).

Recombinant DNA A DNA strand composed of fragments from different sources, which have been joined together using the techniques of genetic engineering: cutting (using restriction enzymes or ligases) and pasting (using polymerase enzymes).

Restriction endonuclease (or restriction enzyme) A class of enzymes that cut DNA by recognizing a specific DNA sequence and cleaving the bonds at that location. Class II enzymes are most often used in genetic engineering.

Restriction fragment length polymorphisms (RFLPs) The occurrence of different restriction enzyme cleavage sites for the same DNA region of different individuals of the same species. Each individual genome thus has a unique set of restriction enzyme–recognition sites and can be used as markers in the sequencing of genomes.

Restriction mapping The use of restriction enzymes along with gel electrophoresis to sequence a DNA fragment. Restriction enzymes of different cutting sites are applied to cloned DNA fragments. Gel electrophoresis separates the cleaved fragments according to size. The sequence of the DNA molecule can be deduced by reading the columns of the gel, each arranged by the particular restriction enzyme cleavage site.

Sequence-tagged site (STS) A short DNA fragment (200–300 base pairs) that is known to be unique and is used as a marker in genetic linkage and physical mapping.

Southern blotting A method for transferring denatured DNA from an agarose gel after gel electrophoresis to a special membrane for hybridization. Used to detect specific sequences based on the occurrence of DNA hybridization.

Structure prediction A set of techniques in bioinformatics that aims to analyze the structure of proteins. Some techniques begin from the DNA sequence, whereas others begin from the protein primary structure of an amino acid chain. Protein sequence and the bonding characteristics of atoms play a part in determining the exact structure of proteins (their secondary structure of motifs and the tertiary structure of the whole protein).

Tissue culture Propagation of plant and animal cells in vitro. Can be used to regenerate whole plants from cultured cells or tissues.

Vector A DNA molecule used to transfer foreign DNA into a host organism. Examples include plasmids, cosmids, and bacteriophages. Viral or phage vectors are often used as delivery vehicles in cloning.

X-ray crystallography Method of visualizing molecules using X-rays to image the three-dimensional arrangement of atoms in a molecule (of DNA or of a protein). The diffraction pattern of X-rays passing through a crystal casts a shadow of the molecule.

Yeast artificial chromosome (YAC) A linear recombinant DNA molecule that replicates in a yeast cell. YACs can contain large DNA fragments (up to 1 million base pairs).

Appendix C: A Brief Chronology of Bioinformatics

1890. The birth of bioinformatics? Herman Hollerith's tabulating machine is used in the U.S. census. The Hollerith tabulating machine uses punched cards akin to the Jacquard Loom (each person is a card), but also uses electricity to read, count, and sort cards. Later, in 1896, Hollerith founds the Tabulating Machine Company, which would later become IBM.

1933. Gel electrophoresis introduced, a standard technique for sequencing DNA.

1938–44. World War II research in mainframe computing advances with the separate work of Alan Turing, John von Neumann, J. Prisper Eckert and John Mauchley, and Vannevar Bush.

1942. Erwin Schrödinger gives a series of lectures under the title "What Is Life?" in which he suggests that hereditary mechanisms operate via a "code script."

1945. The Electronic Numerical Integrator and Computer (ENIAC) is built by Eckert and Mauchley at the Moore School of Electrical Engineering (University of Pennsylvania). It is electronic and digital, makes use of novel storage devices, and is built on von Neumann's unique architecture.

1948. Norbert Wiener publishes *Cybernetics: Control and Communication in the Animal and the Machine.*

1949. Clande Shannon and Warren Weaver publish *A Mathematical Theory of Communication*, derived in part from Shannon's MIT master's thesis, "A Symbolic Analysis of Relay and Switching."

1950s. IBM leads the way in the transition from mainframe military computing to business computing.

1951. Linus Pauling and Corey propose alpha-helix and beta-sheet motifs for proteins, structures that would become instrumental in computer-based protein modeling.

1953. James Watson and Francis Crick publish papers on the structure of DNA, explicitly using the language of information theory and cybernetics. Their research is enabled by Rosalind Frankin's X-ray crystallography imaging of DNA.

1954. John Backus at IBM develops the FORTRAN computer language.

1955. First protein sequence (bovine insulin) published by Fred Sanger.

1958. The Advanced Research Projects Agency (ARPA) is set up by President Eisenhower in response to the Soviet Sputnik launch. Among its aims is the development of advanced communications networks (soon to become the basis for the Internet).

1961. François Jacob and Jacques Monod publish their paper "Genetic Regulatory Mechanisms in the Synthesis of Proteins," in which they apply the model of cybernetics to the workings of the cell. Their *lac operon* model is a modified form of the cybernetic feedback mechanism.

1962. Paul Baran publishes "On Distributed Communications" for RAND, developing the concept of *packet switching* concurrent with Donald Davies (who coined the term). Packet switching would soon become a fundamental concept in the nascent Internet.

1962. Heinrich Matthai and Marshall Nirenberg publish their research "cracking the genetic code," showing how four base pairs combine to form twenty amino acids.

1965. Margaret Dayhoff and colleagues publish *Atlas of Protein Sequence and Structure*, among the first attempts to catalog protein sequences systematically.

1969. ARPA runs the first test of a computer network, ARPAnet, with interface message processors (IMPs) placed at the four nodes (UCLA, Stanford, UC Santa Barbara, and the University of Utah).

1970. The Needleman-Wunsch algorithm for sequence comparison is published and subsequently patented.

1971. Intel announces the 4004 Microprocessor, a "computer on a chip."

1973. First genetic engineering experiments published, incorporating "recombinant DNA."

1973. The Brookhaven Protein Data Bank (PDB) is announced.

1974. Vinton Cerf and Robert Kahn develop the concept of connecting networks of computers into an "internet" and develop the Transmission Control Protocol (TCP).

1975. Two-dimensional electrophoresis and the Southern blot technique are developed.

1977. Techniques for sequencing DNA using restriction enzymes and electrophoresis developed by Sanger and colleagues, known as the "Sanger method."

1980s. The first biotech companies, Genentech and Cetus Corporation, are formed.

1980. The first DNA sequence is published (FX174).

1980. The first patent for a genetically altered microbe is granted in the landmark case *Diamond v. Chakrabarty*. The decision is based in part on the then-contemporary claims for software patents.

1980. IntelliGenetics, the first bioinformatics company, is founded. It markets the PC/GENE software suite.

1981. The Smith-Waterman algorithm for sequence alignment is published and subsequently patented.

1981. IBM introduces its personal computer (PC) to the market.

1982. The Genetics Computer Group (GCG) is formed at the University of Wisconsin Biotechology Center.

1984–85. The first meetings concerning the Human Genome Project (HGP) take place, under the rubric of the U.S. Department of Energy and the University of California. It is formally launched in 1990.

1985. The FASTP algorithm is published and subsequently patented.

1985. Microsoft announces its Windows operating system.

1985. Polymerase chain reaction (PCR) is introduced by researchers working at the Cetus Corporation, including Kary Mullis.

1986. The term *genomics* is used to describe the study of entire genomes.

1986. The European Molecular Biology Laboratory develops the SWISS-PROT protein database.

1987. The use of yeast artificial chromosomes (YACs) as storage and replication vechicles is demonstrated.

1987. The genetic map of *E. coli* is published.

1988. The National Center for Biotechnology Information (NCBI) is established at the National Cancer Institute in the United States.

1988. The first patent for a genetically modified (GM) animal is granted (the OncoMouse or "Harvard Mouse," owned by DuPont).

1988. The first automated gene sequencing machines are developed by Applied Biosystems.

1988. The FASTA sequence alignment algorithm is published, and subsequently patented.

1989. The Oxford Molecular Group is formed, among the first companies dedicated to computer protein modeling.

1990s. A series of bioinformatics startups are formed, including Molecular Applications Group, InforMax, Incyte, Myriad, Institute for Genomic Research (TIGR), Affymetrix, and Lion Bioscience.

1990s. A series of mostly university-based open source bioinformatics initiatives is started, including Bioperl, BioJava, and Open Bioinformatics. The Linux operating system for bioinformatics research used.

1990. The now-standard BLAST algorithm is published. It is subsequently ported for the Web and is accessed daily by thousands of labs worldwide.

1990. The National Human Genome Research Institute (NHGRI) develops GenBank, the main repository for publicly available genome data.

1990. A study estimates that since 1980 more than 150 bioinformatics patents have been granted, with another 400 pending.

1991. Tim Berners-Lee, at the research institute in Geneva, Centre Européenne pour la Recherche Nucléaire (CERN), publishes the protocols that make up the World Wide Web.

1995. The ExPAsy suite of Web-based bioinformatics tools is released by the Swiss Bioinformatics Institute.

1996. The *S. cerevisiae* (yeast) genome is sequenced.

1996. The PROSITE database is released.

1996. Affymetrix produces commercial "GeneChip" microarrays, among the first companies to market microarray devices.

1998. Celera Genomics is formed, with the intention of utilizing new technologies to finish rapidly the sequencing of the human genome.

2000. The U.S. Patent and Trademark Office (PTO) establishes Art Unit 1631 (AU1631) specifically to determine bioinformatics patents. A year later the PTO's *Federal Register Report* announces that subsequent gene-based patents will be required to demonstrate "substantial utility" of their products.

2000. Partnerships between the biotech and information technology sectors are formed: Compaq and Celera, Sun and DoubleTwist, Myriad and Hitachi/Oracle, Apple and Human Genome Sciences, Motorla, Lucent Technologies, and IBM's life science informatics division.

2000. Two surveys estimate that more than 200 bioinformatics databases are in existence, with more than 500 bioinformatics software applications for sale and more than $160 million in software sales in the year 2000.

Appendix D: Biotechnology and Popular Culture; or, Mutants, Replicants, and Zombies

You have seen these types of commercials before. Even before you see the pastoral green fields, laughing grandchildren, confident couples, and sunshine smiles, you already know the commercials are selling you drugs—drugs for the ailments of old age, drugs for a virile sex life, drugs for depression, even drugs for kids with too much sugar and too many channels. The type of commercials made by the pharmaceutical industry has, by the early twenty-first century, become a genre in itself. But something is different about one particular commercial you are seeing now. For one, instead of an elderly man or woman complaining about arthritis, you see Frankenstein's monster. No, literally—he is there, all green, square head complete with bolts, right out of Boris Karloff's performance in the James Whale version from the 1930s. But, unlike Karloff's monster, this one is calm, gentle, talking to you about his arthritic problems ("that was me . . . what a stiff!"). It seems, then, that even monsters need their prescription drugs; this is a monster made by biotechnology, and it needs biotechnology to maintain itself. The monster, luckily, has found a new drug called Osteo Bi-Flex, for osteoarthritis. Osteo Bi-Flex enables the monster to enjoy his life, and we see him gardening, playing the guitar (acoustic) for kids, and doing tai chi in the park. He is a happier monster, leading a much more fulfilling, enriching life—thanks, of course, to Osteo Bi-Flex. No more Karloffian stumbling, grumbling, and terrorizing. The new monster of biotechnology is a bizarre kind of teratological pacifist.

The Osteo Bi-Flex commercial is an example of how layered the popular discourse of biotechnology is; this commercial is an ad for a new drug, but it also deconstructs the genre of the pharmaceutical commercial as well. In it, we see all the stock shots we would expect to see in commercials for Viagra or Celebrex: the initial "crisis" shot of an individual or couple experiencing pain, discomfort, anxiety, embarrassment, followed by a sequence announcing the drug, and concluding with idealized images of nature (gardening—there is always someone gardening), tranquility, and joy. In a sense, the Osteo Bi-Flex commercial demands to be read on several levels. It requires that viewers recognize certain cultural signs (*Frankenstein*, the Karloff version) and that they be able to comprehend the irony in the presentation (almost a mock commercial), but, in the end, it is actually selling a product. It is actually quite a good commercial for this reason alone.

Popular culture is arguably the site in which the ambiguities, anxieties, and tensions of the biotech industry get played out. Of course, the films, games, novels, and comics are often not totally aware that this is happening; or rather, it happens on a level distinct from any authorial "intention." Though a number of cultural theorists have treated popular culture in relation to biotechnology—works by Richard Doyle, Dorothy Nelkin, and Jon Turney come to mind—we still lack a way of understanding the relationship between biotechnology and popular culture that is unique to the concepts, technologies, issues, and knowledges specific to biotech. Cultural studies broadly speaking can, of course, add much to the study of biotechnology in popular culture, but too often such approaches are content to remain at the level of representation that happens anterior to science. Approaches from film studies are strong in their analysis of a particular medium, but it is rare that biotechnology is given concerted attention other than simply being the "content" of a film. The same may be said for the study of science fiction, where biotechnology is often understood to be the technology replacing the positions formerly held by nuclear technologies, information technologies, or nanotechnologies.

What I offer here is not a new approach or anything as ambitious as a methodology, but rather some loose, informal typologies. Clearly, there is much work to be done in sorting out how, for instance, genre is modified as we move from film to video games. But let us assume those problems and take a kind of flyover view of biotechnology in popular culture. I will not bother mentioning some of the more familiar themes in biotechnology-inspired popular culture: the mad scientist, the instrumentality of nature, the gothic in science fiction and horror, and the Frankenstein theme, seen in movies as varied as *Flesh for Frankenstein* (by Andy Warhol), *Young Frankenstein*, *Weird Science*, *Multiplicity*, and *The Hulk*.

Films "about" genetics and biotechnology are usually of two types: those that contain, almost incidentally, genetics and biotechnology, but only to motivate the action of the film, or those that use genetics and biotechnology to raise larger "human" issues. Examples of the former might be *Mimic*, *The Sixth Day*, and *Species*, as well as human-drama serials such as *CSI* or *ER* (though the latter often does raise bioethical issues). Examples of the latter might be Huxley's famous novel *Brave New World* and the films *Blade Runner* (a film that is about many things) and *Gattaca*.

But the intersection of biotechnology and popular culture has produced many other kinds of progeny. First, consider the genre I call *medical horror*. Medical horror includes films, novels, and other media within the horror genre that have medicine, medical practices, or medical technologies as a central part of their plots. A favorite of this genre is the "transplant horror" story, encapsulated in films such as *Eyes Without a Face*, *The Hands of Orlac*, *The Brain That Wouldn't Die*, and, more recently, *The Eye*. In such films, we often see an individual in an accident and in need of a transplant. By chance, the transplanted body part happens to be that of a psychotic killer. Upon transplantation, the body part then takes on a life of its own—or, rather, of its former master. "Parasite horror" is another theme in this genre, in films such as *The Tingler*, a William Castle production that features a parasite along the spinal column that is fed on fear and can be defeated only by screaming. Other variants of this theme include David Cronenberg's early films *Shivers* and *Rabid*, which take tissue transplantation into psychosexual, B-movie eroticism, and the Japanese thriller *Parasite Eve*, based on a novel of the same name, which portrays the apocalyptic awakening of mitochrondrial DNA in humans.

Medical horror owes a great deal to Mary Shelley's *Frankenstein*, but equally so to H. G. Wells' *The Island of Dr. Moreau* (an early narrative about tissue engineering), Algernon Blackwood's *John Silence* series, and Robert Louis Stevenson's *The Strange Case of Dr. Jekyll and Mr. Hyde* (which not only is about pharmaceuticals, but would go well with 1960s documentaries about Albert Hoffman and LSD). In medical horror, the core elements of medical practice are inverted, such that what previously guaranteed health now guarantees the vulnerability of biology. Stories by J. G. Ballard, Michael Blumlein, Thomas Disch, and Octavia Butler, and films such as *The Elephant Man* and *The Kingdom* (as well as the more canonized literary works, such as *Flowers for Alger-*

non and *The Cancer Ward*) take up the dark side of medicine, surgery, and anatomical study within the context of the science fiction and horror genres.

A second genre might be called *biohorror*. Biohorror most often features genetics and biotechnology as experiments gone wrong. The difference, however, is that in biohorror what goes wrong is that everything works perfectly. An engineered virus, for instance, "goes wrong" not because it does not work, but because it works too well. The molecular level takes on a life of its own, with aims that are often counter to those of the human hosts they inhabit. Greg Bear's novels *Blood Music* and *Darwin's Radio* and Joan Slonczewski's *Brain Plague* are variations on this theme, as are Greg Egan's *Teranesia* and Michael Crichton's nanotechnology-inspired *Prey*. Biohorror narratives always involve some sort of extreme but ambivalent transformation. Biohorror is often accompanied by morphological transformation of the human host, as in Cronenberg's version of *The Fly*, or by unrecognizable metamorphosis, as in John Carpenter's remake of *The Thing* (the original, *The Thing from Another World*, is equally good and might be read as the first transgenic biotech film). Biohorror also has social and cultural dimensions, as exhibited in Octavia Butler's "xenogenesis" trilogy and her story "Bloodchild."

The mention of Butler's work raises another focus within biohorror, which is the human-alien hybrid, most often via some sort of virus that follows the logic of alien flesh. *Day of the Triffids* and *Invasion of the Body Snatchers* portray carnivorous plants and clones in a manner not unlike zombie films. Sometimes biohorror results not in the death of the human host, but in a metamorphosis into the posthuman or superhuman. Hence, the predilection of comics (and subsequently films) based on superheros. The *X-Men* mythos is given a particularly genetic twist in the film versions, and *Spider-Man* and *The Hulk* utilize genetics to frame social and psychoanalytic issues of fathers, families, and authority. Biotech encounters pharmaceuticals and reproductive technologies in Cronenberg's *The Brood*, whose unique "psychoplasmics" technologies manifest psychological pathologies as literal growths on the body of the mother figure in the film.

Finally, another genre worth mentioning is the *zombie-epidemic* genre. Actually, this genre combines two already familiar genres: the zombie movie in horror films and the medical thriller novel (represented by the novels of Michael Crichton, Robin Cook, and Richard Preston). The former was established largely in the United States with George Romero's "living dead" trilogy, films that are political satire much more than horror. In Italy, the extensive tradition of horror film there led to Lucio Fulci's *Zombie*, Aldo Lado's *Short Night of Glass Dolls*, and Armando Crispino's *Autopsy*. However, the zombie film is much older than this. In the late 1920s, Bela Lugosi starred in *White Zombie*, followed by Jacques Tournier's *I Walked with a Zombie*, both of which emphasize the colonial context of Haiti and the practices of voodoo. This context has meant that, even for later directors such as Romero, the zombie is indelibly linked to the "silent majority" and slave labor. This connection is further evidenced by mid-twentieth-century films such as Hammer Studio's *Plague of Zombies*.

The other genre that informs the zombie-epidemic story is the medical thriller, or outbreak novel (many of which are made into films). *The Andromeda Strain*, *Coma*, and *Outbreak* are examples in film, and the novels *Vector*, *The Cobra Event*, and *The Hot Zone* are examples in fiction (the last of these is nonfiction, though written in a fictional style). These narratives usually feature either a naturally occurring or top-secret military experiment that "gets loose," starting a chain of reactions that a "virus hunter" or scientist-protagonist must stop, in the process navigating government bureaucracies, civilian victims, and an often conservative military or government bureaucracy or both. However, the outbreak genre itself harkens back to novels such as Daniel Defoe's *A Journal of the Plague Year* and Mary Shelley's underappreciated *The Last Man*. The narrative motifs of contagion, apocalypse, and survival are established in such works.

Although many zombie films are predicated on the supernatural (e.g., Fulci's *The Beyond* and *City of the Walking Dead*), zombie-epidemic films are predicated on science, biology, and epidemiology. I have already mentioned Romero's *The Crazies* and *28 Days Later*, and to them I might add films such as the campy *Omega Man* (starring Charlton Heston as the "last man" and scientist), the Japanese thriller *Cure*, the "virus as serial killer" film *Cabin Fever*, and indie director Georg Koszulinski's film *Blood of the Beast*. A different, but related variant of the zombie-epidemic genre is the cult film *C.H.U.D.* (Cannibalistic Humanoid Underground Dwellers), which combines environmental toxiciology with references to homelessness. In Italy, the tradition of "mondo" exploitation films spawned Ruggero Deodato's disturbing *Cannibal Holocaust*, a critique of both capital consumption and the visual consumption of images. Finally, more "high-brow" examples, such as Albert Camus's novel

The Plague and Ingmar Bergman's film *The Seventh Seal*, are exemplary in their ability to raise a host of complex social, psychological, and even metaphysical issues.

The zombie-epidemic genre is, of course, not limited to film and fiction. In the video game medium, the *Resident Evil* series has explored the zombie-epidemic genre the most, and the film versions build on the action basis of the video games, with minimal plot devoted to the top-secret, underground, corporate biotech firm doing bioweapons research. Other games, such as *Half-Life* and *Doom*, feature creatures that are the result of biotechnology gone horribly wrong, but their significance is limited to their being target practice. Contemporary comics such as *Criminal Macabre*, *Necrowar*, and *Y: The Last Man* continue to explore the epidemic theme in relation to other fields, from medieval history to cosmology.

These are just some of the unique, hybrid genres that have developed in science fiction and horror in relation to biotechnologies. However, a great deal of the studies of biotechnology and popular culture, although extremely valuable in their own right, often presuppose the anteriority of science to science fiction, an assumption that ends up reducing science fiction to a "reaction" or "response" to a previously existing scientific phenomenon. This is the case in a great many examples, as the preponderance of genetics and nanotechnology themes in current films indicate. A more complex cultural studies approach, however, would also have to resist this characterization, for the role of culture is not simply to "mop up" after science has done its job (the sort of "oh, so *that's* what was going on" approach to cultural analysis).

It is not difficult to think of examples of the reverse, when science fiction actually preceded, served as the ground for, or even predicted science (in the most relativistic sense, Ovid practiced transgenics). But perhaps the point should not be one of priority or origins, but of discovering a critical and necessary relationship between biotechnology and culture. Perhaps it is precisely in this zone that we find an ethics and a politics of biotechnology, though not simply the pragmatic kind of bioethics of maxims, but a more exploratory, foundational sort of ethics that ultimately questions foundational concepts such as "life itself."

Popular culture cannot do this, and we would be naive to think that, for instance, Italian B horror films can serve as the ethical model of the future. But this is the point: popular culture is at its best precisely when it is unaware, or even only half-aware, of the issues it is raising. Clearly, if you want to learn about the science of genetics, then watching *The Hulk* is not your best bet. But in the recombinations of pharmaceutical ads, science fiction films, video games, television soap operas, music videos, comics, and reality TV, biotechnology is presented in an incredibly polyvalent manner. Perhaps it can even be argued that popular culture is *the* site in which to understand biotechnology not just as a set of techniques, therapies, and services, but rather as a way of thinking about the body socially, economically, and politically as well.

Notes

Introduction

1. The quotation given in the section heading is from "Weakling Avenger" by Darkthrone, ©2001 Moonfog Productions. Quoted in Girolamo Cardano, "Medical Metaphors in Renaissance Political Theory," *Socius* 4 (April 1994), p. 31.

2. Karl Marx, *Grundrisse: Foundations of the Critique of Political Economy*, trans. Martin Nicolaus (New York: Penguin, 1973), pp. 91, 86.

Chapter 1

1. Anthony Giddens, *The Consequences of Modernity* (Cambridge: Polity, 1990), p. 64.

2. Saskia Sassien, *Globalization and Its Discontents* (New York: New Press, 1998), p. xxv.

3. Malcolm Waters, *Globalization* (London: Routledge, 2001), p. 5.

4. Immanuel Wallerstein, *The Modern World System* (New York: Academic, 1974), p. 15.

5. Manuel Castells, *The Rise of the Network Society* (New York: Blackwell, 1996), p. 67. Castells also notes, in a few unexplored passages, the effects of the information technology paradigm on biology: "Technological convergence increasingly extends to growing interdependence between the biological and microelectronics revolutions,

both materially and methodologically. Thus, decisive advances in biological research, such as the identification of human genes or segments of human DNA, can only proceed because of massive computing power" (p. 63). Although such statements tend to ignore the instrumental role that midcentury cybernetics and its related fields had in the development of a biotech "industry," Castells does note the significance of biotech within globalization. See pp. 47–50.

6. Michael Hardt and Antonio Negri, *Empire* (Cambridge, Mass.: Harvard University Press, 2000), pp. 289–294.

7. Giddens, *The Consequences of Modernity*, p. 21.

8. See Waters, *Globalization*, pp. 17–21.

9. Some, such as Castells (and, arguably, Hardt and Negri), make this the basis for their analyses, whereas others such as Sassien attempt to temper the undue reliance on information technologies by contrasting them or reintegrating them into physical space and geopolitical relationships.

10. Certainly, this is the very point of contention for critics of globalization, for, in many cases, the material conditions of labor, immigration, and tensions within nation-states are far from having been overcome. In addition, the various antiglobalization movements have in many cases appropriated information and media networks to make their own voices heard, creating a kind of "tactical media" out of the very tools that underlie the globalizing process.

11. A common practice since the late 1990s has been to create foreign subsidiary companies to handle separate parts of the product-development process, such as clinical trials or marketing. For an industry perspective, see Jennifer van Brunt, "Borderless Biotech," *Signals Magazine*, December 13, 2000, online at http://www.signalsmag.com.

12. See Eugene Thacker, *Biomedia* (Minneapolis: University of Minnesota Press, 2004).

13. Michael Fortun, "The Human Genome Project: Past, Present, and Future Anterior," in *Science, History, and Social Activism: A Tribute to Everett Mendelsohn*, eds. E. Allen Garland and Roy M. MacLeod (Dordrecht: Kluwer, 2002), p. 355.

14. See Jackie Stevens, "PR for the 'Book of Life,'" *The Nation* (special edition), November 21, 2001.

15. On the institutional politics of the HGP, see Robert Cook-Deegan, *The Gene Wars* (New York: Norton, 1994).

16. See Francis Collins, Michael Morgan, and Ari Patrinos, "The Human Genome Project: Lessons from Large-Scale Biology," *Science* 300 (April 11, 2003): 286–290.

17. For industry reports on bioinformatics, see Oscar Gruss Co., "Trends in Commercial Bioinformatics," *Oscar Gruss Biotechnology Review* (March 13, 2000), available at http://www.oscargruss.com; and Silico Research Limited, *Bioinformatics Platforms*, Research Report/Executive Summary (November 2000), available at http://www.silico-research.com.

18. Georges Canguilhem, "The Concept of Life," in *A Vital Rationalist: Selected Writings*, ed. François Delaporte (New York: Zone, 2000), p. 316.

19. See Lily Kay, *Who Wrote the Book of Life?: A History of the Genetic Code* (Stanford, Calif.: Stanford University Press, 2000), and Evelyn Fox Keller, *Refiguring Life: Metaphors of Twentieth-Century Biology* (New York: Columbia University Press, 1995).

20. On classical information theory, see Claude Shannon and Warren Weaver, *The Mathematical Theory of Communication* (Chicago: University of Illinois Press, 1965). For a summary of information theory in relation to molecular biology, see Kay, *Who Wrote the Book of Life?* pp. 91–102.

21. Michel Foucault, "Candidacy Presentation: Collège de France, 1969," in *Ethics: Subjectivity and Truth: The Essential Works of Foucault, 1954–1984*, vol. 1, trans. Robert Hurley et al., ed. Paul Rabinow (New York: New Press, 1997), pp. 7–8. As Foucault notes, "It [the knowledge of heredity] developed throughout the nineteenth century, starting from breeding techniques, on through attempts to improve species, experiments with intensive cultivation, efforts to combat animal and plant epidemics, and culminating in the establishment of a genetics whose birth date can be placed at the beginning of the twentieth century" (p. 7).

22. Michel Foucault, *"Society Must Be Defended": Lectures at the Collège de France, 1975–1976*, ed. Mauro Bertani and Alessandro Fontana, trans. David Macey (New York: Picador, 2003), p. 243.

23. Michel Foucault, *The History of Sexuality*, vol. 1, *An Introduction*, trans. Robert Hurley (New York: Vintage, 1978), p. 137.

24. Ibid., 1: 245.

25. Foucault, "The Birth of Biopolitics," in *Ethics*, p. 73.

26. Foucault, "*Society Must be Defended*," p. 246.

27. We have seen this in the modern events surrounding "emerging infectious disease," but it already exists in the push toward the computerization of health care ("infomedicine") and the use of information technologies to mediate doctor-patient relationships ("telemedicine").

28. Ian Hacking, "Biopolitics and the Avalanche of Printed Numbers," *Humanities in Society* 5 (1982), p. 281.

29. Ibid., p. 287.

30. I am grateful to Alex Galloway for this encapsulation.

31. Foucault, *The History of Sexuality*, 1: 139.

32. Foucault, "*Society Must be Defended*," pp. 246–247.

33. Michel Foucault, "The Politics of Health in the Eighteenth Century," in *Power: The Essential Works of Foucault, 1954–1984*, vol. 3, trans. Robert Hurley et al., ed. James D. Faubion (New York: New Press, 2000), p. 95. "Security" is as much an economic and cultural notion as it is a military one; the most evident militaristic example currently is the recent scare over bioterrorism (the budgetary boost in the U.S. biowarfare research program). But biopolitics also remains consonant with neoliberalism in its notion of a medical-economic security in the form of health insurance, home care, outpatient services, and the development of biological "banking" institutions (sperm and ova banks, blood banks, tissue banks, etc.).

34. Giorgio Agamben, *Homo Sacer: Sovereign Power and Bare Life*, trans. Daniel Heller-Roazen (Stanford, Calif.: Stanford University Press, 1998), p. 11.

35. Hardt and Negri, *Empire*, pp. 289–294.

36. For example, in a short book review published in *Le Monde* in 1976 (just after the publication of *Discipline and Punish*), Foucault criticizes what we might now call "pop science" books, which, in a monumental sweep, move from the minutiae of the mol-

ecule to the structure of human societies. See Michel Foucault, "Bio-histoire et bio-politique," in *Dits et écrits 1954–1988*, ed. Daniel Defert, François Ewald, and Jacques Lagrange (Paris: Gallimard, 1994), pp. 94–97. In particular, Foucault notes that we should be cautious in too easily moving from the molecule to the masses, as if the line between biology and culture were really a matter of homogeneity and scale. He notes that diversity and diversification is the key to understanding contemporary biopolitics, that "the species must not be defined by a prototype but by a set of variations" (p. 96, my translation). In this short article Foucault also makes an explicit connection to genetics. In genetics, the population is not just defined by their manifest, visible characteristics, but, "molecular biology has permitted the repair of factors dependent on the immunological structure and the enzymatic equipment of cells—characteristics whose condition is rigorously genetic" (p. 96, my translation).

37. As Ian Hacking has shown in his thorough historical study, statistics is well known to have grown out of biology as a particular field of application (biology as statecraft, as it were). Likewise, scientific method was central to Lamarck in the very concept of a science of life, or *bio-logy*, and the ability to quantifiably account for the biological was the central promise of the life sciences and biology. Add to this another, equally relevant development: that of the beginnings of digital computers and software. In order to examine fully the changes Foucault broadly points to during the late eighteenth through the nineteenth century, we would have to take into account the work of Charles Babbage and Ada Lovelace. Perhaps, then, the real nexus for biopolitics is not just in the development of statistics or in the life sciences, but in the biotechnical nexus between statistics and biology, informatics and genetics, Babbage and Darwin, Lovelace and Lamarck.

38. Karl Marx, *Economic and Philosophic Manuscripts of 1844*, trans. Martin Milligan (New York: Prometheus, 1988), p. 71.

39. Ibid., p. 71. This basic formulation remains in Marx's later works, but it is largely subsumed under the discussion of commodities and labor power. However, in the first volume of *Capital*, for instance, Marx further qualifies the estrangement of labor by noting how it must emerge from the treatment of the body as property. See *Capital*, vol. 1 (New York: Penguin, 1990), p. 271.

40. Marx, *Economic and Philosophic Manuscripts*, p. 71, emphasis in original.

41. Ibid., p. 72.

42. Ibid., p. 76.

43. Karl Marx, *Grundrisse: Foundations of the Critique of Political Economy*, trans. Martin Nicolaus (New York: Penguin, 1973), p. 272, emphasis in original.

44. Marx, *Capital*, Vol. 1: 290.

45. In this sense, Marx relies a great deal on Hegel—Hegel the philosopher but also Hegel the biologist. It is not difficult to see Hegel's organicism behind Marx's emphasis on the "species being," and indeed the sections of the 1844 manuscripts on estranged labor can be read side by side with the sections on Hegel's introduction to *Phenomenology of Spirit*.

46. Marx, *Economic and Philosophic Manuscripts*, p. 81 (italics removed).

47. Marx further notes that productive activity is estranged because it is external to the human subject (it does not directly benefit the subject), because within the capitalist system it is necessary or "coerced" labor (labor as an immediate "means of life"), and because it belongs to someone else (the capitalist) and is thus further rendered as property. See *Economic and Philosophic Manuscripts*, p. 74.

48. Ibid., pp. 77–78, emphasis in original.

49. Nick Dyer-Witheford's work on the biopolitical dimensions of Marx's notion of "species being" is an important contribution in this area. Dyer-Witheford makes a helpful distinction between "human commodities" (consumers-subjects), "chosen peoples" (fundamentalism), and "commoners" (collective dissent movements). See Dyer-Witheford's paper "1844/2004/2044: The Return of Species Being" submitted to *Historical Materialism* (Jan. 2004). However, although Dyer-Witheford reads the "species being" in the broader context of the antiglobalization movement, my use of Marx's term remains rooted in the way that the biotech industry recontextualizes biology as a type of labor power. My main point of interest is the strange and frustrating "nonhuman" character of biomaterial labor.

50. Marx, *Economic and Philosophic Manuscripts*, p. 75, emphasis in original.

51. Marx, *Capital*, 1: 283. Marx makes a similar statement elsewhere in *Capital*: "We mean by labor-power, or labor-capacity, the aggregate of those mental and physical capabilities existing in the physical form, the living personality, of a human being, capabilities which he sets in motion whenever he produces a use-value of any kind" (p. 270). And again: "Labor-power exists only as a capacity of the living individual. Its production consequently presupposes his existence" (p. 274).

52. In the 1844 manuscripts Marx notes: "Admittedly animals also produce. They build themselves nests, dwellings, like the bees, beavers, ants, etc. But an animal only produces what it immediately needs for itself or its young. It produces one-sidedly, while man produces universally" (*Economic and Philosophic Manuscripts*, p. 77). And again, though somewhat more ambiguously, in *Capital*: "We presuppose labor in a form in which it is an exclusively human characteristic. A spider conducts operations which resemble those of the weaver, and a bee would put many a human architect to shame by the construction of its honeycomb cells. But what distinguishes the worst architect from the best of bees is that the architect builds the cell in his mind before he constructs it in wax. At the end of every labor process, a result emerges which had already been conceived by the worker at the beginning, hence already existed ideally" (1: 283–284). There is much to explore in such passages regarding the contemporary research into the "self-organization" and complexity of ant swarming, wasp nest building, bird flocking, and so on.

53. Marx, *Economic and Philosophic Manuscripts*, p. 76, emphasis in original.

54. Ibid., p. 77, emphasis in original.

55. These two meanings of the term are encapsulated by Marx: "just as nature provides labor with the *means of life* in the sense that labor cannot live without objects on which to operate, on the other hand, it also provides the *means of life* in the more restricted sense—i.e., the means for the physical subsistence of the *worker* himself." Ibid., p. 72, emphasis in original.

56. Ibid., pp. 75–76, emphasis in original. In volume 1 of *Capital*, Marx speaks of the instruments of labor in a way that recalls his earlier formulation of the "inorganic body," in effect placing the inorganic body of the 1844 manuscripts within the context of capitalism: "Leaving out of consideration such ready-made means of subsistence as fruits, in gathering which a man's bodily organs alone serve as the instruments of his labor, the object the worker directly takes possession of is not the object of labor but its instrument. Thus nature becomes one of the organs of his activity, which he annexes to his own bodily organs" (1: 285).

57. Marx, *Economic and Philosophic Manuscripts*, p. 70.

58. Marx, *Grundrisse*, p. 84.

59. Foucault, "*Society Must be Defended*," p. 243.

60. Ibid., p. 243.

61. Foucault, "The Politics of Health in the Eighteenth Century," in *Ethics*, pp. 95–96.

62. Paolo Virno, *A Grammar of the Multitude*, trans. Isabella Bertoletti, James Cascaito, and Andrea Casson (New York: Semiotext[e], 2004), p. 82. Virno continues, stating that "capitalists are interested in the life of the worker, in the body of that worker, only for an indirect reason: this life, this body, are what contains the faculty, the potential, the *dynamis*" (pp. 82–83). The concept of "biomaterial labor" discussed more fully in chapters 3 and 4 differs from Virno's interpretation of Foucauldian biopolitics, principally in that it suggests that the biotech industry operates beyond post-Fordism and is able effectively to separate the prior indissociability of life and labor. Genes, enzymes, and cell perform a specific type of biological labor in the lab, apart from the body.

63. Even in the early, so-called humanist Marx, the human subject exists in a conflicted relation to nature. On the one hand, in the human-nature relation, production is fully integrated with "life activity" insofar as all production is direct use value. On the other hand, the very act of appropriation, of working upon nature, immediately objectifies nature, thereby objectifying the human subject's "inorganic body." At some points, Marx seems to want to preserve some essential, even extraeconomic relation between human and nature, but at other points (especially in his commentaries on Hegel) he acknowledges that this "knowledge of estrangement" is in fact part of the human condition.

64. The phrase *living dead labor* is an homage to the subgenre of zombie-epidemic films, which, in many ways, are the most astute critique of the biotech industry. See appendix IV.

65. For more on the Mo-cell line, see Lori Andrews and Dorothy Nelkin, *Body Bazaar: The Market for Human Tissue in the Biotechnology Age* (New York: Crown, 2001), pp. 25–31. Also see E. Richard Gold, *Body Parts: Property Rights and the Ownership of Human Biological Materials* (Washington, D.C.: Georgetown University Press, 1996), pp. 23–33.

66. For more on the OncoMouse, see Donna Haraway, *Modest_Witness@Second_Millennium.FemaleMan©_Meets_OncoMouse™: Feminism and Technoscience* (New York: Routledge, 1997), pp. 87–85. On transgenics generally, see J. A. H. Murray, ed., *Transgenesis* (New York: Wiley and Sons, 1992).

67. For more on mammalian bioreactors, see Andrew Kimbrell, *The Human Body Shop: The Cloning, Engineering, and Marketing of Life* (Washington, D.C.: Gateway, 1997), pp. 207–225. On medical transgenics generally, see R. B. Church, ed., *Transgenic Models in Medicine and Agriculture* (New York: Wiley-Liss, 1990).

68. In fact, biological estrangement extends to human beings in its more conventional context as well. The controversy surrounding the patenting of cell lines is an example. When a biopsy is taken from a patient and then cultured into a cell line and patented, the patient's biology continues to "work" independently of the patient and appears as a kind of body-outside-the-body (in a strange way, a biotechnical instance of an "out of body experience").

Chapter 2

1. See Erwin Schrödinger, *What Is Life?* (Cambridge: University of Cambridge Press, 1967), pp. 20–22.

2. See Francis Crick's papers published during the 1960s, such as "The Genetic Code," *Scientific American* 207 (1962): 66–75; "The Recent Excitement in the Coding Problem," *Progress in Nucleic Acids Research* 1 (1963): 163–217; and "Towards the Genetic Code," *Discovery* 23.3 (1962): 8–16.

3. See their landmark paper, François Jacob and Jacques Monod, "Genetic Regulatory Mechanisms in the Synthesis of Proteins," *Journal of Molecular Biology* (1961): 318–59. Also see Jacob's *The Logic of Life: A History of Heredity* (New York: Pantheon, 1974), pp. 267–298.

4. See their paper, Heinrich Matthai, Oliver W. Jones, Robert G. Martin, and Marshall Nirenberg, "Characteristics and Composition of RNA Coding Unit," *Proceedings of the National Academy of Sciences* 48 (1962): 1580–1588. Also see Marshall Nirenberg, "The Genetic Code II," *Scientific American* 208 (1963): 80–94.

5. See Lily Kay, *Who Wrote the Book of Life? A History of the Genetic Code* (Stanford, Calif.: Stanford University Press, 2000); and Hans-Jörg Rheinberger, *Toward a History of Epistemic Things: Synthesizing Proteins in the Test Tube* (Stanford, Calif.: Stanford University Press, 1997).

6. See Claude Shannon's Ph.D. dissertation, "An Algebra for Theoretical Genetics," Department of Mathematics, Massachusetts Institute of Technology, April 15, 1940. See also John von Neumann's lectures, republished as *The Computer and the Brain* (New

Haven, Conn.: Yale University Press, 2000), pp. 68–70. Von Neumann, along with Alan Turing, was partially responsible for introducing biological concepts into computer science, thereby providing the discourse of artificial intelligence with a biological dimension.

7. As Judith Butler notes, "the anatomical is only 'given' through its signification, and yet it appears to exceed that signification, to provide the elusive referent in relation to which the variability of signification performs. Always already caught up in the signifying chain by which sexual difference is negotiated, the anatomical is never given outside its terms and yet it is also that which exceeds and compels that signifying chain, that reiteration of difference, an insistent and inexhaustible demand." Judith Butler, *Bodies That Matter* (New York: Routledge, 1993), p. 90.

8. Bioinformatics is one of the more fascinating subdisciplines within biotech, spawning innovative, research-based approaches involving genetic algorithms and distributed computing. Computer companies such as Compaq and IBM have been involved, the latter having spent $100 million in starting its life science initiative around 2000–2001. In addition, in science industry publications such as *Nature Biotechnology* and *Biospace.com*, there is a growing demand for individuals with both molecular biology and computer programming backgrounds for bioinformatics jobs.

9. For more on bioinformatics, see D. Benton, "Bioinformatics: Principles and Potential of a New Multidisciplinary Tool," *Trends in Biotechnology* 14.8 (August 1996): 261–272; Diane Gershon, "Bioinformatics in a Post-Genomics Age," *Nature* 389 (September 27, 1997): 417–418; Brendan Horton, "The Breakdown on Bioinformatics," *Nature* 388 (August 7, 1997): 603–605; Ken Howard, "The Bioinformatics Gold Rush," *Scientific American* (July 2000): 58–63; N. M. Luscombe, D. Greenbaum, and M. Gerstein, "What Is Bioinformatics? A Proposed Definition and Overview of the Field," *Methods of Informatic Medicine* 40 (2001): 346–358; Aris Persidis, "Bioinformatics," *Nature Biotechnology* 17 (August 1999): 828–830; and Paolo Saviotti et al., "The Changing Marketplace of Bioinformatics," *Nature Biotechnology* 18 (December 2000): 1247–1249.

10. The immediate technical predecessors to bioinformatics databases were the first attempts to "enumerate" genetic and protein molecules as informational sequences. For early examples of this approach applied to protein classes, see Margaret Dayhoff et al., *Atlas of Protein Sequence and Structure*, vol. 1 (Silver Spring, Md.: National Biomedical Research Foundation, 1965). For an application to individual protein molecules, see R. W. Holley and D. M. Powers "The Base Sequence of Yeast Alanine

Transfer RNA," *Science* 147 (1965): 1462–1465; and Fredrick Sanger and E. O. P. Thompson, "The Amino-Acid Sequence in the Glycyl Chain of Insulin," *Biochemical Journal* 53 (1953): 353–366; 366–374.

11. The first computer databases of genetic and protein data were established in the 1980s and 1990s, as personal computing made database management (and, later, online services) more affordable for university-based research labs. See E. E. Abola et al., "Protein Data Bank," in *Crystallographic Databases*, ed. F. H. Allen et al. (Cambridge: Data Commission of the International Union of Crystallography, 1987), pp. 107–132. Also see F. C. Bernstein, "Protein Data Bank: A Computer-Based Archival File for Macromolecular Structures," *Journal of Molecular Biology* 112 (1977): 535–542.

12. See Persidis, "Bioinformatics."

13. See Jon Agar, "History of Science on the World Wide Web," *British Journal for the History of Science* 29 (1996): 223–227.

14. BLAST has become somewhat of a standard in bioinformatics research, providing a comprehensive cross-database search that includes all data associated with the public genome project as well as bibliographic data on related research (not unlike a library search) and data from other biological databases such as the Protein Data Bank. See S. F. Altschul et al., "Basic Local Alignment Search Tool," *Journal of Molecular Biology* 215 (1990): 403–410.

15. On the development of lab technologies for biotech, see John Hodgson, "Gene Sequencing's Industrial Revolution," *IEEE Spectrum* (November 2000): 36–42. On PCR, see Paul Rabinow, *Making PCR: A Story of Biotechnology* (Chicago: University of Chicago Press, 1996).

16. Although early online databases were public domain, biotech corporations' serious interest brought about privatized databases, which can be accessed through company subscriptions. Many of the databases discussed here exist exclusively online because this approach facilitates more efficient uploading of new data into a centralized information resource. For examples of research, see A. Bairoch and B. Broeckmann, "The SWISS-PROT Protein Sequence Data Bank," *Nucleic Acids Research* 19 (1991): 2247–2249; and Michael Krawczak et al., "Human Gene Mutation Database—A Biomedical Information and Research Source," *Human Mutation* 15 (2000): 45–51.

17. The issue of cross-platform compatibility and standardization of file formats has become pressing in the bioinformatics industry. Although software companies want

to protect their interests, a number of researcher-led groups have attempted to initiate protocols for bioinformatics research. The effort to establish these "ontology" standards seems to involve more than technical configuration; it points to the central problem in bioinformatics in managing "information" across both "wet" and "dry" media or platforms (in vitro and in silico). See Clare Samson, "Finding a New Language for Bioinformatics," *Nature Biotechnology* 15 (November 1997): 1253–1256.

18. See the reports from Oscar Gruss Co., "Trends in Commercial Bioinformatics," *Oscar Gruss Biotechnology Review* (March 13, 2000), available at http://www.oscargruss.com; and Silico Research Limited, *Bioinformatics Platforms*, Research Report/Executive Summary (November 2000), available at http://www.silico-research.com.

19. See Pharmaceutical Research and Manufacturers Association (PhRMA), *Pharmaceutical Industry Profile 2003* (Washington, D.C.: PhRMA, 2003).

20. Ibid.

21. Ibid. PhRMA is one of the leading pharmaceutical industry organizations worldwide. Of course, its reports reflect the political and health-care-related ideologies of the pharmaceutical industry, and it is thus also a powerful lobbying body surrounding issues such as Medicare.

22. See Philip Dean, Paul Gane, and Edward Zanders, "Pharmacogenomics and Drug Design," in *Pharmacogenomics: The Search for Individualized Therapies*, ed. Julio Licino and Ma-Li Wong (Weinheim, Germany: Wiley-VCH, 2002), pp. 143–158.

23. For an example, see Klaus Lindpaintner, "The Importance of Being Modest: Reflections on the Pharmacogenetics of Abacavir," *Pharmacogenomics* 3.6 (2002): 835–838.

24. The study, published in November of 2000, can be accessed at http://www.genewatch.org. It was also published on *The Guardian* Web site at http://www.guardianunlimited.co.uk.

25. Urs Meyer, "Introduction to Pharmacogenomics: Promises, Opportunities, and Limitations," in *Pharmacogenomics: The Search for Individualized Therapies*, ed. Julio Licino and Ma-Li Wong (Weinheim, Germany: Wiley-VCH, 2002), p. 3. For more on pharmacogenomics, see D. S. Bailey, A. Bondar, and L. M. Furness, "Pharmacogenomics: It's Not Just Pharmacogenetics," *Current Opinion in Biotechnology* 9 (1998): 595–601; Thomas Bumol and August Watanabe, "Genetic Information, Genomic

Technologies, and the Future of Drug Discovery," *Journal of the American Medical Association* 285 (2001): 551–555; Brian Claus and Dennis Underwood, "Discovery Informatics: Its Evolving Role in Drug Discovery," *Drug Discovery Today* 7.18 (September 2002): 957–966; W. E. Evans and M. V. Relling, "Pharmacogenomics: Translating Functional Genomics into Rational Therapeutics," *Science* 286.5439 (1999): 487–491; G. S. Ginsburg and J. J. McCarthy, "Personalized Medicine: Revolutionizing Drug Discovery and Patient Care," *Trends in Biotechnology* 19.12 (2001): 491–496; M. W. Linder and R. Valdes Jr., "Pharmacogenetics in the Practice of Laboratory Medicine," *Molecular Diagnosis* 4 (1999): 365–379; Roy Pettipher and Lon Cardon, "The Application of Genetics to the Discovery of Better Medicines," *Pharmacogenomics* 3.2 (2002): 257–263; and A. D. Roses, "Pharmacogenetics and the Practice of Medicine," *Nature* 405 (2000): 857–865.

26. The phrase has also been used recently by Donna Haraway, Sarah Franklin, and Nikolas Rose. The use here certainly builds on their work, but is also meant to refer specifically to this dual political and ontological aspect of fields such as bioinformatics and pharmacogenomics, which integrate biology and information in novel ways.

27. Nikolas Rose, "The Politics of Life Itself," *Theory, Culture, and Society* 18.6 (2001), p. 13. Rose notes how this "molecular politics" is also related to the reformulation of "risk" in the genetic context and, furthermore, how this genetic risk impacts the way individuals frame their own health and lifestyles.

28. Sarah Franklin, "Life Itself: Global Nature and the Genetic Imaginary," online at the Department of Sociology, Lancaster University, http://www.comp.lancs.ac.uk/sociology/soc048sf.html. As Franklin states, "the very concept of 'genetic information' is a collapse of matter and message at the heart of the contemporary life sciences." Franklin's focus is primarily the "genetic imaginary," or the complex social process through which genetic "life itself" is imagined, such as seen in the film, merchandise, spin-off documentaries, and museum exhibits surrounding *Jurassic Park*.

29. See S. L. Teitelbaum, "Bone Resorption by Osteoclasts," *Science* 289 (September 1, 2000): 1504; and D. S. Yamashita and R. A. Dodds, "Cathepsin K and the Design of Inhibitors of Cathepsin K," *Current Pharmaceutical Design* 6.1 (January 2000): 1–24.

30. See the press releases at the Human Genome Sciences Web site, http://www.hgsci.com.

31. Interestingly enough, this claim was made not only in Human Genome Sciences' press releases, but also in an advertisement for Apple computers, available at

http://www.apple.com. Michael Fannon, Vice President for Human Genome Sciences, states that "the Power Macintosh platform gives us the computational power we need, coupled with maintainability and ease of use that, to this day, have never been replicated on other platforms."

32. See the press releases from Celera Genomics, available at http://www.celera.com.

33. PhRMA, *Pharmaceutical Industry Profile 2003*, p. 15.

34. U.S. PTO no. 5,501,969 and U.S. PTO no. 5,861,29, respectively. These patents can be viewed online at http://www.uspto.gov.

35. For more on the techniques of pharmacogenomics, see Thomas Lengauer, *Bioinformatics: From Genomes to Drugs* (Weinheim, Germany: Wiley-VCH, 2002). Also see the articles in Julio Licino and Ma-Li Wong, eds., *Pharmacogenomics: The Search for Individualized Therapies* (Weinheim, Germany: Wiley-VCH, 2002).

36. Georges Canguilhem, "The Concept of Life," in *A Vital Rationalist*, ed. François, Delaporte, trans. Arthur Goldhammer (New York: Zone, 2000), p. 303.

37. Ibid. Canguilhem's proposition needs to be qualified, of course, for it can quite easily be argued that Plato's theory of forms and the faculty of reason in the *Timaeus*—in its analogizing between anatomy and the Platonic idea—is equally valid as a candidate for the philosophy of biology. However, Canguilhem extends his analyses of Aristotle to point to the way in which both the study of the concept and the study of organic life were circumscribed within a more rigorous, analytical framework. "It was Aristotle the naturalist who based his system for classifying animals on structure and mode of reproduction, and it was the same Aristotle who used that system as a model for his logic" (p. 303).

38. For a review of this phenomenon, see my article "Shattered Body, Shattered Self," *Afterimage* 29.5 (March–April 2002).

39. "Le concept et la vie" was first published in *Revue philosophique de Louvain* 64 (May 1966): 193–223.

40. Canguilhem, "The Concept of Life," p. 303.

41. Ibid., pp. 316–317.

42. Ibid., p. 316.

43. As Canguilhem notes in a separate essay, "it is not only the history of anatomy and physiology that begins with Aristotle but also the history of what was long called 'natural history,' including the classification of living things, their orderly arrangement in a table of similarities and differences, study of their kinship through morphological comparison, and, finally, study of the compatibility of different modes of existence." *Ideology and Rationality in the Life Sciences*, trans. Arthur Goldhammer (Cambridge, Mass.: MIT Press, 1988), p. 133.

44. "It must then be the case that the soul is substance as the *form* of a natural body which potentially has life, and since this substance is actuality, soul will be the actuality of such a body." See Aristotle, *De anima* (On the soul), trans. Hugh Lawson-Tancred (New York: Penguin, 1986), book II, section 1, p. 157.

45. The other famous example Aristotle mentions is the analogy of the eye: "For if the eye was an animal, then sight would be its soul, being the substance of the eye that is in accordance with the account of it. . . . So just as pupil and sight *are* the eye, so, in our case, soul and body *are* the animal." *De anima*, book II, section 1, p. 158.

46. Georges Canguilhem, "The Epistemology of Biology," in *A Vital Rationalist*, p. 80, 81.

47. Ibid., p. 68. Canguilhem cites Georges Cuvier's notion of "modes of life," which is instructive in the context of contemporary bioinformatics and computational biology: "This flux [of biological life] is composed of molecules, which change individually yet remain always the same type. Indeed, the actual matter that constitutes a living body will soon have dispersed, yet that matter serves as the repository of a force that will compel future matter to move in the same direction." Georges Cuvier, *Rapport historique sur les progress des sciences naturelles depuis 1789 jusqu'à ce jour* (Paris: De L'imprimerie impériale, 1810), p. 70.

48. Georges Canguilhem, "The Question of Normality in the History of Biological Thought," in *Ideology and Rationality in the Life Sciences*, p. 128.

49. Canguilhem, "The Epistemology of Biology," p. 67.

50. Canguilhem, "The Question of Normality," p. 141.

51. Canguilhem, "The Epistemology of Biology," p. 69. The passage is from Bichat's *Recherches physiologiques sur la vie et la mort.*

52. The concept of "animation" has a wide range of evocative associations in the age of computer animation and computer graphics interface. It also, interestingly enough, serves as the main criteria for "life" in many science fiction films, when the conventional criteria of anatomical structure or material composition fail; consider the Cold War "alien invasion" films such as *Day of the Triffids* or *Invasion of the Body Snatchers*, or more contemporary films such as *The Abyss*, *Alien*, or *The Thing*.

53. Georges Canguilhem, "Vie," *Encyclopedia universalis* (Paris: Encyclopedia Universalis France, 1973), 16: 764–769.

54. Canguilhem's approach looks for two things methodologically: "thematic conservation" across disparate disciplines and paradigms (e.g., Aristotelian hylomorphism in the work of eighteenth-century naturalism and nineteenth-century physiology), and what Canguilhem calls "scientific ideology," or the ways in which discursive exchanges in the sciences both inform those sciences and take them astray into sociology, politics, economics, and culture.

55. Canguilhem, "The Epistemology of Biology," p. 87.

56. Ibid., p. 88.

57. See Stuart Kauffman, *The Origins of Order: Self-Organization and Selection in Evolution* (New York: Oxford University Press, 1993), pp. 298–312.

58. Canguilhem, "The Epistemology of Biology," p. 88.

59. Georges Canguilhem, *The Normal and the Pathological*, trans. Carolyn Fawcett (New York: Zone, 1989), p. 278.

60. Even prior to these developments, molecular biology research during the 1960s had already emphasized the network-based, systemswide activity of genes and proteins; François Jacob and Jacques Monod's famous research on genetic regulatory mechanisms is an important example in this regard.

61. In an everyday context, this definition of information is operative in the handling of e-mail; a portion of a network utility ensures that the e-mail of 4 kb sent in London arrives as an e-mail of 4 kb in Seattle.

62. Claude Shannon and Warren Weaver, *The Mathematical Theory of Communication* (Chicago: University of Illinois Press, 1965), p. 8. Shannon reiterates this point when he notes that "the fundamental problem of communication is that of reproducing at one point either exactly or approximately a message selected at another point" (p. 31). In information theory terms, "semantic aspects . . . are irrelevant to the engineering problem" (p. 31).

63. Evelyn Fox Keller, *Refiguring Life: Metaphors of Twentieth-Century Biology* (New York: Columbia University Press, 1995). As Fox Keller notes, "the notion of genetical information that Watson and Crick invoked was not literal but metaphoric" (p. 19).

64. By the time of Watson and Crick's much-lauded 1953 papers, the language of informatics in molecular biology was already quite developed, such that the two could note how, in a molecular form such as DNA, "many different permutations are possible, and it therefore seems likely that the precise sequence of the bases is the code which carries the genetical information." See James Watson and Francis Crick, "General Implications of the Structure of Deoxyribonucleic Acid," *Nature* 171 (1953): 964–967. Fox Keller thus argues that, contra Shannon and Weaver, for Watson and Crick, "if 'genetical information' is to have anything to do with life, it must involve meaning." *Refiguring Life*, p. 94.

65. The quotation given in the section heading is from "Madrigal III" by Ulver, ©1997 Magic Arts Publishing. Quoted in Henry Cornelius, "Occult Medicine and the Political Context of Enlightenment Mathematics," *Medical Sociology* 21.1 (1994), p. 78.

66. The prevalence of the central dogma is one of the curious instances in the history of molecular biology. Any cursory understanding of genes and proteins cannot but help to recognize the inherently complex nature of genetic function. Although Jacob and Monod's research did contribute to the understanding of mRNA and its cousins, it was, more important, a statement concerning the nonlinear, multiagential nature of gene function. Indeed, during the same period in which some researchers were working on "cracking the genetic code," biologists such as C. H. Waddington, Richard Lewontin, and Stuart Kauffman were performing studies showing how the complexity of genetic function required new, nonreductive approaches in biology. Yet it is difficult not to see the specter of the central dogma still very much alive in the biotech industry, with its emphasis on pills and "silver bullet" therapies. This emphasis makes sense economically—it would be difficult to market a therapy based on complexity and much easier to market a drug—but in many cases it does not make sense medically, as the low rate of current positive drug response and ADRs attest.

87. Ibid., pp. 19–20.

88. Ibid., p. 19. Canguilhem continues: "The cell is a milieu for intracellular elements; it lives in an interior milieu that is either on the scale of the organ or the organism, which organism itself lives in a milieu that is for it, in a sense, what the organism is for its component parts."

89. Richard Lewontin, *The Triple Helix: Gene, Organism, Environment* (Cambridge, Mass.: Harvard University Press, 2000), p. 38.

90. Ibid., p. 100.

91. Canguilhem, "The Living and Its Milieu," p. 21.

Chapter 3

1. See the press release "White House Press Release on Human Genome Project," March 14, 2000, White House Library Web site, http://www.whitehouse.gov.

2. Most molecular biologists distinguish between "junk DNA" and "noncoding" regions. The latter are regions that do not directly code for a protein, but that may play a part, however indirectly, in protein synthesis or, as stated, in the maintenance of the structural and sequential integrity of DNA itself. The former is often thought to be of archaeological or evolutionary interest, kind of like a genetic "fossil" buried in our genomes. Other hypothesis concerning introns (the junk DNA) suggest that they play a role only in the development of the embryo and are subsequently inactivated.

3. See Ken Howard, "The Bioinformatics Gold Rush," *Scientific American* (July 2000): 58–63. Also see the Celera press release "Celera Genomics Completes Sequencing Phase of the Genome from One Human Being," April 6, 2000, available at http://www.celera.com.

4. See Howard, "The Bioinformatics Gold Rush."

5. Francis Collins, Michael Morgan, and Aristides Patrinos, "The Human Genome Project: Lessons from Large-Scale Biology," *Science* (April 11, 2003), p. 290.

6. Ibid.

7. Samuel Broder, G. Subramanian, and Craig Venter, "The Human Genome," in *Pharmacogenomics: The Search for Individualized Therapies*, ed. Julio Licino and Ma-Li Wong (Weinheim, Germany: Wiley-VCH, 2002), pp. 9–34.

8. Broder, Subramanian, and Venter state that the "analysis of genomic sequence provides several clues towards predicting pathogenic potential of a microbe. The phenomenon of lateral transfer of genetic material in microbes is one of great interest and has been shown to contribute to genomic diversity in free-living bacteria and to a lesser extent in obligate bacteria. . . . These observations have substantial implications for the development of new anti-microbial agents, and the selection of the best anti-microbial agents for any host. This is a template for conceptualizing individualized medicine of the future." Ibid., p. 19.

9. Ibid., p. 29.

10. In the years following the IHGSC and Celera publications, there have been contentious debates concerning access to genome data and efforts to open the sharing of information and knowledge within the scientific community.

11. Adam Smith, *The Wealth of Nations* (New York: Bantam, 2003), book 4, p. 537.

12. As Marx notes, the aim of the liberal economists is to "present production—see e.g. Mill—as distinct from distribution etc., as encased in eternal natural laws independent of history, at which opportunity *bourgeois* relations are then quietly smuggled in as the inviolable natural laws on which society in the abstract is founded." *Grundrisse: Foundations of the Critique of Political Economy*, trans. Martin Nicolaus (New York: Penguin, 1973), p. 87.

13. Ibid., p. 83.

14. Ibid., p. 84.

15. Both the HGP and the HGDP adopted this technique. See chapter 4 for more.

16. See Francis Crick, "The Genetic Code," *Scientific American* 207 (1962): 66–75. Also see Lily Kay, *Who Wrote the Book of Life? A History of the Genetic Code* (Stanford, Calif.: Stanford University Press, 2000), pp. 128–144.

17. Erwin Schrödinger, *What Is Life?* (Cambridge: University of Cambridge Press, 1967), p. 21. Schrödinger also states that "the chromosome structures are at the same time instrumental in bringing about the development they foreshadow" (p. 22).

18. Ibid., p. 22.

19. See Norbert Wiener, *Cybernetics: Or Control and Communication in the Animal and the Machine* (Cambridge, Mass.: MIT Press, 1996 [orig. 1948]); and Claude Shannon and Warren Weaver, *The Mathematical Theory of Communication* (Chicago: University of Illinois Press, 1965). Wiener was concerned primarily with information as related to systemic regulation (through feedback loops) and with the larger social implications of cybernetics and the grouping of humans and machines under the categories of information and negative feedback. Shannon, working at Bell Labs, was concerned primarily with a technical question: how to transmit a message at point A to point B with the greatest accuracy and lowest probability of error, contributing in part to developments in telecommunications that are evident in today's "wired" condition of global networks.

20. For Shannon, "noise" is a type of information, albeit an unsuccessful type, whereas for Wiener noise is the opposite of information. See Wiener, *Cybernetics*, pp. 60–94, and 155–168; Shannon and Weaver, *Mathematical Theory of Communication*, pp. 8–16, and 31–35.

21. Katherine Hayles, *How We Became Posthuman: Virtual Bodies in Cybernetics, Literature, and Informatics* (Chicago: University of Chicago Press, 1999), p. 2.

22. Ibid., pp. 1–13.

23. For their original research papers on DNA's structure, see James Watson and Francis Crick, "A Structure of Deoxyribonucleic Acid," *Nature* 171 (1953): 737–738; and "General Implications of the Structure of Deoxyribonucleic Acid," *Nature* 171 (1953): 964–967. See also Kay, *Who Wrote the Book of Life?* pp. 128–179.

24. The most explicit example is the "central dogma" of molecular biology in the 1950s and 1960s. See Kay, *Who Wrote the Book of Life?* pp. 173–175.

25. For examples of research in systems biology, see Hiroaki Kitano, ed., *Foundations of Systems Biology* (Cambridge, Mass.: MIT Press, 2001), as well as information about the work pioneered by Leroy Hood at the Institute for Systems Biology, available at http://www.systemsbiology.org.

26. See also Ruth Hubbard and Elijah Wald, *Exploding the Gene Myth* (Boston: Beacon, 1997). As they state, "the language that geneticists use often carries considerable ideological baggage. Molecular biologists, as well as the press, use verbs like 'control,' 'program,' or 'determine' when speaking about what genes or DNA do. These are all inappropriate because they assign far too active a role to DNA. The fact is that DNA doesn't 'do' anything; it is a remarkably inert molecule. It just sits in our cells and waits for other molecules to interact with it" (p. 11).

27. Susan Oyama, *The Ontogeny of Information: Developmental Systems and Evolution* (Durham, N.C.: Duke University Press, 2000), p. 14.

28. I use the term *genetic reductionism* here to cover a range of critiques and analyses of molecular genetics and biotech, all of which question the reductive science implicit in current fields such as genomics. For a wide range of approaches, see Evelyn Fox Keller, *Refiguring Life: Metaphors of Twentieth-Century Biology* (New York: Columbia Univeristy Press, 1995); Richard Lewontin, *Biology as Ideology: The Doctrine of DNA* (New York: Harper-Collins, 1992); and Dorothy Nelkin and Susan Lindee, *The DNA Mystique: The Gene as a Cultural Icon* (New York: W. H. Freeman, 1995).

29. Marx, *Grundrisse*, p. 86, italics removed. Living labor, for Marx, is implicit in all of labor's forms, though in qualitatively different ways. In a sense, the challenge posed to capital is how to maximize the living labor in waged labor. But, as Marx notes, in the process of quantifying and objectifying living labor, something of living labor is unavoidably lost. The moment living labor transforms itself into labor power for sale (the moment living labor produces only exchange values in commodities), something of living labor escapes. Capital must somehow revive living labor in wage labor, must somehow make the dead live: "Objectified labor ceases to exist in a dead state as an external, indifferent form on the substance, because it is itself again posited in a moment of living labor; as a relation of living labor to itself in an objective material, as the *objectivity* of living labor" (p. 360, emphasis in original).

30. Ibid., p. 360, emphasis in original.

31. "Labor is the living, form-giving fire; it is the transitoriness of things, their temporality, as their formation by living time." Ibid., p. 361.

32. Antonio Negri, *Marx beyond Marx*, trans. and introduced by Harry Cleaver, Michael Ryan, and Maurizio Viano (Brooklyn: Autonomedia/Pluto, 1991), pp. 67–68, emphasis in orginal.

33. Nick Dyer-Witheford, *Cyber-Marx: Cycles and Circuits of Struggle in High-Technology Capitalism* (Chicago: University of Illinois Press, 1999), p. 65.

34. Marx, *Grundrisse*, p. 693.

35. Ibid.

36. Ibid., pp. 690–706.

37. See Geoffrey Bowker and Susan Leigh Star, *Sorting Things Out: Classification and Its Consequences* (Cambridge, Mass.: MIT Press, 1999).

38. Friedrich Kittler, "Gramophone, Film, Typewriter," in *Literature, Media, Information Systems*, ed. John Johnson (Amsterdam: G&B Arts, 1997), pp. 28–49.

39. See C. J. Date, *An Introduction to Database Systems,* 8th ed. (New York: Addison-Wesley, 2003) p. 240. See also Gary G. Bitter, ed., *Macmillan Encyclopedia of Computers* (New York: Macmillan, 1992); and Peter Rob and Carlos Coronel, *Database Systems: Design, Implementation, and Management* (Belmont, Mass.: Wadsworth, 1993).

40. For early accounts on the development of bioinformatics databases, see M. A. Andrade and C. Sander, "Bioinformatics: From Genome Data to Biological Knowledge," *Current Opinion in Biotechnology* 8.6 (December 1997): 675–683. See also Howard, "The Bioinformatics Gold Rush."

41. See Charles Cantor, "The Challenges to Technology and Informatics," in *The Code of Codes: Scientific and Social Issues in the Human Genome Project*, ed. Daniel Kevles and Leroy Hood (Cambridge, Mass.: Harvard University Press, 1992), pp. 98–111.

42. Joan Fujimura and Michael Fortun, "Constructing Knowledge across Social Worlds: The Case of DNA Sequence Databases in Molecular Biology," in *Naked Science: Anthropological Inquiry into Boundaries, Power, and Knowledge*, ed. Laura Nader (New York: Routledge, 1996), p. 170.

43. An example of this tension is in bioinformatics applied to databasing. On the one hand, the consensus view is that "genes" actually do very little and are but one component among many that contribute to given biochemical processes. On the other, a great many of the database tools in bioinformatics focus on or are predicated on the centrality of genes and DNA as their starting point. For examples of this tension from

the perspective of computer science, see Pierre Baldi, *The Shattered Self* (Cambridge, Mass.: MIT Press, 2000).

44. Inscribing technologies also form a unique relationship to bodies and to death. Whereas Kittler describes literature as evoking a zone of fantasy that extends from it, mechanical media are themselves constituted by the phantasmic, by the tension between realism or simulation, and the understood logic of representation. Thus, there are ghost voices on the airwaves, ghost images in the shadow play of film, and psychic "automatic writing." As Kittler states, "media always already provide the appearance of specters." "Gramophone, Film, Typewriter," p. 41.

45. Ibid., p. 46.

46. Ibid., p. 32.

47. Marx, *Grundrisse*, p. 99. "Thus production, distribution, exchange and consumption form a regular syllogism; production is the generality, distribution and exchange the particularity, and consumption the singularity in which the whole is joined together" (p. 89). Further, Marx notes how, in classical political economy, production is seen to be driven by "general natural laws," whereas distribution among the classes is a result of "social accident" (p. 89).

48. "That exchange and consumption cannot be predominant is self-evident. Likewise, distribution as distribution of products; while as distribution of the agents of production it is itself a moment of production. A definite production thus determines a definite consumption, distribution, and exchange as well as *definite relations between these different moments*. Admittedly, however, *in its one-sided form*, production is itself determined by other moments." Marx, *Grundrisse*, p. 99, emphasis in original.

49. The productive aspects of consumption and distribution are further elaborated by Marx in the *Grundrisse*. Whereas consumption seems to be the mere endpoint of the production process, production "produces consumption (i) by creating the material for it; (ii) by determining the manner of consumption; and (iii) by creating the products, initially posited by it as objects, in the form of a need felt by the consumer" (p. 92). Much of these propositions has been elaborated by thinkers such as Jean Baudrillard, Guy Debord, and Herbert Marcuse, albeit in significantly different ways. Likewise, whereas distribution seems to be the mere allocation of goods, "before distribution can be the distribution of products, it is: (i) the distribution of the instruments of production, and (ii), which is a further specification of the same

relation, the distribution of the members of the society among the different kinds of production" (p. 96).

50. Both the Celera and public-consortium projects constructed their databases from genetic samples of selected individuals, taken as representative of the population as a whole. It was often noted that Celera chose individuals of varying ethnic backgrounds to ensure a comprehensive sampling. For an account of the process, see J. Craig Venter, "The Sequence of the Human Genome," *Science* 291 (February 16, 2001): 1304–1351.

51. The HGDP is now housed at the Morris Institute for Population Studies at Stanford University. Information on the controversy sparked by the HGDP can be found at the Rural Advancement Foundation International (RAFI) archives online at http://www.rafi.org.

52. See Vandana Shiva, *Biopiracy: The Plunder of Nature and Knowledge* (Toronto: Between the Lines Press, 1997).

53. Clive Trotman, "Introns-Early: Slipping Lately?" *Trends in Genetics* 14.4 (April 1998): 132–134. Science fiction writer Greg Bear has taken up the implications on the evolutionary, "introns-early" theory in his novel *Darwin's Radio* (New York: Ballantine 1999), suggesting that the "fossilized" DNA is there waiting to be reactivated to initiate another phase shift in human evolution.

54. The quotation given in the section heading is from "Raining Murder" by Darkthrone, ©2001 Moonfog Productions. Quoted in Arnaud Villeneuve, "Medicine, Materiality, and Spectacle in Industrialized Paris," in *Decadent Biology: Science Studies and the 19th Century*, ed. Claude Louis (Berkeley: University of California Press, 1993), p. 189.

55. Bataille's particular take on political economy has been discussed at length in other studies. Suffice it to say that considering his interests in the anthropological significance of expenditure, ritual, and uselessness, Bataille approaches political economy from the vantage point of excess rather than utility. His starting point is therefore abundance, and thus the central problematic of political economy becomes how to deal with excess, or how to expend. For further analysis of Bataille's political economy, see Denis Hollier, *Against Architecture* (Cambridge, Mass.: MIT Press, 1995); and Julian Pefanis, *Heterology and the Postmodern: Bataille, Baudrillard, and Lyotard* (Durham, N.C.: Duke University Press, 1991).

56. See the essays "The Notion of Expenditure" and "Base Materialism and Gnosticism" in Georges Bataille, *Visions of Excess, Selected Writings, 1927–1939*, ed. Allan Stoekl, trans. Allan Stoekl, Carl R. Lovitt, and Donald M. Leslie Jr. (Minneapolis: University of Minnesota Press, 1985).

57. Bataille, "The Notion of Expenditure," pp. 141, 142.

58. Steven Shaviro, *Connected, or What it Means to Live in the Network Society* (Minneapolis: University of Minnesota Press), pp. 220–221.

59. See Georges Bataille, *The Accursed Share*, Vol. 1, trans. Robert Hurley (New York: Zone, 1991).

60. Ibid., 1: 21. Bataille reiterates this position more directly when he states that "on the surface of the globe, for living matter in general, energy is always in excess; the question is always posed in terms of extravagance. The choice is limited to how the wealth is to be squandered" (p. 23, italics removed).

61. Lev Manovich, "Database as a Symbolic Form," *Nettime* list (December 14, 1998), http://www.nettime.org. A fuller account is given in Manovich's *The Language of New Media* (Cambridge, Mass.: MIT Press, 2001), pp. 218–236.

62. Marx, *Grundrisse*, p. 693.

63. Ibid., p. 706.

64. Ibid., pp. 703–704.

65. See Brian Goodwin, *How the Leopard Changed Its Spots: The Evolution of Complexity* (New York: Touchstone, 1994); and Stuart Kauffman, *At Home in the Universe: The Search for the Laws of Self-Organization and Complexity* (Oxford: Oxford University Press, 1995).

66. See Evelyn Fox Keller, *The Century of the Gene* (Cambridge, Mass.: Harvard University Press, 2000).

Chapter 4

1. Many press releases and news items can be found at the Rural Advancement Foundation International (RAFI) archives Web site at http://www.rafi.org. Also see

Patricia Kahn, "Genetic Diversity Project Tries Again," *Science* 266 (November 4, 1994): 720–722.

2. The idea for the HGDP came to Cavalli-Sforza as a postdoc at Cambridge University under the eugenicist Ronald Fisher, who was then applying mathematical techniques to studying Mendelian laws in human characteristics such as height, weight, and intelligence (e.g., IQ scores). Fisher's book *The Genetical Theory of Natural Selection*, published in 1930, reiterated late-nineteenth-century fears about the degeneration and deterioration of the population from the overbreeding of the lower classes. Cavalli-Sforza soon became a leading figure in population genetics. In the 1980s, he and his colleagues collected DNA samples in Africa to use statistics to study human genetic differences in select populations. For an excellent analysis of the HGDP, see Jennifer Reardon, "The Human Genome Diversity Project: A Case Study in Co-production," *Social Studies of Science* 31.3 (2001): 357–388.

3. See Debra Harry (no, not that one), *Biocolonialism: A New Threat to Indigenous Peoples*, a report from the Indigenous Peoples Council on Biocolonialism (1998), available at http://www.foel.org/LINK/LINK93/biocolonialism.html.

4. See Vandana Shiva, *Biopiracy: The Plunder of Nature and Knowledge* (Toronto: Between the Lines, 1997), pp. 65–85. Also see the GeneWatch report *Privatising Knowledge, Patenting Genes: The Race to Control Genetic Information*, GeneWatch Briefing no. 11 (June 2000), available at http://www.genewatch.org.

5. See Reardon, "The Human Genome Diversity Project."

6. The former HGDP is now based at Stanford University's Morris Institute for Population Studies.

7. Other organizations instrumental in lobbying against biopiracy include the Foundation for Economic Trends (FET) and the Institute for Science in Society (I-SIS).

8. On bioinformatics, see Ken Howard, "The Bioinformatics Gold Rush," *Scientific American* (July 2000): 58–63. Also see chapter 1 in this book.

9. On pharmacogenomics, see Julio Licino and Ma-Li Wong, eds., *Pharmacogenomics: The Search for Individualized Therapies* (Weinheim, Germany: Wiley-VCH, 2002). Also see chapter 1 in this book.

10. For instance, Affymetrix's "GeneChip" technology, Incyte's "LifeSeq" databases, Perkin-Elmer's automated genetic sequencing computers.

11. On bioinformatics and post-genomics, see Diane Gershon, "Bioinformatics in a Post-Genomics Age," *Nature* 389 (September 27, 1997): 417–418.

12. For reportage on the human genome projects, see the news articles "Genetic Code of Human Life Is Cracked by Scientists," *New York Times*, June 26, 2000; "Special Issue: Genome—The Race to Decode the Human Body," *Newsweek* (April 10, 2000); and "Special Issue: Cracking the Code," *Time* (July 3, 2000).

13. On developments in population genetics, see Aravinda Chakravarti, "Population Genetics—Making Sense out of Sequence," *Nature Genetics* 21 (January 1999): 56–60. Also see Steve Olson, "The Genetic Archaeology of Race," *The Atlantic Online* (April 2001), http://www.theatlantic.com.

14. For example, the two most common genetically based components of difference include SNPs, or the minute sequence differences from one individual to another, and HAPs, or the physical arrangement of SNPs on a chromosome. See Nicholas Wade, "In the Hunt for Useful Genes, a Lot Depends on Snips," *New York Times*, August 11, 1998, available at http://www.nytimes.com.

15. See Vicki Brower, "Mining the Genetic Riches of Human Populations," *Nature Biotechnology* 16 (April 1998): 337–340; and Gary Taubes, "Your Genetic Destiny for Sale," *MIT Technology Review* (April 2001): 41–46.

16. Both the Gene Trust and the First Genetic Fund borrow the rhetoric of investment banking to contextualize genetic research as long-term investments, benefitting not just the individual, but the species as a whole. GenBank is one of the first online databases to be established by the U.S. government–funded HGP. The Gene Trust Web site makes this explicit in its Web-based advertising (e.g., the DNA you donate now could save lives in the future).

17. A Web site dedicated to HAP research is Hapcentral, http://www.hapcentral.com. Along with the myriad of genomics databases now online, a centralized, U.S. government–funded SNP database exists at the Whitehead Institute, http://www.genome.wi.mit.edu/SNP/human/index.html.

18. See Michael Fortun, "Towards Genomic Solidarity: Lessons from Iceland and Estonia," *Open Democracy* (October 7, 2003), available at http://www.opendemocracy.net.

19. See Taubes, "Your Genetic Destiny for Sale."

20. See the UK BioBank Web site at http://www.ukbiobank.ac.uk.

21. WMA, "Declaration on Ethical Considerations Regarding Health Databases," adopted by the WMA General Assembly (2002), available at http://www.wma.net/e/policy/d1.htm.

22. Michael Fortun, "Experiments in Ethnography and Its Performance," on the Mannvernd Web site at http://www.mannvernd.is.

23. Ibid.

24. Michel Foucault, *The History of Sexuality*, vol. 1, *An Introduction*, trans. Robert Hurley (New York: Vintage, 1978), pp. 141–142.

25. Ibid., 1: 147, 148 (italics removed).

26. Much of this analysis is predicated on textbooks in population genetics, including A. J. Ammerman and Luigi Luca Cavalli-Sforza, *The Neolithic Transition and the Genetics of Populations in Europe* (Princeton, N.J.: Princeton University Press, 1984); Luigi Luca Cavalli-Sforza, *The History and Geography of Human Genes* (Princeton, N.J.: Princeton University Press, 1994); and Luigi Luca Cavalli-Sforza and F. Cavalli-Sforza, *The Great Human Diasporas* (Menlo Park, Calif.: Addison-Wesley, 1995).

27. Luigi Luca Cavalli-Sforza, *Genes, Peoples, and Languages* (New York: North Point, 2000), p. 25.

28. Quoted in Reardon, "The Human Genome Diversity Project," p. 362. The full reference is Theodosius Dobzhansky, *Genetics and the Origin of Species* (New York: Columbia University Press, 1951 [orig. 1937]), p. 138.

29. See Reardon, "The Human Genome Diversity Project," pp. 361–362. Also see Daniel Kevles, *In the Name of Eugenics: Genetics and the Uses of Human Heredity* (Cambridge, Mass.: Harvard University Press, 1995), pp. 164–175.

30. See Elaine Daston and Katherine Park, *Wonders and the Order of Nature, 1150–1750* (New York: Zone, 1998), pp. 25–39.

31. See Zakiya Hanafi, *The Monster in the Machine: Magic, Medicine, and the Marvelous in the Time of the Scientific Revolution* (Durham, N.C.: Duke University Press, 2000). See also Daston and Park, *Wonders and the Order of Nature*, pp. 201–215.

32. See Cavalli-Sforza, *The History and Geography of Human Genes.*

33. Cavalli-Sforza, *Genes, Peoples, and Languages*, p. 20.

34. See Dudley Wilson, *Signs and Portents: Monstrous Births from the Middle Ages to the Enlightenment* (New York: Routledge, 1993). See Daston and Park, *Wonders and the Order of Nature*, pp. 177–190.

35. See G. H. Hardy, "Mendelian Proportions in a Mixed Population," *Science* 28 (1908): 49–50. Also see the sections in Daniel Hartl and Andrew Clark, *Principles of Population Genetics* (Sunderland, Mass.: Sinauer Associates, 1997). The Hardy-Weinberg principle is by no means universal to all population genetics approaches, but it is being taken here as emblematic of the kind of complicated interweaving of statistics and culture that constitutes the field.

36. See Cavalli-Sforza, *Genes, Peoples, and Languages*, pp. 66–73.

37. Foucault, *History of Sexuality*, 1: 137.

38. Ibid., 1: 137, 139.

39. Ibid., 1: 139.

40. Ibid., 1: 139, emphasis in original.

41. Michel Foucault, *"Society Must Be Defended": Lectures at the Collège de France, 1975–76*, trans. David Macey, ed. Mauro Bertani and Alessandro Fontana (New York: Picador, 2003), p. 243.

42. "The great eighteenth-century demographic upswing in Western Europe, the necessity for coordinating and integrating it into the apparatus of production, and the urgency of controlling it with finer and more adequate power mechanisms cause 'population,' with its numerical variables of space and chronology, longevity and health, to emerge not only as a problem but as an object of surveillance, analysis, intervention, modifications, and so on." Michel Foucault, "The Politics of Health in the

Eighteenth Century," in *Power: The Essential Works of Michel Foucault 1954–1984*, vol. 3, trans. Robert Hurley et al., ed. James Faubion (New York: New Press, 2000), p. 95.

43. Foucault, *History of Sexuality*, 1: 142.

44. Malthus notes that "it follows necessarily that the average rate of the actual increase of population over the greatest part of the globe, obeying the same laws as the increase of food, must be totally of a different character from the rate at which it would increase *if unchecked*. The great question, then, which remains to be considered, is the manner in which this constant and necessary check upon population practically operates." Thomas Malthus, "A Summary View of the Principle of Population," in *An Essay on the Principle of Population* (New York: Penguin, 1985 [orig. 1798]), p. 242.

45. Regarding threats to the population, Malthus observes that "a foresight of the difficulties attending the rearing of a family acts as a preventive check; and the actual distresses of some of the lower classes, by which they are disabled from giving the proper food and attention to their children, act as a positive check." *An Essay on the Principle of Population*, p. 89.

46. Thomas Malthus, *Principles of Political Economy* (New York: Augustus M. Kelly, 1964 [2d ed. 1836]), p. 313.

47. Malthus, *An Essay on the Principle of Population*, pp. 313–314.

48. Ian Hacking, *The Taming of Chance* (Cambridge: Cambridge University Press, 1990), p. 1.

49. Ibid., p. 52.

50. Ibid., p. 1.

51. Ibid., p. 5.

52. Foucault, "*Society Must be Defended*," p. 246.

53. See Michel Foucault, "Security, Territory, and Population," in *Ethics: Subjectivity and Truth: The Essential Works of Michel Foucault 1954–1984*, vol. 1, trans. Robert Hurley et al., ed. Paul Rabinow (New York: New Press, 1997), pp. 67–72.

54. Foucault, *"Society Must be Defended,"* p. 246.

55. Ibid., p. 255.

56. David Arnold, *Colonizing the Body: State Medicine and Epidemic Disease in 19th Century India* (Berkeley: University of California Press, 1993), p. 8.

57. More information on the IHSD can be found at the deCODE Web site at http://www.decode.com.

58. See the Mannvernd Web site at http://www.mannvernd.is.

59. On the concept of "endocolonialism," see Paul Virilio, *The Art of the Motor* (Minneapolis: University of Minnesota Press, 1995), pp. 99–133.

60. See Kari Steffanson and Jeffrey Gulcher, "The Icelandic Healthcare Database and Informed Consent," *New England Journal of Medicine* 342.24 (June 15, 2000). Other articles are available through the deCODE Web site (see note 57).

61. Accounting for approximately 0.1 percent of the total genome (or roughly one million base-pair variations per individual), these SNPs are thought to contribute to a range of phenotypic characteristics, from the physical markers that make one person different from another to the susceptibility to single base pair mutation conditions (such as sickle-cell anaemia or diabetes).

62. For an example of a HAP database, see Genaissance's Web site at http://www.genaissance.com. For the SNP database, see http://www.genome.wi.mit.edu/SNP/human.

63. The quotation given in the section heading is from "Madrigal IV" by Ulver, ©1997 Magic Arts Publishing. Quoted in G. B. Della Porta, "Vitalism and the Role of Physicians in Early Modern Expansionism," *Renaissance Studies* 4 (2002), p. 19.

64. See Harry, "Biocolonialism." Also see Jeremy Rifkin, *The Biotech Century: Harnessing the Gene and Remaking the World* (New York: Tarcher/Putnam, 1998).

65. On postcolonialism and technoscience, see Donna Haraway, *Modest_Witness@Second_Millennium.FemaleMan©_Meets_OncoMouse™: Feminism and Technoscience* (New York: Routledge, 1997); and Sandra Harding, *Is Science*

Multicultural? Postcolonialisms, Feminisms, and Epistemologies (Bloomington: Indiana University Press, 1998).

66. See Vandana Shiva, "Biodiversity, Biotechnology, and Profits," in *Biodiversity: Social and Ecological Perspectives*, ed. Vandana Shiva (New Jersey: Zed Books, 1991), pp. 43–58.

67. Ibid., 44–45.

68. Troy Duster, *Backdoor to Eugenics* (New York: Routledge, 2003), p. 40. Duster notes two significant trends in modern genetic screening practices: first, the fact that "genetic screening programs, already in place throughout most of the United States, can be distinguished from all previous health screening in the degree to which 'risk-populations' to be identified have frequently been linked to ethnicity and race" (p. 5), and, second, "state and national registries for information received from newborn genetic screening programs are already in place, collecting data on the chromosome and genetic trait status of millions of infants" (p. 5).

69. See Evelyn Fox Keller, *Reflections on Gender and Science* (New Haven, Conn.: Yale University Press, 1995). Also see Haraway, *Modest_Witness*.

70. Frantz Fanon, "Medicine and Colonialism," in *A Dying Colonialism*, trans. Haakon Chevalier (New York: Grove, 1965), pp. 121, 125.

71. Fanon paraphrases this more familiar view of colonial intervention: "Colonialism obviously throws all the elements of native society into confusion. The dominant group arrives with its values and imposes them with such violence that the very life of the colonized can manifest itself only defensively, in a more or less clandestine way" (ibid., p. 130). This view has, of course, been complicated more recently by Homi Bhabba's psychoanalytic reading of the colonial encounter.

72. Ibid., pp. 134, 139. Fanon also notes that "in centers of colonization the doctor is nearly always a landowner as well" (p. 133).

73. Ibid., p. 122. In his psychological analysis, Fanon notes the attitude of the colonial doctor that produces this debt: "This is what we have done for the people of this country; this country owes us everything; were it not for us, there would be no country" (p. 122).

74. Ibid., pp. 140–141.

75. Ibid., p. 142.

76. Ibid., p. 140.

77. In light of recent trends to emphasize the role of information technology and computers in biotech, Fanon's comments are especially prescient: "All the efforts exerted by the doctor, by his team of nurses, to modify this state of thing encounter, not a systematic opposition, but a 'vanishing' on the part of the patient." Ibid., p. 129.

78. Ibid., pp. 144–145.

79. Fortun, "Towards Genomic Solidarity."

80. Fanon, "Medicine and Colonialism," p. 142.

Chapter 5

1. See S. N. Cohen, A. C. Y. Chang, H. W. Boyer, and R. B. Helling, "Construction of Biologically Functional Bacterial Plasmids in Vitro," *Proceedings of the National Academy of Sciences* 70 (1973): 3240–3244; and J. F. Morrow, S. N. Cohen, A. C. Y. Chang, H. W. Boyer, H. M. Goodman, and R. B. Helling, "Replication and Transportation of Eukaryotic DNA in *Escherichia coli*," *Proceedings of the National Academy of Sciences* 71 (1974): 1743–1747.

2. For a current, though uncritical, analysis of the history of biotechnology as an industry, see Cynthia Robbins-Roth, *From Alchemy to IPO: The Business of Biotechnology* (New York: Perseus, 2000). On the history of biotechnology from an industrial standpoint, see Robert Bud, "Biotechnology in the Twentieth Century," *Social Studies of Science* 21 (1991): 415–457.

3. On the notion of translation, see Bruno Latour, *We Have Never Been Modern* (Cambridge, Mass.: Harvard University Press, 1993), pp. 39–43.

4. Although the recombinant DNA research during the 1970s instigated public outcry at scientists "playing God," and although such public anxieties led to some formalized, bioethical debate (for instance, the Asilomar conferences held in California in 1973 and 1975), left unexamined were the direct connections between research science, governmental sanctioning, and bioscience business (patenting, marketing, investment). For more on the social and political controversies brought up by

recombinant DNA, see Susan Wright, *Molecular Politics: Developing American and British Regulatory Policy for Genetic Engineering, 1972–82* (Chicago: University of Chicago Press, 1994).

5. Daniel Bell, *The Coming of the Post-industrial Society* (New York: Basic, 1999), p. 126.

6. Ibid., p. 127.

7. Ibid.

8. Ibid., pp. 127, 128.

9. Manuel Castells, *The Rise of the Network Society* (New York: Blackwell, 1996), p. 67.

10. Castells, for instance, includes a few descriptive pages on biotechnology, but they are subsumed under a broader chapter on changes in microelectronics and information technologies during the 1970s; Castells thus takes the biotech industry as a subset of microelectronics, of which it is not. See ibid., pp. 47–51.

11. Karl Marx. *Capital*, trans. Ben Fowkes (New York: Penguin, 1990 [1867]), 1: 493 n. 4.

12. Ibid., 1: 283, 285, 517.

13. See Karl Marx, *Grundrisse: Foundations of the Critique of Political Economy*, trans. Martin Nicolaus (New York: Penguin, 1973), pp. 690–706.

14. Marx, *Capital*, 1: 494.

15. "The machine, therefore, is a mechanism that, after being set in motion, performs with its tools the operation as the worker formerly did with similar tools. Whether the motive power is derived from man, or in turn from a machine, makes no difference here." Ibid., 1: 495.

16. Ibid., 1: 497.

17. Marx, *Grundrisse*, p. 705.

18. Marx, *Capital*, 1: 501.

19. "As soon as a machine executes, without man's help, all the movements required to elaborate a raw material, and needs only supplementary assistance from the worker, we have an automatic system of machinery, capable of constant improvement in its details . . . a mechanical monster whose body fills whole factories, and whose demonic power, at first hidden by the slow and measured motions of its gigantic members, finally bursts forth in the fast and feverish whirl of its countless working organs." Marx, *Capital*, 1: 503.

20. Marx, *Grundrisse*, p. 692, italics removed.

21. Ibid., p. 693.

22. Marx, *Capital*, 1: 270, 274.

23. Ibid., 1: 271, 272. What changes in each step is the role that human labor power plays in the production process. For Marx, human labor power is affected by technology in two primary ways. First, it is affected by the relationship between modes of production and natural resources, or between the body and natural forces. There is, in a sense, a physics to industrialism, encapsulated in the technologies of the steam engine or the weaving loom. Water, steam, animals, and a range of other natural elements serve as the raw motive force for production. These forces begin to replace direct human labor power. The "inorganic body" Marx speaks of in the 1844 manuscripts becomes an "objective organization of production." Second, human labor power is affected by the mechanical forces that displace skilled labor or manual labor in the utilization of natural forces: value becomes transformed from a use value to the value of a product for exchange and labor becomes differentiated between labor to produce a product, and labor to produce machines (to produce products).

24. Marx, *Grundrisse*, p. 693.

25. Marx, *Capital*, 1: 274.

26. "Capital is dead labor which, vampire-like, lives only by sucking living labor, and lives the more, the more labor it sucks." Ibid., 1: 342.

27. "Owing to its conversion into an automaton, the instrument of labor confronts the worker during the labor process in the shape of capital, dead labor, which dominates and soaks up living labor-power." Ibid., 1: 548.

28. But we should remember that, for Marx, living labor is always the capacity for labor power, the potential for labor to enter into the labor-capital relation and therefore become "objectified labor." Thus, there is some blurring between living labor and labor power, for in capitalism the former is always determined by the latter. Living labor, in capitalism, seems to have its telos in labor power. Why, then, make the distinction at all? Because, quite simply, living labor is not labor power.

29. Antonio Negri, *Marx beyond Marx*, trans. Harry Cleaver, Michael Ryan, and Maurizio Viano (Brooklyn: Autonomedia/Pluto, 1991), p. 113.

30. See Harry Cleaver, *Reading Capital Politically* (San Francisco: AK, 2000), pp. 65–71.

31. Marx, *Grundrisse*, p. 266.

32. Nick Dyer-Witheford, *Cyber-Marx: Cycles and Circuits of Struggle in High-Technology Capitalism* (Chicago: University of Illinois Press, 1999), p. 65.

33. Negri, *Marx beyond Marx*, p. 67.

34. Ibid., p. 70.

35. Raniero Panzieri, "The Capitalist Use of Machinery: Marx versus the Objectivists," posted online at http://www.geocities.com/cordobakaf/panzieri.html. Previously published in translation in *Working Class Autonomy and the Crisis*, ed. Ed Emery (London: Red Notes Collective, 1979).

36. Ibid.

37. Maurizio Lazzarato, "Immaterial Labor," in *Radical Thought in Italy*, ed. Paolo Virno and Michael Hardt (Minneapolis: University of Minnesota Press, 1996), p. 133. Hardt and Negri echo, for the most part, Lazzarato's definition when they state that "since the production of services results in no material and durable good, we define the labor involved in this production as immaterial labor—that is, labor that produces an immaterial good, such as a service, a cultural product, knowledge, or communication." See Michael Hardt and Antonio Negri, *Empire* (Cambridge, Mass.: Harvard University Press, 2000), p. 290.

38. Lazzarato, "Immaterial Labor," p. 142; and Hardt and Negri, *Empire*, p. 30.

39. Lazzarato, "Immaterial Labor," p. 145.

40. Ibid., p. 135.

41. Hardt and Negri, *Empire*, p. 298.

42. Lazzarato, "Immaterial Labor," p. 145.

43. Negri, *Marx beyond Marx*, p. 72.

44. See chapters 1 through 3 in Robert Bud, *The Uses of Life: A History of Biotechnology* (Cambridge: Cambridge University Press, 1993).

45. For a critique of this reliance on informatics, see Katherine Hayles, *How We Became Posthuman: Virtual Bodies in Cybernetics, Literature, and Informatics* (Chicago: University of Chicago Press, 1999), pp. 1–25.

46. The quotation given in the section heading is from "Madrigal IV" by Ulver, ©1997 Magic Arts Publishing. Quoted in T. B. von Hohenheim, "The 'Nature' of Power Relationships in Early Modern Occult Science," *Cultural Anthropology* 13 (1988), p. 35.

47. For more information on the HGP, see the Web site http://www.ornl.gov/hgmis. See also Daniel Kevles and Leroy Hood, eds., *The Code of Codes: Scientific and Social Issues in the Human Genome Project* (Cambridge, Mass.: Harvard University Press, 1992).

48. Each of these companies' Web sties also contains press releases and links to relevant news articles. But the situation should be explicated in all its complexity. That is, although in one sense the bottom line for biotech startups is economics, it would be inaccurate to create of this situation a dichotomy in which a greedy biotech corporation invades the public good of the federal genome project. The practice of "patent then publish" has been a motto of university-based research for some time, and major pharmaceutical companies have a long-standing tradition of developing products based on academic research. In addition, we need to also ask how "public" the HGP really is. Like the Manhattan Project before it, the HGP is a project of Western "big science," carried out by laboratories and research institutes mostly in the United States and Europe. Within the United States, speculations on the availability of genetic drugs restrict such expensive treatments to a very small sector of the health-care-paying public. And, as most researchers concede, we are very far away from a medicine utopia

in which genetically designed drugs and gene therapies are commonplace medical practices.

49. See Dan Schiller, *Digital Capitalism: Networking the Global Market System* (Cambridge, Mass.: MIT Press, 1999).

50. See Michael Dawson and John Bellamy Foster, "Virtual Capitalism," in *Capitalism and the Information Age: The Political Economy of the Global Communication Revolution*, ed. Robert McChesney, et al. (New York: Monthly Review Press, 1998), pp. 51–69.

51. See Catherine Waldby, *The Visible Human Project: Informatic Bodies and Posthuman Medicine* (New York: Routledge, 2000). Waldby's concept of biovalue operates in the interstices between technique and knowledge production in the ongoing management of biomedical bodies.

52. Cited in Aris Persidis, "Bioinformatics," *Nature Biotechnology* 17 (August 1999): 828–830.

53. For more on the field of bioinformatics from a technical (yet readable) standpoint, see Andreas Baxevanis, et al., ed., *Bioinformatics: A Practical Guide to the Analysis of Genes and Proteins* (New York: Wiley-Liss, 2001).

54. See Fredric Jameson, "Culture and Finance Capital," in *The Cultural Turn: Selected Writings on the Postmodern, 1983–1998* (London: Verso, 1998), pp. 136–161.

55. For example, see the industry news portals Biospace, at http://www.biospace.com, and GenomeWeb, at http://www.genomeweb.com.

56. For a transcript, see "White House Press Release on Human Genome Project," March 14, 2000, White House Library Web site at http://www.whitehouse.gov.

57. The Gene Trust Web site is at http://www.dna.com. It is noteworthy that both James Watson and Jim Clark (of Netscape renown) are on the board of the Gene Trust, exemplifying in their dual presence the biological and informatic aspects of much biotech research.

58. See Jean Baudrillard, *Simulations* (New York: Semiotext[e], 1983).

59. On DNA chips or microarrays, see Andrew Marshall and John Hodgson, "DNA chips: An Array of Possibilities," *Nature Biotechnology* 16 (January 1998): 27–31; and

Michael Ramsey, "The Burgeoning Power of the Shrinking Laboratory," *Nature Biotechnology* 17 (November 1999): 1061–1062.

60. For a summary of this concept, see Cleaver, *Reading Capital Politically*, pp. 66–69.

61. See Lily Kay, *Who Wrote the Book of Life? A History of the Genetic Code* (Stanford, Calif.: Stanford University Press, 2000). Evelyn Fox Keller and Richard Lewontin, as well as others, have for many years posed critical challenges to the DNA-centric view in genetics and biotechnology. See Evelyn Fox Keller, *The Century of the Gene* (Cambridge, Mass.: Harvard University Press, 2000); Richard Lewontin, *Biology as Ideology: The Doctrine of DNA* (New York: Harper-Collins, 1992); Lynn Margulis and Dorian Sagan, *What Is Life?* (New York: Simon and Schuster, 1995); Susan Oyama, *The Ontogeny of Information: Developmental Systems and Evolution* (Durham, N.C.: Duke University Press, 2000).

62. See Manuel De Landa, "Non-organic Life," in *Zone 6: Incorporations*, ed. Jonathan Crary and Sanford Kwinter (New York: Zone, 1992), pp. 128–167. For popular accounts of chaos theory and the sciences of complexity, see James Gleick, *Chaos* (New York: Penguin, 1988); and Fritjof Capra, *The Web of Life: A New Understanding of Living Systems* (New York: Doubleday, 1997).

63. De Landa, "Non-organic Life," p. 133.

Chapter 6

1. Much has been written about Romero's zombie films from the perspective of film and cultural studies. For my purposes here, it is interesting to note how the zombies in film are portrayed biopolitically as a kind of mob rule. However, the ideological content of this zombie/mob changes throughout Romero's films. For instance, *Night of the Living Dead* configures zombies as a kind of lynch mob (for the protagonist), whereas *Dawn of the Dead* (which takes place in a shopping mall) portrays zombies as the new consumers of the 1980s. *Day of the Dead* presents perhaps the most complex representation of the zombies—at once masses of immigrants, but also the civilian populace "fenced out" of military governance.

2. "Convention on the Prohibition of the Development, Production, and Stockpiling of Bacteriological (Biological) and Toxic Weapons and on Their Destruction," Article I.2. Available online at the Federation of American Scientists (FAS) Web site, http://www.fas.org. For more on international bioweapons policy and regulation, see

Malcolm Dando, *The New Biological Weapons: Threat, Proliferation, and Control* (Boulder, Colo.: Lynne Rienner, 2001).

3. Just a few years prior to the making of *The Crazies*, a not-yet-known author named Michael Crichton published a thriller called *The Andromeda Strain*, which was subsequently made into a technically innovative film, in effect jump-starting the subgenre of the "outbreak" film.

4. For a more informal comment on these topics, see my article "Body Horror Is Back (Because It Never Left)," *Mute Magazine* online (September 24, 2003), at http://www.metamute.com.

5. The full title is the Public Health Security and Bioterrorism Preparedness and Response Act of 2002 (H.R. 3448), available online via the White House Web site, http://www.whitehouse.gov.

6. See, for instance, the press release "President's Budget Includes $274 Million to Further Improve Nation's Bio-surveillance Capabilities," January 29, 2004, at the U.S. Department of Health and Human Services Web site, http://www.hhs.gov.

7. Giorgio Agamben has analyzed the way in which the structure of the exception pervades sovereignty and its double, "bare life." For Agamben, sovereign power is defined by its relation to the exception, not by its ability to enforce a rule. The exception—not the contract—is therefore the basis of sovereign power, in which the exception, or state of emergency, becomes the rule by which power is exercised. See Giorgio Agamben, *Homo Sacer: Sovereign Power and Bare Life*, trans. Daniel Heller-Roazen (Stanford, Calif.: Stanford University Press, 1998).

8. For example, see WHO, "Global Public Health Response to Natural Occurrence, Accidental Release or Deliberate Use of Biological and Chemical Agents or Radionuclear Material That Affect Health," Fifty-fifth World Health Assembly, agenda item 13.15, WHA55.16 (May 18, 2002), available at http://www.who.int.

9. Further information about the CDC's disease surveillance systems such as the NEDSS can be found at the CDC Web site, http://www.cdc.gov.

10. See BMA, *Biotechnology Weapons and Humanity* (Amsterdam: Harwood Academic, 1999), pp. 53–68. There is also a large body of U.S. military reports from the 1970s and 1980s on the use of genetic engineering as a bioweapon. For example, see U.S.

Army, *Biological Warfare Threat Study*, USA CMLS Threat Section (Washington, D.C.: Government Printing Office, October 6, 1983).

11. Richard Preston, *The Hot Zone* (New York: Anchor, 1995), p. 19. Such descriptions—which also often appear in reputable magazines and newspapers—serve to heighten the level of body anxiety regarding biowar. They overlap a great deal with their fictional counterparts in the novels of Michael Crichton, Robin Cook, and others.

12. See BMA, *Biotechnology Weapons and Humanity*, pp. 30–33.

13. One result has been tighter controls on biological substances used in science research. The U.S. Bioterrorism Act (Title II, Subtitle A, sections 201–204) outlines protocols for regulating the circulation of potentially hazardous biological substances. These tighter controls have led, some argue, to negative restrictions on life science research. See Daniel Kevles, "Biotech's Big Chill," *MIT Technology Review* (August 2003): 40–50.

14. The *Nova* special was reported by Judith Miller, Stephen Engelberg, and William Broad, the same authors who published the book *Germs: Biological Weapons and America's Secret War* (New York: Touchstone, 2001). What is surprising is that although the book spends a considerable amount of time discussing the U.S. complicity in the biowar problem, the *Nova* special devotes nearly all of its time to discussing the Soviet bioweapons program of the 1980s, making almost no mention of the U.S. programs.

15. For a fascinating account of "biowar" in antiquity, see Adrienne Mayor, *Greek Fire, Poison Arrows, and Scorpion Bombs: Biological and Chemical Warfare in the Ancient World* (New York: Overlook Duckworth, 2003), as well as James Poupard and Linda Miller, "History of Biological Warfare: Catapults to Capsomeres," *Annals of the New York Academy of Sciences* 666 (1992): 9–19. Also see Mark Wheelis, "Biological Warfare before 1914," in *Biological and Toxic Weapons: Research, Development, and Use from the Middle Ages to 1945*, ed. Erhard Geissler, and John Ellis van Courtland Moon (Oxford and Stockholm: Oxford University Press and Stockholm International Peace Research Institute, 1999),pp. 8–34.

16. See chapters 3 and 5 in Mayor, *Greek Fire, Poison Arrows, and Scorpion Bombs*.

17. See Thucydides, *History of the Peloponnesian War*, trans. Rex Warner (New York: Penguin, 1972), pp. 151–156.

18. Ibid., p. 155.

19. On the Black Death, see Philip Ziegler, *The Black Death* (New York: Harper, 1971). Also see Robert Gottfried, *The Black Death: Natural and Human Disaster in Medieval Europe* (New York: Free Press, 1983).

20. Gabriele de Mussis, "The Arrival of the Plague," in *The Black Death*, ed. and trans. Rosemary Horrox (Manchester: Manchester University Press, 1994), pp. 14–26.

21. As might be expected, the dominant interpretation of the Black Death in Europe was that of divine retribution. The excerpt from de Mussis's chronicle reads thus: "The dying Tartars, stunned and stupefied by the immensity of the disaster brought about by the disease, and realizing that they had no hope of escape, lost interest in the siege. But they ordered corpses to be placed in catapults and lobbed into the city in the hope that the intolerable stench would kill everyone inside. . . . And soon the rotting corpses tainted the air and poisoned the water supply, and the stench was so overwhelming that hardly one in several thousand was in a position to flee the remains of the Tartar army. Moreover one infected man could carry the poison to others, and infect people and places with the disease by look alone." Ibid., p. 17.

22. This monitoring includes surveillance of imported goods and checks for acts of sabotage within processing plants and distribution of food inside the United States.

23. See Miller, Engleberg, and Broad, *Germs*, pp. 18–33.

24. For news updates, see the Organic Consumers Association at http://organicconsumers.org/madcow.htm.

25. See Mark Wheelis, "Biological Sabotage in World War I," in *Biological and Toxin Weapons: Research, Development, and Use from the Middle Ages to 1945*, pp. 35–62.

26. See Friedrich Frischknecht, "The History of Biological Warfare," *European Molecular Biology Organization Report* 4 (2003): 547–52. Also see BMA, *Biotechnology, Weapons, and Humanity*, pp. 22–30.

27. See Wheelis, "Biological Sabotage in World War I."

28. See Olivier Lepick, "French Activities Related to Biological Warfare, 1919–45," in *Biological and Toxin Weapons: Research, Development, and Use from the Middle Ages to 1945*, pp. 70–90.

29. The text of the Geneva Protocol is available from a number of online sources, including Bradford University's Department of Peace Studies at http://www.bradford.ac.uk/acad/sbtwc/keytext/genprot.htm.

30. See Sheldon Harris, "The Japanese Biological Warfare Programme: An Overview," in *Biological and Toxin Weapons: Research, Development, and Use from the Middle Ages to 1945*, pp. 127–152.

31. See Sheldon Harris, *Factories of Death: Japanese Biological Warfare, 1932–45 and the American Cover-up* (New York: Routledge, 2002).

32. See Leonard Cole, *Clouds of Secrecy: The Army's Germ Warfare Tests over Populated Areas* (Lanham, M.d.: Rowman and Littlefield, 1988).

33. See Gradon Carter and Graham Pearson, "British Biological Warfare and Biological Defense, 1925–45," in *Biological and Toxin Weapons: Research, Development, and Use from the Middle Ages to 1945*, pp. 185–188.

34. See Cole, *Clouds of Secrecy*. Also see B. J. Bernstein, "The Birth of the U.S. Biological Warfare Program," *Scientific American* 255 (1987): 94–99.

35. See Cole, *Clouds of Secrecy*.

36. See Mark Wheelis and Malcolm Dando, "On the Brink: Biodefense, Biotechnology, and the Future of Weapons Control," *Chemical and Biological Weapons Convention Bulletin* 58 (December 2002): 3–7.

37. See Frischknecht, "The History of Biological Warfare," pp. S48–49. See also former Soviet bioweapons researcher Ken Alibek's book *Biohazard* (New York: Random House, 1999).

38. See BMA, *Biotechnology, Weapons, and Humanity*, pp. 56–60.

39. The most comprehensive history of eugenics is Daniel Kevles's *In the Name of Eugenics: Genetics and the Uses of Human Heredity* (Cambridge, Mass.: Harvard University Press, 1995).

40. On the impact of American eugenics on gender, sexuality, and race, see Nancy Ordover, *American Eugenics* (Minneapolis: University of Minnesota Press, 2002).

41. See Kevles, *In the Name of Eugenics*, pp. 113–22.

42. See Daniel Kevles, "Out of Eugenics: The Historical Politics of the Human Genome," *The Code of Codes: Scientific and Social Issues in the Human Genome Project*, ed. Daniel Kevles and Leroy Hood (Cambridge, Mass.: Harvard University Press, 1992), pp. 3–37.

43. For a critique of the new eugenics, see Critical Art Ensemble, "Eugenics: The Second Wave," *Ctheory* (1998), available at http://www.ctheory.net.

44. See Miller, Engleberg, and Broad, *Germs*, pp. 199–202.

45. See Poupard and Miller, "History of Biological Warfare," pp. 11–13, and Wheelis, "Biological Warfare before 1914," pp. 17–27.

46. Cited in Poupard and Miller, "History of Biological Warfare," p. 12.

47. David Arnold, *Colonizing the Body: State Medicine and Epidemic Disease in 19th Century India* (Berkeley: University of California Press, 1993), p. 9.

48. See Roy Porter, *The Greatest Benefit to Mankind: A Medical History of Humanity* (New York: Norton, 1997), pp. 462–492. Particular examples of the former are Columbus in Haiti, Cortez in central Mexico, the Spanish conquest of Paraguay. Examples of the latter are the British in sub-Saharan Africa and the French in the Congo and West Africa.

49. Ibid., p. 481.

50. For an analysis from the industry perspective, see Jennifer van Brunt, "Who's Going to Pay for Global Health?" *Signals Magazine* (October 24, 2003), available at http://www.signalsmag.com.

51. There are many contemporary accounts by medical missionaries, but for historical views see Meera Abraham, *Religion, Caste, and Gender: Missionaries and Nursing History in South India* (Sandvika, Norway: BI Publications, 1996); and Michael Gelfand, *Christian Doctor and Nurse: The History of Medical Missions in South Africa from 1799–1976* (Dereham, United Kingdom: Aitken, 1984).

52. See Frantz Fanon, "Medicine and Colonialism," in *A Dying Colonialism*, trans. Haakon Chevalier (New York: Grove, 1965), pp. 121–147.

53. See Paul Virilio and Sylvère Lotringer, *Pure War* (New York: Semiotext[e], 1997), pp. 91–102.

54. For an overview, see the anthology John Arquilla and David Ronfeldt, eds., *In Athena's Camp: Preparing for Conflict in the Information Age* (Santa Monica, Calif.: RAND, 1997).

55. See Gerfried Stocker and Christine Schöpf, eds., *InfoWar: Ars Electronica 1998* (Linz, Austria: Springer, 1998).

56. See Virilio and Lotringer, *Pure War*, pp. 94–95.

57. The WHO's Global Outbreak and Response Network, inaugurated in 2000, played an instrumental role in managing the 2003 SARS outbreak.

58. Michel Foucault, *The History of Sexuality, vol. 1, An Introduction*, trans. Robert Hurler (New York: Vintage, 1978), p. 142.

59. Michel Foucault, "Security, Territory, and Population," in *Ethics: Subjectivity and Truth: The Essential Works of Michel Foucault 1954–1984, vol. 1*, trans. Robert Hurley et al., ed. Paul Rabinow (New York: New Press, 1997), p. 71.

60. Foucault, "The Birth of Biopolitics," in *Ethics*, p. 73.

61. Michel Foucault, *"Society Must Be Defended": Lectures at the Collège de France, 1975–76*, trans. David Macey, ed. Mauro Bertani and Alessandro Fontana (New York: Picador, 2003), p. 266.

62. Agamben, *Homo Sacer*, pp. 14–17.

63. Foucault, *"Society Must be Defended,"* p. 26.

64. Ibid., p. 27.

65. Ibid., p. 37.

66. See Foucault, "The Birth of Biopolitics."

67. Foucault, *"Society Must be Defended,"* p. 39.

68. Ibid., p. 249.

69. Ibid., pp. 239–240.

70. "Sovereign power's effect on life is exercised only when the sovereign can kill. The very essence of the right of life and death is actually the right to kill: it is at this moment when the sovereign can kill that he exercises his right over life." Ibid., p. 240.

71. "Thus life, not death, becomes the focus of a new kind of power. . . . It is the power to 'make' live and 'let' die. The right of sovereignty was the right to take life or let live. And then this new right is established: the right to make live and to let die." Ibid., p. 241.

72. Ibid., p. 254.

73. Ibid., p. 60.

74. Ibid., p. 246.

75. For a historical study of hospital technologies, see Bettyann Holtzmann Kevles, *Naked to the Bone: Medical Imaging in the Twentieth Century* (Reading, United Kingdom: Addison-Wesley, 1997).

76. Scott Montgomery, "Codes and Combat in Biomedical Discourse," *Science as Culture* 2.3 (1991), p. 374.

77. Ibid., p. 375.

78. Foucault, "*Society Must be Defended*", pp. 246, 252.

79. Ibid., p. 137.

80. Ibid., pp. 92–96.

81. For an overview see FAS, "Analysis of the Anthrax Attacks," available at http://www.fas.org.

82. See Barbara Hatch Rosenberg, "Anthrax Attacks Pushed Open an Ominous Door," *Los Angeles Times* (September 22, 2002).

83. See FAS, "Analysis of the Anthrax Attacks."

84. See Judith Miller, Stephen Engelberg, and William Broad, "U.S. Germ Warfare Research Pushes Treaty Limits," *New York Times*, September 4 2001.

85. See the press release "Bayer to Supply Government by Year-End with 100 Million Tablets for $95 Million," available at http://www.news.bayer.com.

86. In 2001, there was also a patent-access litigation lawsuit against Bayer and two other companies for allegedly paying competitors to keep them from introducing generic versions of Cipro on the market. It is estimated that Cipro sales for Bayer average $1 billion yearly.

87. See Paul Virilio and Sylvère Lotringer, *Crepuscular Dawn*, trans. Mike Taormina (New York: Semiotext[e], 2002), pp. 135–136.

88. Ibid., p. 135. Virilio's three types of bombs roughly correspond to three technical revolutions in speed: "So, these are the three characteristics for the three revolutions. The revolution in *physical transportation* came first: movement and acceleration up to supersonic speeds. The revolution in *transmissions*, which comes second, is the revolution of live transmission. It is the cybernetic revolution. It is the ability to reach the light barrier, in other words, the speed of electromagnetic waves in every field, not only television and tele-audition, but also tele-operation. Finally, the revolution in *transplants*, the last revolution, introduces this technology of transmission inside the body by means of certain techniques. After the revolution in transportation and the revolution in transmissions, now with the twenty-first century begins the revolution in intra-organic transplants" (p. 99).

89. Ibid., p. 136.

90. Ibid., pp. 154 (italics removed), p. 146.

91. The quotation given in the section heading is from "Sin Origin" by Darkthrone, ©2001 Moonfog Productions. Quoted in F. Jollivet Castelot, "Nosology, Necrology, and Medicine in Medieval Plague Poetry," *Medicine, Literature, and Language* 2.2 (1999), p. 32.

92. See Wheelis and Dando, "On the Brink." Also see their article "Back to Bioweapons?" *Bulletin of the Atomic Scientists* 59.1 (January–February 2003): 40–46.

93. See Plato, *The Republic*, trans. Allan Bloom (New York: Basic, 1991), book V, sections 457–461.

94. See Emily Martin, *Flexible Bodies: The Role of Immunity in American Culture from the Days of Polio to the Age of AIDS* (Boston: Beacon, 1994), pp. 127–142.

95. See Jennifer Brower and Peter Chalk, *The Global Threat of New and Reemerging Infectious Diseases* (Santa Monica, Calif.: RAND, 2003).

96. See the news articles by Brian Krebs, "'Good' Worm Fixes Infected Computers," *Washington Post* online, August 18, 2003, at http://www.washingtonpost.com, and Paul Roberts, "Blaster Worm Continues to Spread," *PC World*, August 12, 2003, at http://www.pcworld.com.

Chapter 7

1. See Catherine Arnst and John Carey, "Biotech Bodies," *Business Week* (July 1998): 56–63; and Nancy Parenteau, "The Organogenesis Story," *Scientific American* (April 1999): 83–84. For more on tissue engineering in general, see Lawrence Bonassar and Joseph Vacanti, "Tissue Engineering: The First Decade and Beyond," *Journal of Cellular Biochemistry* 30–31 (1998): 297–303; and David Mooney and Antonios Mikos, "Growing New Organs," *Scientific American* (April 1999): 60–65.

2. In modern embryology, after conception the embryo begins cellular multiplication and division until at a certain stage the cells begin to "differentiate"—that is, what were once general embryonic cells (embryonic stem cells) begin to specify their structure, shape, and function. Thus, those stem cells that differentiate to become neural cells are often branch shaped, whereas those stem cells that go on to become muscle cells are often spindle shaped, and so forth. For further information on stem cell research and the latest news on policy changes, see the NIH Web site at http://www.nih.gov.

3. See Robert Lanza, Robert Langer, and W. L. Chick, eds., *Principles in Tissue Engineering* (New York: Landes, 1997); and Charles Patrick, Antonios Mikos, and Larry McIntire, eds., *Frontiers in Tissue Engineering* (New York: Pergamon, 1998).

4. Tissue engineering pioneers Robert Langer and Joseph Vacanti have defined tissue engineering as "an interdisciplinary field that applies the principles of engineering and the life sciences toward the development of biological substitutes that restore,

maintain, or improve tissue function." See Robert Langer and Joseph Vacanti, "Tissue Engineering," *Science* 260 (1993), p. 920. In this and other articles by tissue engineering researchers, common themes are how to apply the data generated in genomics (and proteomics) toward a more sophisticated understanding of cell structure and function, and the need for further research in biomaterials (or various types of scaffolding structures in which to "seed" cells). The implication is that the combination of development-enhancing biomaterials and a thorough knowledge of the genetic basis of cell structure and function would pave the way for the ability to (re)generate tissues and even organs in vivo.

5. See Georges Canguilhem, *The Normal and the Pathological*, trans. Carolyn Fawcett (New York: Zone, 1989).

6. On cyberspace and the problems of corporeality or "the meat," see William Gibson's *Neuromancer* (New York: Ace, 1984). Case, the protagonist and hacker in Gibson's now-classic work of cyberpunk science fiction, contrasts the infinite complexity and luminosity of the cyberspatial matrix with the base, low necessities of the flesh. In this sense, *Neuromancer* can be seen as one of the key participants in a branch of technoculture that tends more often than not toward a technophilia laced with transcendental implications, something also found, for example, in the robotics and artificial intelligence work of Hans Moravec.

7. Bonassar and Vacanti, "Tissue Engineering: The First Decade and Beyond," p. 297.

8. See Aris Persidis, "Tissue Engineering," *Nature Biotechnology* 17 (May 1999): 508–510.

9. Bonassar and Vacanti define *cells and materials* (the latter also referred to as the *scaffold* or *matrix*) as the critical components for tissue engineering because of their interaction: "The cellular component is necessary for the generation of new tissue through production of extracellular matrix and is responsible for the long-term maintenance of this matrix. The scaffold material provides mechanical stability of the construct in the short term and provides a template for the three-dimensional organization for the developing tissue." Bonassar and Vacanti, "Tissue Engineering: The First Decade and Beyond," p. 299. What is interesting about this description from the point of view of ontogeny is how Bonassar and Vacanti split form and function between the scaffold/matrix and the cells. Even though the biodegradable scaffold eventually disintegrates as the new tissue or organ literally takes shape, Bonassar and Vacanti suggest that the cells themselves—although functional—cannot realize their own form. The

larger question this raises is whether both form and function are latent in the cell itself or whether the phenomenon of cellular function arises as an interaction of many biomolecular systems acting in concert.

10. Research for this chapter was done between 1998 and 2001 and initially written in 2001. As Bonassar and Vacanti state, "When used currently, the term tissue engineering has come to imply some combination of cells, scaffold material, and bioactive peptides used to guide the repair or formation of tissue." Bonassar and Vacanti, "Tissue Engineering: The First Decade and Beyond," p. 298. For more on the techniques of tissue engineering and developments in specific areas of research, consult Lanza, Lauger, and Chick, *Principles in Tissue Engineering*; and Particle, Mikos, and McIntire, *Frontiers in Tissue Engineering*.

11. See Langer and Vacanti, "Tissue Engineering," and Mooney and Mikos, "Growing New Organs."

12. For research on universal donor cells, see R. J. Armstrong and C. N. Svendsen, "Neural Stem Cells: From Cell Biology to Cell Replacement," *Cell Transplantation* 9.2 (2000): 139–152; N. Ende et al., "Pooled Umbilical Cord Blood as a Possible Universal Donor for Marrow Reconstitution and Use in Nuclear Accidents," *Life Sciences* 69.13 (2001): 1531–1539; and Roger Pedersen, "Embryonic Stem Cells for Medicine," *Scientific American* (April 1999): 68–73.

13. See Robert Langer, "Tissue Engineering: Status and Challenges," *E-biomed* 1 (March 7, 2000): 5–6, and Mooney and Mikos, "Growing New Organs."

14. See Mooney and Mikos, "Growing New Organs"; and M. Stading and R. Langer, "Mechanical Shear Properties of Cell-Polymer Cartilage Constructs," *Tissue Engineering* 5.3 (1999): 241–250.

15. For research on biodegradable polymer scaffolds, see V. Hasirci et al., "Versatility of Biodegradable Biopolymers: Degradability and an in Vivo Application," *Journal of Biotechnology* 86.2 (2001): 135–150; D. von Heimburg, et al., "Influence of Different Biodegradable Carriers on the in Vivo Behavior of Human Adipose Precursor Cells," *Plastic Reconstructive Surgery* 108.2 (2001): 411–420; and P. X. Ma and R. Zhang, "Microtubular Architecture of Biodegradable Polymer Scaffolds," *Journal of Biomedical Materials Research* 56.4 (2001): 469–477.

16. For research on bioreactors in tissue engineering, see E. J. Doolin et al., "Effects of Microgravity on Growing Cultured Skin Constructs," *Tissue Engineering* 5.6 (1999):

573–582, and S. P. Hoerstrup et al., "New Pulsatile Bioreactor for in Vitro Formation of Tissue Engineered Heart Valves," *Tissue Engineering* 6.1 (2000): 75–79.

17. For research on growth factors in tissue engineering, see H. von Recum, et al., "Growth Factor and Matrix Molecules Preserve Cell Function on Thermally Responsive Culture Surfaces," *Tissue Engineering* 5.3 (1999): 251–265; Y. Tabata et al. "De Novo Formation of Adipose Tissue by Controlled Release of Basic Fibroblast Growth Factor," *Tissue Engineering* 6.3 (2000): 279–289; and L. Weiser et al. "Effect of Serum and Platelet-Derived Growth Factor on Chondrocytes Grown in Collagen Gels," *Tissue Engineering* 5.6 (1999): 533–544.

18. See Mooney and Mikos, "Growing New Organs"; and Patrick, Mikos, and McIntyre, eds., *Frontiers in Tissue Engineering*.

19. See A. I. Caplan, "Stem Cell Delivery Vehicle," *Biomaterials* 11 (1990): 44–46; and J. Gao et al., "Tissue-Engineered Fabrication of an Osteochondral Composite Graft Using Rat Bone Marrow–Derived Mesenchymal Stem Cells," *Tissue Engineering* 7.4 (2001): 363–371. Stem cells are discussed in more detail in chapter 8, where the relation of the concept of "regeneration" is related to technical modes of configuring biological time. Although stem cell research is often mentioned in the media apart from tissue engineering, the two fields have for some time been converging, especially as research into the ability to control cell differentiation offers tissue engineering more concrete ways of directing cell aggregates into viable living tissue structures. For more on this topic, see "Special Report: Tissue Engineering," *Scientific American* (April 1999), pp. 86–89. For our purposes here, I discuss stem cells only as they relate to tissue engineering—principally as another option for cell sourcing in the initial steps of tissue engineering techniques. For general information on stem cells and stem cell research (as well as for related news concerning policy, etc.), consult the NIH Web site at http://www.nih.gov.

20. See especially the landmark article by James Thomson et al., "Embryonic Stem Cell Lines Derived from Human Blastocysts," *Science* 282 (1998): 1145–1147.

21. See C. R. Bjornson et al., "Turning Brain into Blood: A Hematopoietic Fate Adopted by Adult Neural Stem Cells in Vivo," *Science* 283 (1999): 534–537.

22. See Pedersen, "Embryonic Stem Cells for Medicine."

23. For example, see Terrence Deacon et al., "Blastula-Stage Stem Cells Can Differentiate into Dopaminergic and Serotonergic Neurons after Transplantation," *Experimental Neurology* 149 (1998): 28–41.

24. See Bonassar and Vacanti, "Tissue Engineering: The First Decade and Beyond."

25. Ibid., p. 299.

26. Pedersen, "Embryonic Stem Cells for Medicine," p. 72.

27. Robert Langer and Joseph Vacanti, "Tissue Engineering: The Challenges Ahead," *Scientific American* (April 1999), p. 86.

28. David Stocum, "Regenerative Biology and Medicine in the 21st Century," *E-biomed* 1 (March 7, 2000), p. 17.

29. See Caplan, "Stem Cell Delivery Vehicle"; and Mooney and Mikos, "Growing New Organs."

30. See Langer, "Tissue Engineering: Status and Challenges."

31. See Mooney and Mikos, "Growing New Organs." Also see the research in Lonnie Shea et al., "DNA Delivery from Polymer Matrices for Tissue Engineering," *Nature Biotechnology* 17 (June 1999): 551–554.

32. Of course, the gap between the hypothetical and the experimental is constantly changing with a field as new as tissue engineering. Nevertheless, that relationship is not simply one of hypotheses that are later fulfilled in experiment; alternative approaches or unexpected research results often take the field in new directions. An example, mentioned previously, is the use of stem cells in tissue engineering, where early techniques for synthesizing bioartificial tissues were based on the capacity of biomaterials science to engineer nonorganic living tissue equivalents. For an early experimental application of tissue engineering, see I. V. Yannas et al., "Design of an Artificial Skin I: Basic Design Principles," *Journal of Biomedicine Materials* 14 (1980): 65–81. For examples of speculation on the future of tissue engineering, see the special issues of *Scientific American* (April 1999) and *MIT Technology Review* (April 2001).

33. See Bonassar and Vacanti, "Tissue Engineering: The First Decade and Beyond"; and Persidis, "Tissue Engineering."

34. Canguilhem, *The Normal and the Pathological*, pp. 54–61.

35. Ibid., pp. 65–86.

36. Ibid., p. 51.

37. Ibid., pp. 91–101.

38. Ibid., p. 91.

39. Ibid., p. 89.

40. For a popular-press overview on disease research and genetic medicine, see "Special Issue: The Future of Medicine," *Time* (January 11, 1999). For example, the basis of much gene therapy is that in many genetically based disorders, the body is inhibited by a lack of a certain gene (which would play a central part in the production of needed proteins for the body's sustenance). The role of gene therapy is then to intervene by providing the needed molecules, which would then be taken up and incorporated by the body on the molecular level, thereby restoring it to an equilibrium state. Although this theory was successful early on, it has since run into a number of complications, primary among them a lack of consideration of the network of molecular elements and processes involved, a lack that exists because of a centralized focus on the individual's DNA as the causal source. That is, gene therapy has been operating on a model similar to Bernard's, in which the problem, on the genetic level, is the excess or deprivation of a given gene or DNA sequence. Much is at stake in this therapeutic notion of medicine, including the historically specific, scientifically engaged, and dynamic production of a set of criteria that will constitute a medical-biological norm in a particular scientific-medical context (e.g., medical genetics, drug design, pathobiology, and so on).

41. See Arthur Kroker, *Spasm: Virtual Reality, Android Music, Electric Flesh* (New York: St. Martin's, 1993), pp. 36–46.

42. See Marshall McLuhan, *Understanding Media* (Cambridge, Mass.: MIT Press, 1995), pp. 7–13.

43. Michel Foucault, "The Birth of Biopolitics," in *Ethics: Subjectivity and Truth: The Essential Works of Michel Foucault 1954–1984*, Vol. 1, trans. Robert Hurley et al., ed. Paul Rabinow (New York: New Press, 1997), p. 73.

44. See chapter 3 for the role of biopolitics in genomics and population genomics.

45. The primary challenges for tissue engineering research lie in this relationship between a highly individualized medicine and a universalized medicine. For

individualized tissue engineering, the incomplete knowledge about the genetic and molecular factors in cellular development, differentiation, and decay has been a significant obstacle in fulfilling the vision of growing an organ on demand from a patient-specific cell sample. Likewise, tissue engineering's goal of establishing a modular approach to cell and tissue transplant using either universal donor cells (immunorepressed stem cells) or off-the-shelf organs (tissue structures generated from universal donor cells) faces the obstacles surrounding the maintenance of viable cell and tissue structures in environments that may compromise immunorepression. For comments on individual versus universal models in tissue engineering, see Langer and Vacanti, "Tissue Engineering: The Challenges Ahead," pp. 86–89; Pedersen, "Embryonic Stem Cells for Medicine," pp. 68–73; and Sophie Petit-Zeman, "Regenerative Medicine," *Nature Biotechnology* 19 (March 2001): 201–206.

46. The quotation given in the section heading is from "Rust" by Darkthrone, ©2003 Moonfog Productions. Quoted in Donnolo Sabbati, "Alchemical Metaphors in Early Modern Surgical Treatises," *Journal of Modern Languages* 24 (1999), p. 6.

47. See Jean Baudrillard, *Simulations* (New York: Semiotext[e], 1983), pp. 38–49; and *Seduction* (New York: St. Martin's, 1990), pp. 37–52.

48. On the historical-social aspects of anatomy, see Jonathan Sawday, *The Body Emblazoned: Dissection and the Human Body in Renaissance Culture* (New York: Routledge, 1995). On the role of hospital technologies in modern medicine's enframing of the body, see Joel Howell, *Technology in the Hospital: Transforming Patient Care in the Early Twentieth Century* (Baltimore: Johns Hopkins University Press, 1995).

49. Kroker, *Spasm*, pp. 36–45.

50. See Bruno Latour, *We Have Never Been Modern* (Cambridge, Mass.: Harvard University Press, 1993), pp. 10–12.

51. Ibid., pp. 34–43. Also see Latour, *Pandora's Hope: Essays on the Reality of Science Studies* (Cambridge, Mass.: Harvard University Press, 1999), pp. 1–24.

52. Latour, *We Have Never Been Modern*, p. 81.

53. See Latour, *Pandora's Hope*, pp. 122–123.

54. See Mooney and Mikos, "Growing New Organs," p. 65.

Chapter 8

1. See Boyce Rensberger, *Life Itself: Exploring the Realm of the Living Cell* (New York: Oxford University Press, 1996), pp. 62–80.

2. Quoted in Sophie Petit-Zeman, "Regenerative Medicine," *Nature Biotechnology* 19 (March 2001): 201–206.

3. Lawrence Bonassar and Joseph Vacanti, "Tissue Engineering: The First Decade and Beyond," *Journal of Cellular Biochemistry* 30–31 (1998), p. 297.

4. David Mooney and Antonios Mikos, "Growing New Organs," *Scientific American* (April 1999), p. 65.

5. See Stephen Hall, "The Recycled Generation," *New York Times Magazine*, January 30, 2000.

6. See Petit-Zeman, "Regenerative Medicine." See also David Stocum, "Regenerative Biology and Medicine in the 21st Century," *E-biomed* 1 (March 7, 2000): 17–20.

7. See Robert Langer and Joseph Vacanti, "Tissue Engineering," *Science* 260 (1993): 920–926.

8. Mooney and Mikos, "Growing New Organs."

9. See Catherine Arnst and Johnathan Carey, "Biotech Bodies," *BusinessWeek*, July 27, 1988, pp. 56–63. Also see Nancy Parenteau, "The Organogenesis Story," *Scientific American* (April 1999): 83–84.

10. See the news article "Bionic Bodies Special: Test Tube Thumb," *BBC News* online, November 4, 1998, at http://news.bbc.co.uk.

11. For a survey of techniques in regenerative medicine, see Charles Patrick, Antonios Mikos, and Larry McIntire, eds., *Frontiers in Tissue Engineering* (New York: Pergamon, 1998).

12. See Mooney and Mikos, "Growing New Organs."

13. See Roger Pedersen, "Embryonic Stem Cells for Medicine," *Scientific American* (April 1999): 68–73. Two central articles that describe stem cells are Michael

Shamblott, "Derivation of Pluripotent Stem Cells from Cultured Human Primordial Germ Cells," *Proceedings of the National Academy of Sciences* 95 (1998): 13726–13731; and James Thomson et al., "Embryonic Stem Cell Lines Derived from Human Blastocysts," *Science* 282 (1998): 1145–1147. For background information on stem cells see the NIH stem cells primer at http://www.nih.gov. Also see the testimony by Lori Andrews before the President's Council on Bioethics, "Stem Cell Research: Current Law and Policy with Emphasis on the States," Session 4, July 24, 2003, at http://www.bioethics.gov.

14. See Gretchen Vogel, "Harnessing the Power of Stem Cells," *Science* 283 (March 1999): 1432–1434.

15. On multipotency in stem cells, see "Derivation of Pluripotent Stem Cells," Shamblott et al., and Thomson et al., "Embryonic Stem Cell Lines," as well as the background information at the NIH Web site.

16. See Bonassar and Vacanti, "Tissue Engineering: The First Decade and Beyond"; and Mooney and Mikos, "Growing New Organs."

17. For a survey of stem cell research in regenerative medicine, see the following research articles: C. R. Bjornson et al., "Turning Brain into Blood: A Hematopoietic Fate Adopted by Adult Neural Stem Cells in Vivo," *Science* 283 (1999):534–537; L. D. K. Buttery et al., "Differentiation of Osteoblasts and in Vitro Bone Formation from Murine Embryonic Stem Cells," *Tissue Engineering* 1.7 (February 2001): 89–100; Robert Phillips et al., "The Genetic Program of Hematopoietic Stem Cells," *Science* (June 2, 2000): 1635–1640; Maya Schuldiner et al., "Effects of Eight Growth Factors on the Differentiation of Cells Derived from Human Embryonic Stem Cells," *Proceedings of the National Academy of Sciences* 97 (2000): 11307–11312; Dale Woodbury et al., "Adult Rat and Human Bone Marrow Stromal Cells Differentiate into Neurons," *Journal of Neuroscience Research* 61.4 (August 2000): 2–7.

18. See also Robert Langer, "Tissue Engineering: Status and Challenges," *E-biomed* 1 (March 7, 2000): 5–6; and Petit-Zeeman, "Regenerative Medicine."

19. See Bonassar and Vacanti, "Tissue Engineering: The First Decade and Beyond."

20. See Nicholas Wade, "Teaching the Body to Heal Itself," *New York Times* online, November 7, 2000, at http://www.nytimes.com.

21. For example, in the cellular basis of aging or cancer cures. See Hall, "The Recycled Generation."

22. See Kathryn Brown, "The Human Genome Business Today," *Scientific American* (July 2000): 50–59.

23. For example, see the industry news articles "Gero-technology Companies Turn Back the Hands of Time," *Biospace.com*, March 28, 2000, at http://www.biospace.com, and "Merger Makes Curis a Contender in Regenerative Medicine," *Biospace.com*, March 31, 2000, at http://www.biospace.com.

24. Aristotle, *De generatione et corruptione* (On generation and corruption), trans. Harold H. Joachim, in *The Basic Works of Aristotle*, ed. Richard McKeon (New York: Modern Library, 2001), book I, chap. 2, p. 478.

25. Ibid., book I, chap. 3, p. 479.

26. Somites are segmented blocks of tissue in the animal embryo that in vertebrates give rise to the vertebral column, ribs, and striated muscle.

27. Brian Goodwin and Ricard Solé, *Signs of Life: How Complexity Pervades Biology* (New York: Basic, 2000), p. 61.

28. Ibid., pp. 61 62.

29. Susan Squier, "Life and Death at Strangeways: The Tissue Culture Point of View," in *Biotechnology and Culture: Bodies, Anxieties, and Ethics*, ed. Paul Brodwin (Seattle: University of Washington Press, 2000), p. 28.

30. Ibid., p. 28.

31. Catherine Waldby, "Stem Cells, Tissue Cultures, and the Production of Biovalue," *Health: An Interdisciplinary Journal for the Social Study of Health, Illness, and Medicine* 6.3 (2002), p. 310.

32. Aristotle, *Politica* (Politics), trans. Benjmain Jowett, in *The Basic Works of Aristotle*, ed. McKeon, book I, chap. 9, p. 1139. Aristotle had already identified the difference between what Marx would later call use value and exchange value. The latter, for Aristotle, constitutes the art of accumulating wealth in itself. See book I, chap. 9 of the *Politics*.

33. Paul Virilio and Sylvère Lotringer, *Pure War*, trans. Mark Polizzotti (New York: Semiotext[e], 1997), p. 136.

34. See ibid., pp. 35–36. Also see *The Virilio Reader*, ed. James Der Derian (London: Blackwell, 1998), pp. 117–133.

35. See Virilio and Lotringer, *Pure War*, pp. 64–65.

36. Examples of the former include the cyborg integration of human and machine in McLuhan's "mechanical bride" (the automobile), and examples of the latter include the pervasive sensory coverage of Debord's "society of the spectacle." For further examples, see Paul Virilio, *The Art of the Motor*, trans. Julie Rose (Minneapolis: University of Minnesota Press, 1995), pp. 99–132.

37. Paul Virilio, *The Information Bomb*, trans. Chris Turner (London: Verso, 2000), pp. 2–3.

38. Karl Marx, *Capital*, trans. Ben Fowkes (New York: Penguin, 1990 [orig. 1867]), 1: 336.

39. Ibid., 1: 339.

40. Ibid.

41. Antonio Negri, "The Constitution of Time," in *Time for Revolution*, trans. Matteo Mandarini (New York: Continuum, 2003), p. 29.

42. Ibid.

43. Ibid., p. 36, emphasis in original.

44. Michael Hardt and Antonio Negri, *Empire* (Cambridge, Mass.: Harvard, University Press, 2000), pp. 289–294. See also Maurizio Lazzarato, "Immaterial Labor," in *Radical Thought in Italy*, ed. Paolo Virno and Michael Hardt (Minneapolis: University of Minnesota, 1996), pp. 133–147.

45. Hardt and Negri, *Empire*, pp. 354–359.

46. The quotation given in the section heading is from "I, Voidhanger" by Darkthrone, © 2001 Moonfog Productions. Quoted in J. Bohme, "Biology, Necrology,

and the Relation between Magic and Medicine in the Early Renaissance," *Early Modern Studies Quarterly* 2.2 (1989), 34–42.

47. Negri, *Time*, p. 102.

Chapter 9

1. Donna Haraway, "The Promises of Monsters: A Regenerative Politics of Inappropriate/d Others," in *Cultural Studies*, ed. Lawrence Grossberg, Cary Nelson, and Paula Treichler (New York: Routledge, 1992), p. 296.

2. Ibid., p. 298.

3. Haraway suggests that in order to begin refiguring nature, two movements must occur: first, the separatist divisions between the human and technological must become sites of critique and complexification (which also implies problematizing rationalist, progressive histories of technology and agentially biased views on technology as tool). Second, in the multitudinous zones wherein nature and culture intersect, traditional notions of the agential, autonomous subject need to be resituated within a field that is at once larger (a network) and more specified (discrete, contextual elements).

4. The point here is both political and ontological. In the ontological question, these entities, when denaturalized and not accepted within their normative contexts, are indeed monstrous, cyborg creations, often literal fusions of genetic flesh and computer circuits. The issue here is not phenomenological—How do we as subjects relate to these creations? How do we as subjects fill these passive empty objects with our sense experience?—or postmodernist (the body is not disappearing or a determinstic product of language). The issue is critical—How is subjectivity both installed in one sector of a situation (e.g., the patient) and liquidated from another (the body-object)? When we consider a cyborg entity such as the DNA chip, the point is not to recuperate it into models of the modern subject, but rather to ask how it networks itself, as a discrete, natural-technical, material-semiotic unit, into the processes of technoscience.

5. See the essays "Regimes, Pathways, Subjects" and "Entering the Post-media Era," in Félix Guattari, *Soft Subversions*, ed. Sylvère Lotringer, trans. David Sweet and Chet Wiener (New York: Semiotext[e], 1996), pp. 112–131 and 106–112. On Guattari's approaches to the question of subjectification, see *Molecular Revolution*, trans. Rosemary Sheed (New York: Penguin, 1984).

6. Guattari's perspective is contextualized by the present age of information technologies. In more detail, the three "paths" include: (1) Pathways of power, where the mode of exercise is from the outside (outside-in). Its main object is land (territorializing), and the historical example is early-modern Christianity's organization of knowledge in monastic and religious institutions ("Age of European Christianity"). (2) Pathways of knowledge, where the mode of exercise is from within the subject (inside-out). Its main object is the map (deterritorialization), and the historical examples are the Enlightenment and the Industrial Revolution in the emergence of modern capitalism ("Age of Capitalist Deterritorialization of Modes of Knowledge and Technique"). (3) Pathways of self-reference, where the mode of exercise is self-transformation (infolding and outfolding). Its main object is the self that becomes other (change, process, transformation), and the historical examples are the liberatory possibilities in information technologies ("Age of Planetary Computerization"). See Guattari, "Regimes, Pathways, Subjects," pp. 114–117.

7. Ibid., p. 124.

8. As Guattari states, "our project, on the contrary, is to attempt to rethink these three necessarily interwoven paths/voices. No engagement with the creative phyla of the third path/voice is tenable unless new existential territories are concurrently established." Ibid., p. 125.

9. See ibid., pp. 125–126. The two characteristics Guattari emphasizes in relation to postmedia subjectivities are "processual" character and its "singularization." By "processual," he means that human-media subjectivities should resist calcification into either well-worn dichotomies (subject/object, human user/nonhuman tool), or the standardization characteristics of the information technology economy (proprietary ownership, obsolescence, restricted access). By "singularization," he means that part of what subjectivity should mean is not the right to be the same (your essence, your identity, your statistical data), but rather the right to be singular, different, and always different from your self.

10. Haraway, "The Promises of Monsters," p. 298.

11. The Tissue Culture and Art Project Web site is at http://www.tca.uwa.edu.au.

12. The SymbioticA Web site is at http://www.symbiotica.uwa.edu.au.

13. If this question can be addressed, then perhaps Haraway's program for cyborgs, monsters, and material-semiotic nodes can provide a theoretical framework to discuss

"bioinformatic bodies." This framework would mean considering discrete elements—such as DNA chips, automated sequencers, online biological databases, tissue-engineered organs, and so on—as material-semiotic nodes that play a central role in the means whereby these bioscientific discourses and practices are refiguring the "natural" body.

14. As artists Oron Catts and Ionat Zurr note on the SymbioticA Web site, "the Semi-Living ear was constructed out of degradable/absorbable biopolymers and seeded with human chondrocytes (cartilage tissue). The ear confronts the viewer with the notion of the body as fragmented and as consisting of communities of cells (tissues/organs) collaborating together. These communities can be sustained alive and grow separated and independent to the body with the support of an artificial support system. The ear belongs to a new class of object/being in the continuum of life—that of the Semi-Living. The Semi-Living are constructed of living and non-living materials, and are new autonomous entities located in the fuzzy border between the living/non-living, grown/constructed, born/manufactured, and object/subject."

15. The property of one's own body is constitutive of modern political theory; in Locke, for instance, ownership over one's own body is a citizen's fundamental right. The topic takes a different turn in Marx, however, for whom this very right serves as the basis for the selling of labor power in waged labor.

16. See E. O. Wilson, *Sociobiology: The New Synthesis* (Cambridge, Mass.: Harvard University Press, 2000).

17. See Ruth Hubbard and Elijah Wald, *Exploding the Gene Myth* (Boston: Beacon, 1997); Richard Lewontin, *Biology as Ideology: The Doctrine of DNA* (New York: HarperCollins, 1992); and Jeremy Rifkin, *The Biotech Century: Harnessing the Gene and Remaking the World* (New York: Tarcher/Putnam, 1998).

18. The SubRosa Web site is at http://www.cyberfeminism.net. As the group states in its manifesto (available online), "SubRosa is a reproducible cyberfeminist cell of cultural researchers committed to combining art, activism, and politics to explore and critique the intersections of the new information and biotechnologies in women's bodies, lives, and work."

19. The *Expo EmmaGenics* brochure includes advertisements for devices such as the "Palm XY" (a wireless "mate finder"), the "Zygote Monitor" (for televising embryonic development for home videos), and even recipes for human caviar.

20. See Paul Rabinow, "Artificiality and Enlightenment: From Sociobiology to Biosociality," in *The Science Studies Reader*, ed. Mario Biagioli (New York: Routledge, 1999), pp. 407–417.

21. The Biotech Hobbyist Web site is at http://www.biotechhobbyist.org.

22. The CAE Web site is at http://www.critical-art.net.

23. See CAE, *Flesh Machine: Cyborgs, Designer Babies, and New Eugenic Consciousness* (Brooklyn: Autonomedia, 1998), pp. 3–6.

24. CAE, *The Molecular Invasion* (Brooklyn: Autonomedia, 2002), pp. 59–76.

25. Ibid., p. 3.

26. Ibid., p. 111, emphasis in original.

27. The quotation given in the section heading is from "Madrigal IV" by Ulver, © 1997 Magic Arts Publishing. Quoted in Robert Flud, "The Non-Modern Interstitiality of the Animal in the Mythos of Norwegian Black Metal," *Journal of Popular Culture Studies* 31 (October 2000): 12–26.

28. Michel Foucault, *The History of Sexuality*, Vol. 1, *An Introduction*, trans. Robert Hurley (New York: Vintage, 1978), p. 159.

29. David Garcia and Geert Lovink, "The ABC of Tactical Media," posted to Nettime, May 16, 1997, at http://www.nettime.org. The Next 5 Minutes festival is dedicated to exploring tactical media practices and promotes "tactical media labs" worldwide, at http://www.n5m.org.

30. Alex Galloway, *Protocol: How Control Exists after Decentralization* (Cambridge, Mass.: MIT Press, 2004), p. 175. One of Galloway's examples is computer viruses, which exploit the very property that enable networks to exist: their horizontality and connectedness, coupled with the near-monopolistic nature of Windows operating systems.

31. For instance, see Albert-László Barabási and Zoltán Oltavi, "Network Biology: Understanding the Cell's Functional Organization," *Nature Reviews* 5 (February 2004): 101–113; Rita Colwell, "Future Directions in Biocomplexity," *Complexity* 6.4 (2000): 21–22; H. Jeong, "The Large-Scale Organization of Metabolic Networks," *Nature* 407

(October 5, 2000): 651–654; Stuart Kauffman, *At Home in the Universe: The Search for the Laws of Self-Organization and Complexity* (Oxford: Oxford University Press, 1995); Pierre Legrain, "Protein-Protein Interaction Maps: A Lead Towards Cellular Functions," *Trends in Genetics* 17 (2001): 346–352. For more, see chapter 6 of my book *Biomedia* (Minneapolis: University of Minnesota Press, 2004). Such alternative approaches can be grouped broadly into three categories: network science approaches (which study networks broadly and take biological networks as a special case), systems biology approaches (which are rooted more in the life sciences and high-technology sectors), and biocomplexity approaches (which are grounded in both theoretical and empirical research).

Index

DATE DUE
